AF566861

SP
LLC

Affectionately dedicated
to
my grand children

Aarnav
Kiran
Rohan

Nematode Disease Complexes in Horticultural Crops

P. Parvatha Reddy

Former Director
Indian Institute of Horticultural Research
Bangalore - 560 094, India

2011

STUDIUM PRESS LLC

P.O. Box 722200, Houston, Texas 77072, USA
Tel: (281) 776-8950, Fax: (281) 776-8951
e-mail: studiumpress@gmail.com
Website: http://www.studiumpress.in

Nematode Disease Complexes in Horticultural Crops

ISBN: 1-933699-80-9

Published by:

STUDIUM PRESS, LLC

P.O. Box-722200, Houston, Texas-77072, USA

Tel. 281-776-8950; Fax: 281-776-8951

E-mail: *studiumpress@gmail.com*

Printed at Thomson Press (India) Ltd.

PREFACE

Plants are constantly exposed to numerous organisms, many of which are common components of the soil biosphere. It thus seems appropriate to consider interrelationships among such organisms and the ultimate effects of these complexes upon host plants. It is realistic to assume that because a host is infected by one pathogen its response to additional invaders will be altered. These alterations may have significant influence upon disease development within a particular host, epidemiology of all pathogens involved, and ultimately, on disease management. Plant parasitic nematodes are usually a part of the soil microfauna, and should be considered when disease complexes are studied.

Nematode interactions are important biological phenomena and of great significance in horticulture. It is a fascinating subject which is multidisciplinary in nature and concerns any scientist involved with plant health. Nature does not work with pure cultures and that many plant diseases are influenced by associated organisms. Different parasites on the same plant interact, which result in disease complexes, and these interactions may lead to susceptibility by predisposition or resistance through preinduction of resistance against a particular parasite. Plant parasitic nematodes often play a major role in disease interactions. The fungus-nematode and the bacterium-nematode interactions are numerous, and weakly parasitic fungal and bacterial parasites can cause considerable damage once they gain entry into plant roots in the presence of feeding nematodes. Some species of nematodes belonging to five genera *Xiphinema, Longidorus, Paralongidorus, Trichodorus* and *Paratrichodorus* have a unique and important role as pathogens of plants, because they also transmit certain viruses to their host plants.

The information on interaction of nematodes with fungi, bacteria, viruses, rhizobia, arbuscular mycorrhizal fungi, other nematodes and insects in causing disease complexes in horticultural crops (fruits, vegetables, plantation, spice, tuber, ornamental, medicinal and aromatic crops) is very much scattered. There is no book at present which comprehensively and exclusively deals with the above aspects on horticultural crops. The present book is divided into two parts. The first part deals with the principles of nematode interactions with wilt-inducing and root rot fungi, arbuscular mycorrhizal fungi, plant viruses, bacterial wilt, rhizobia, nematodes, insects and management of these disease complexes. The second part deals with crop-wise disease complexes and their management in horticultural crops. The book is illustrated with excellent quality photographs enhancing the quality of publication. The book is written in lucid style, easy to understand language along with adoptable recommendations for enhancing the productivity.

In view of greater emphasis being given by the Government of India to horticulture by establishing horticulture mission and exclusive State Horticultural Universities, there is an urgent need for good text books to teach courses on different disciplines of horticulture. In this context, the present book is of immense help to all those concerned with interaction of nematodes with other micro-organisms in horticultural crops.

This book can serve as a practical guide to practicing farmers of horticultural crops. Further, it is a useful reference to policy makers, research and extension workers and students. The material can also be used for teaching post-graduate and under-graduate courses. Suggestions to improve the contents of the book are most welcome (E-mail: reddy_parvatha@yahoo.com).

I profusely thank Dr. M.S. Rao, Principal Scientist (Nematology), Division of Entomology and Nematology, Indian Institute of Horticultural Research, Bangalore, who has kindly made available some of the material included in this book. The publisher, Studium Press LLC, New Delhi, deserves commendation for their professional contribution.

July 30, 2010

P. Parvatha Reddy
34, UAS Layout, Ist Main, 7th Cross
Sanjay Nagar, Bangalore-560 094.

CONTENTS

Section I

Principles of Nematode Interactions

Chapter 1
INTRODUCTION

Plants are constantly exposed to numerous organisms, many of which are common components of the soil biosphere. It thus seems appropriate to consider interrelationships among such organisms and the ultimate effects of these complexes upon host plants. It is realistic to assume that because a host is infected by one pathogen its response to additional invaders will be altered. These alterations may have significant influence upon disease development within a particular host, epidemiology of all pathogens involved, and ultimately, on disease management. Plant parasitic nematodes are usually a part of the soil microfauna, and should be considered when disease complexes are studied.

The presidential address of Fawcett (1931) to the American Phytopathological Society at Cleveland, Ohio, USA, was a landmark in history of plant pathology. His assertion that nature does not work with pure cultures alone, but most frequently with associations; impelled plant pathologists to study the effect of inoculation of plants with known mixtures of microorganisms on the development of disease. The tremendous input of information since then has brought several new facts to light and has led many to question the dogma of specific etiology of Koch's postulates (Wallace, 1978).

Under natural conditions a plant is a potential host to various micro-organisms and they can influence each other by occupying the same habitat. Infection by one pathogen can alter host response to subsequent infection by another pathogen. Different parasites on the same plant interact, which results in disease complexes, and these interactions may lead to susceptibility by predisposition or resistance through preinduction of resistance against a particular parasite (Sidhu and Webster, 1981).

Nematode interactions are important biological phenomena and of great significance in horticulture. It is a fascinating subject which is multidisciplinary by nature and concerns any scientist involved with plant health. In nature plants are rarely exposed to the influence of a single pathogen. Nature does not work with pure cultures and that many plant diseases are influenced by associated organisms. Plant parasitic nematodes often play a major role in disease interactions. Interaction involving nematodes is important because they contribute substantially to variability in crop growth (Zadoks and Sachein, 1979). It is possible that some dramatic crop losses involving nematodes are due to the interaction between several determinants that exacerbate the effects (Wallace, 1983). Nematodes interact with different groups of plant pathogens and root symbionts. It seems reasonable to expect that infection by one pathogen may alter the host response to subsequent infection by another (Taylor, 1990). The fungus-nematode and the bacterium-nematode interactions are numerous, and weakly parasitic fungal and bacterial parasites can cause considerable damage once they gain entry into plant roots in the presence of feeding nematodes.

Ecological associations between nematodes and other organisms have been recognized and an account of the biology of associations of nematodes can be grouped into the following:

- Etiological associations often causing disease complexes.
- Competitive associations.
- Nematodes as vectors of plant viruses and other plant pathogens.
- Nematodes as parasites and predators.
- Nematodes as prey and host to other predators and parasites.

1.1. INTERACTION WITH BACTERIA

Nematodes may predispose plants to bacterial diseases. Nematodes in association with bacteria induce different symptoms in plants from those found when either pathogen is present alone. The association between gall-inducing nematodes and a species of bacteria results in toxin production in the host plant. Bacteria can enter only plants through wounds or through natural openings or by vectors. The role of nematodes in relation to bacterial pathogens has usually been regarded as providing wounds in the roots through which bacteria may enter the plant. This appeared to be the case in the association between *Ralstonia solanacearum* and the root-knot nematodes *Meloidogyne incognita* and *Agrobacterium tumefaciens* associated with *M. hapla*.

Nematodes can act as vectors by carrying bacteria on their body surface. In this case an obligate etiological relationship is established between nematode and bacterium on the manifestation of the resulting disease (Taylor, 1990). In the association between *Rhodococcus fascians* and *Aphelenchoides ritzemabosi* there is evidence that the bacterium attaches specifically to the cuticle of the nematode and is thus transported to the strawberry host plant (Hawn, 1971; Riley *et al.*, 1988). Both pathogens together induce various grades of leaf deformation. Strawberry plants attacked by the nematode alone do not show leaf deformations and bacterial infection alone induces no severe disease.

Bacterial wilt caused by *Ralstonia solanacearum* is more severe in resistant cultivars of tomato and eggplant in the presence of *M. incognita*. The combination of the two pathogens suppresses the survival rate of wilt-resistant tomato plants to 33-36%. Both capsicum and eggplant are prone to many soil borne diseases among which the bacterial wilt (*R. solanacearum*) in combination with root-knot nematode (*M. incognita*) takes heavy toll every year all over the world (Naik, 2003).

In other associations ecto-and endoparasitic nematodes can facilitate bacterial invasion by providing wounds in the host root (Poinar and Hansen, 1986).

1.2. INTERACTION WITH FUNGI

Much experimental evidence indicates a biological interaction between nematodes and certain soil-borne fungi. In most interactions the nematodes are not essential for establishment and development of fungal pathogens. However, the nematodes usually assist and enhance the pathogenicity mechanism of the fungus towards modifications in the host plants. For many years it was thought that nematode wounds facilitate the entry of fungi.

M. incognita and *Rhizocotnia bataticola* or *Sclerotium rolfsii* when inoculated simultaneously in soil, reduce the germination of seeds in okra, brinjal and tomato (Chhabra and Sharma, 1981; Shukla and Swarup, 1970). The incidence and severity of root rot of brinjal, tomato and okra, caused by the soil-borne fungi such as *R. solani, R. bataticola* and *Phomopsis vexans* was increased in the presence of *M. incognita* (Hazarika and Roy, 1974; Chahal and Chhabra, 1984; Chhabra *et al.*, 1977; Sharma *et al.*, 1980).

Evidence indicates that interactions between Fusarium and root-knot and cyst nematodes are biological and physiological rather than mechanical in nature (Roy *et al.*, 1989). Giant cells and syncytia induced by some endoparasites act as transfer cells and their activity are directly related to the physiological stage of the nematode. Most synergistic interactions between these endoparasitic nematodes and Fusarium occur after three to four weeks of nematode infection. Three to four weeks represents the time for reaching the maximum level of concentration of total protein and other metabolites in transfer cells induced by the nematodes. Many fungi benefit from this enriched medium and Fusarium hyphae in root-knot galls are more extensive than those in non-infected roots. Moreover, root exudates reflect the biochemical and physiological changes induced by nematode infection, and hence the development of fungi in roots is influenced by the presence of additional metabolic products in the rhizosphere and on the roots. Root exudates known to attract the motile stage of fungal pathogens (Zentmyer, 1961) represent a source of nutrients for the soil microflora and may be a stimulus for the germination of dormant spores. Thus changes induced by a nematode in the root exudates may be the first stage in the synergistic interaction between nematode and fungi (Taylor, 1990).

Plant parasitic nematodes are known to predispose some plants to fungal pathogens (Bergeson, 1972; Sidhu and Webster, 1977; Mai and Abawi, 1987) and this type of interaction was genetically evaluated in the *M. incognita-Fusarium oxysporum* f. sp. *lycopersici* disease complex on tomato host. Endoparasitic nematodes such as root-knot and cyst nematodes have long been known as primary pathogens for their ability to predispose plants to infection by secondary pathogens such as several species of *Fusarium, Phytophthora* and *Rhizoctonia*.

Migratory nematodes also represent a predisposing factor to infection by certain fungi. Mechanical wounding of the root favours the entry of other pathogens. For example, *Pratylenchus* sp. is reported to interact synergistically with *Verticillium* and *Belonolaimus* sp. and *Trichodorus* sp. with Fusarium (Faulkner *et al.*, 1970). The mechanism of synergism with migratory nematodes is different from that with sedentary endoparasites, because only necrosis and other processes of destruction develop rather than root physiological changes.

The kinds of effects observed in the nematode-fungus etiological relationships are:

- Fungus disease aggravated.
- Host growth suffered.
- Resistance to fungus reduced.
- Fungus suppressed by nematode.
- Nematode suppressed by fungus.
- Susceptibility to fungus increased (Norton, 1978).

1.3. INTERACTION WITH VIRUSES

Some species of nematodes belonging to the families Longidoridae and Trichodoridae have a unique and important role as pathogens of plants, because they also transmit certain viruses to their host plants. Hewitt *et al.* (1958) provided first experimental evidence that *Xiphinema index* acted as vector of grapevine fanleaf virus. In the Longidoridae, three genera – *Xiphinema, Longidorus* and *Paralongidorus* – transmit the nepoviruses, which are polyhedral, isodiametric particles about 28 nm in diameter. Nepoviruses are reported to infect a wide host range, from wild plants to annual and perennial crops. Nepoviruses are pathogens of economic importance, as they affect and damage major crops. Grapevine fan leaf and strawberry and raspberry ring spot viruses are transmitted by the Longidoridae. *X. index* is distributed worldwide and is closely related to its natural host, grapevine. All vectors of grapevine viruses are longidorids and 12 viruses of grapevine are known or suspected to be nematode transmitted. Grapevine degeneration is characterized by deformation and discolouration of the leaves, reduced vigour and poor fruit setting. The crop may be drastically affected, with average loss of above 60% (Rudel, 1985). In Great Britain and Central Europe, strawberry and raspberry crops are stunted and often die, when mixed infection of nematode vectors and related nepoviruses are present (Murant *et al.*, 1990).

In the Trichodoridae, 13 species of *Trichodorus* and *Paratrichodorus* are known to transmit rod-shaped, short and long (45-115 nm and 180-210 nm) particles classified as tobraviruses. Tobacco rattle and pea early browning viruses, transmitted by the Trichodoridae, infect a wide range of wild and cultivated plants. Tobacco rattle induce diseases of economically important bulbous flower crops in Europe and potato crops in Europe and USA. Browning disease has spread in Europe and induces severe stunting and premature death in large areas of pea crops (Taylor, 1980).

1.4. INTERACTION WITH OTHER NEMATODES

Interaction between different species of nematodes is usually antagonistic rather than beneficial. Two species can cohabit in the same host, but competition is severe when both species have similar feeding behaviour. Interaction between ectoparasites can be stimulatory, but competitive advantage increases as the host-parasite relationship becomes more complex. Combined infection of ecto- and endo-parasitic nematodes are, in most cases, suppressive for ectoparasites as sedentary endoparasites are more advanced parasites. In any case two or more species of nematodes may interact and the outcome of the interaction is density and time dependent. The combined effect usually leads to increase of the disease, which becomes more severe when two or more species of nematodes interact with another pathogen (Eisenback, 1985).

Nematodes influence the plant response to other pathogens and are influenced by associated microorganisms. Nematode attack sometimes increases the effect of disease caused by other organisms. Thus plant breeders should be aware of the possibility of nematode attack as the pathological additive effect of two pathogens, when they develop cultivars with resistance to fungi and bacteria.

1.5. INTERACTION WITH RHIZOBIA

Plant parasitic nematodes affect growth indirectly, by interfering with the plant's symbiotic relationship with rhizobia. Nodules from nematode-infected plants had lower fresh weight, specific nitrogenase and leghaemoglobin content than nodules from uninfected plants. The nematode effect depends on the nematode population density and can be reversed by removing the pathogen from the root system (Ko *et al.*, 1984).

1.6. INTERACTION WITH ARBUSCULAR MYCORRHIZAL FUNGI (AMF)

Another system in which nematodes seem to be important is the mycorrhizal association between certain fungi and plant roots. AMF are obligate symbionts and benefit plants by increasing the flow of essential elements into roots, by increasing the root surface for uptake of soil constituents, through the hyphal network at the exterior of the root, or through vesicles and arbuscles (branching structures) within cortical cells. Generally, AMF have a stimulating effect on plant growth by increasing shoot, root and seed weights as well as pod and seed numbers (Carling and Brown, 1980).

Nematode infection can damage the roots sufficiently to block the symbiotic plant-fungal associations. Field observations have indicated that root-knot nematodes cause a reduction in mycorrhizal level in the plant system (Jatala, 1986). AMF usually protect plants from pathogenic fungi and nematodes.

Sikora (1979) suggested that the presence of the mycorrhizal fungus in the plant root system prior to nematode attack may exert its influence on nematodes by altering root attractiveness, reduction in juvenile penetration, impeding juvenile development and retarding giant cell formation. These effects probably result from complex physiological changes associated with the mycorrhizal infection, and are not caused by direct competition between the two organisms. Attraction could be adversely influenced by changes in the chemicals of root exudates, by hardening of the root tissue due to increased lignin level in the endo- and exodermis of mycorrhizal plants. Increased phenolics and decreased auxin levels in the roots may impede juvenile growth and giant cell formation (Dehne, 1982). In general, mycorrhiza-plant parasitic nematode interactions result in a less deleterious effect of nematodes on plants. Roots of carrot and tomato colonized by AMF resulted in a decreased population of *Meloidogyne hapla* and *M. incognita*, respectively (Sikora, 1979). Nematode suppression might be the result of competition for plant nutrients between AMF and nematodes. AMF are very dependent on host photosynthates during the early stages of root colonization when fungal structures are forming. This period of AMF establishment coincides with the entry and selection of feeding site by the nematode juveniles. Insufficient nutrients for nematodes could result in decreased reproduction of the early generation. Conversely *M. arenaria* population on grape plants increased in the presence of AMF. Population densities were greater in mycorrhizal than in non-mycorrhizal plants (Atilano *et al.*, 1981).

1.7. INTERACTION WITH INSECTS

Nematodes parasitize many groups of insects and play an important role in natural control of these pests (Nickle and Welch, 1984). In some cases, the nematodes invade insects without causing them damage and also are transported for long distances. *Bursaphelenchus cocophilus*, which causes red ring disease in coconut and other palms, is vectored by an insect *Rhynchophorus palmarum.*

1.8. MECHANISMS OF INTERACTION

Numerous interactions of plant-parasitic nematodes with the plant-pathogenic fungi, viruses, bacteria, nematodes and possibly, rickettsia-like microorganisms have been described. The mechanisms operating in these interactions are largely unknown. Plant-parasitic nematodes are frequently vectors of plant-pathogenic viruses or bacteria, or, occasionally, of pathogenic fungi. Nematodes may aid ingress of plant-pathogenic fungi and bacteria by simple, physical wounding of the host plant. Plant-parasitic nematodes may also predispose hosts or modify expression of their resistance to plant pathogens by altering the host physiology. Altered nutritional status, and elicitation or suppression of host toxins, antifungal metabolites and growth factors are mechanisms by which nematodes may affect predisposition. Disease development in complex diseases may be controlled by changes in rhizosphere flora mediated by the nutritional quality and quantity of exudates from nematode-parasitized roots which enhance or suppress growth of organisms antagonistic to plant pathogens. Such exudates may also overcome fungistasis. By limiting host root development, nematodes may induce drought stress in the host, a factor thought to influence development of some plant diseases. By these mechanisms, nematodes also exert influence on their own reproduction and cohabitation in host plants. To the present time, work in this area has been largely descriptive. Research clarifying the mechanisms of interaction between nematodes and other plant pathogens and the interplay of these mechanisms in the development of complex diseases will open interesting new means of disease control. It may also lead to a better understanding of the factors important in regulating basic ecological interactions and processes among soilborne parasites in the rhizosphere.

Chapter 2

INTERACTION OF NEMATODES WITH WILT-INDUCING FUNGI

By far the greatest number of complexes reported involved plant parasitic nematodes along with wilt-inducing fungi as the other component. The root-knot nematodes accelerate and increase wilt symptom development and ultimately increase in death rate of cultivars of many crops. The presence of root-knot nematodes was reported to increase the severity of wilt disease in many horticultural crops such as carnation (Ferraz and Lear, 1976); tuberose (Rao *et al.*, 2001, 2002); watermelon (Yen *et al.*, 1998) and tomato (Bhagawati and Goswami, 2000). The combination of nematode and wilt-inducing fungi often results in a synergistic interaction wherein the crop loss is greater than expected from either pathogen alone or an additive effect of the two together. A cultivar susceptible to the interaction can result in a total crop failure. The problem for the farmer is compounded by such factors as saprophytic capability, wide host range and long-term survival of the pathogens. Thus, productivity of land for a highly valuable crop is impaired for many years.

2.1. NEMATODE - FUSARIUM WILT DISEASE INTERACTIONS

Disease complexes involving *Fusarium oxysporum* and plant nematodes have had a long history of investigation. A considerable number of *F. oxysporum* formae speciales (approximately 70) are known to interact with nematodes (Armstrong and Armstrong, 1981). The most studied formae speciales is *F. oxysporum* f. sp. *lycopersici*, the causal agent of wilt in tomato. Interactions between *Meloidogyne* spp. and *Fusarium* wilt pathogens have been studied and documented in several host crops, including beans (France and Abawi, 1994); tomatoes (Abawi and Barker, 1984; Suleman *et al.*, 1997); coffee (Bertrand *et al.*, 2000); peas (Siddiqui and Mahmood, 1999); and bananas (Jonathan and Rajendran, 1998).

2.1.1. Root-knot Nematode - Fusarium Wilt Disease Interactions

2.1.1.1. *Vegetable crops*

(i) *Tomato*: Jenkins and Coursen (1957) induced wilting in *Fusarium* wilt-resistant tomato variety 'Chesapeake' only when root-knot nematodes were present along with fungal inoculum. Furthermore, when *M. hapla* was combined with the fungus, only 60% of the plants wilted, whereas *M. incognita acrita* promoted wilt in 100% of the plants.

Bhagawati and Goswami (2000) reported the interaction of *M. incognita* and *F. oxysporum* f. sp. *lycopersici* on tomato (*Lycopersicon esculentum*). It was found that when either or both pathogens were inoculated simultaneously or nematode was inoculated 10 days prior to inoculation of the fungus, the symptoms were visible by 20 days as against no such symptoms in plants where fungus was inoculated 10 days prior to nematodes. The intensity of wilt after 40-60 days of inoculation was significantly higher when simultaneously inoculated or prior inoculation of nematodes compared to prior inoculation of fungus. The number of galls and egg masses and nematode population in soil was significantly reduced when simultaneously inoculated or prior inoculation of nematodes.

Goode and McGuire (1967) have pointed to another possibly important aspect of the Fusarium-root-knot interrelationship. They found that nematode infection enables certain races of *F. oxysporum* f. sp. *lycopersici* to attack tomato varieties ordinarily resistant to those races and suggested that the fungus mutates within the host.

(ii) *Brinjal*: Brinjal was not susceptible to *F. oxysporum* unless *M. incognita* was also present (Smits and Noguera, 1982).

(iii) *French bean*: *M. incognita* in association with soil-borne fungus such as *F. solani* was responsible for wilt disease complex in French bean. In the presence of nematode, the wilt due to *F. solani* increased considerably (Singh *et al.*, 1981).

(iv) *Cauliflower*: Maximum wilt score due to *F. oxysporum* f. sp. *conglutinans* (4.0) and wilt symptom index (8.0) was observed in cauliflower plants which received *M. incognita* 14 days before the inoculation of wilt fungus (Pathak and Keshar, 2004).

2.1.1.2. *Ornamental crops*

(i) *Crossandra*: *M. incognita* interact synergistically with wilt fungi *F. oxysporum* and *F. solani* and results in premature death of crossandra plants (Khan and Reddy, 1992). *F. solani* and *F. oxysporum* are invariably associated with the lesions caused by *P. delattrei* in crossandra (Khan and Reddy, 1992; Srinivasan and Muthurksihanan, 1976). Srinivasan (1974) found that wilt symptoms appeared earlier when nematodes were added two weeks prior to *F. solani*.

(ii) *Carnation*: Ferraz and Lear (1976) studied the interaction of four plant parasitic nematodes (*Circonemoides curvatum, Pratylenchus dianthus, Rotylenchus robustus* and *Meliodogyne hapla*) and *F. oxysporum* f. sp. *dianthi* in carnations. They reported that plants wilted earlier and more severe symptoms were present with the nematode-fungus combination than with the fungus alone. All nematodes except *R. robustus* showed synergistic interaction with *Fusarium.*

In an experiment carried out to study the role of root-knot nematode, *M. incognita* in predisposing carnation (*Dianthus caryophyllus*) to *Fusarium* wilt, it was observed that the appearance of the wilt symptoms were accelerated when *M. incognita* was inoculated two weeks prior to *F. oxysporum* f. sp. *dianthi*. The rate of wilting was observed to be 2.1 and 4.6, respectively during 12th and 25th week of observation, while the root galling index (RGI) values were recorded to be 4.15 and 4.75, respectively. It was observed that maximum mortality was recorded when *M. incognita* was inoculated 2 weeks prior to *F. oxysporum* f. sp. *dianthi* which was 33.4% and 79.2%, respectively, at 12th and 25th week of observation. The plant growth was also reduced significantly due to prior inoculation of *M. incognita* (Shylaja, 2005).

(iii) *Tuberose*: In an experiment undertaken to assess the impact of *Fusarium* wilt-root knot disease complex on the flower yield in tuberose, it was observed that the disease complex drastically reduced the yield of flowers in *P. tuberosa,* thereby bringing down the production of flowers in this commercial crop. It was found that the damage was maximum in presence of both pathogens *viz; M. incognita* and *F. oxysporum* f. sp. *dianthi*, when compared to the presence of either one of these pathogens (Shylaja, 2005).

2.1.1.3. *Medicinal crops*

(i) *Dioscorea*: A positive interaction between *M. incognita* and *F. oxysporum* f. sp. *dioscorea* occurs in *Dioscorea rotundata* cv. Habanero.

2.1.1.4. *Spice crops*

(i) *Ginger*: Rotting of ginger roots by *F. solani* became more severe in the presence of *M. incognita* (Doshi and Mathur, 1987).

2.1.2. Burrowing Nematode – Fusarium Wilt Disease Interactions

2.1.2.1. *Fruit crops*

(i) *Banana*: The incidence of Panama wilt in banana caused by *Fusarium oxysporum* f. sp. *cubense* was doubled in the presence of *R. similis*. The onset of wilt was also considerably shortened. *R. similis* caused the breakdown of resistance to panama wilt in "Robusta" bananas.

Stover (1966) suggested that *F. solani* may contribute to the extension of lesions on banana roots caused by *R. similis* and even to the death of roots.

(ii) *Citrus*: Combined infection of grapefruit seedlings by *R. similis* and *Fusarium* spp. resulted in more severe damage than from the nematode or fungus alone (Feder and Feldmesser, 1961).

2.1.3. Citrus Nematode – Fusarium Wilt Disease Interactions

2.1.3.1. *Fruit crops*

(i) *Citrus*: Under greenhouse conditions, the citrus nematode, *Tylenchulus semipenetrans* apparently interacts with *F. solani* to reduce citrus seedling growth, and the effect of the pathogen combination is noticeably greater than that of either pathogen alone. This may be important in subsequent root rot of citrus (Van Gundy and Tsao, 1963). In the presence of *T. semipenetrans,* significantly more citrus plants become infected by *F. oxysporum* and *F. solani* (O'Bannon *et al*., 1967).

2.1.4. Stunt Nematode – Fusarium Wilt Disease Interactions

2.1.4.1. *Vegetable crops*

(i) *Pea*: *Tylenchorhynchus claytoni* increases wilt more rapidly on pea roots infected with *F. oxysporum* f. sp. *pisi* race 1 than in roots free from the fungus (Davis and Jenkins, 1963).

2.1.5. Lance Nematode – Fusarium Wilt Disease Interactions

2.1.5.1. *Vegetable crops*

(i) *Peas*: A nematode-fungus relationship occurs in peas. The chief symptoms of the "early yellowing" disease, root rot, is dependent upon the presence of both *Hoplolaimus uniformis* and *F. oxysporum* f. sp. *pisi* race 3 (Labruyere *et al*., 1959).

Table 2.1: Nematode – Fusarium wilt disease interactions

Crop	*Nematode*	*Fusarium spp.*	*Reference(s)*
Banana	*M. incognita acrita*	*F. o.* f. sp. *cubense*	Loos, 1959
	R. similis	*F. o.* f. sp. *cubense*	Newhall, 1958
Sweet orange	*Tylenchulus semipenetrans*	*F. solani*	Van Gundy & Tsao, 1963
Grapefruit	*R. similis*	*Fusarium* spp.	Feder & Feldmesser, 1961
Papaya	*Meloidogyne* spp.	*Fusarium* spp.	Saeed *et al*., 1972
Tomato	*M. incognita acrita* & *M. hapla*	*F. o.* f. sp. *lycopersici*	Jenkins & Coursen, 1957
	M. incognita & *M. javanica*	*F. o.* f. sp. *lycopersici*	Bowman & Bloom, 1966; Harrison & Young, 1941; Chahal & Chhabra, 1984; Hasan & Khan, 1985
	M. incognita	*F. o.* f. sp. *lycopersici*	Bhagawati & Goswami, 2000
	P. coffeae & *P. penetrans*	*F. o.* f. sp. *lycopersici*	Hirano & Kawamura, 1972
Eggplant	*M. incognita*	*F. oxysporum*	Smits & Noguera, 1982
Pea	*M. incognita acrita* & *M. incognita*	*F. o.* f. sp. *pisi*	Davis & Jenkins, 1963

Table 2.1: (*Contd...*)

Table 2.1: (*Contd...*)

Crop	*Nematode*	*Fusarium spp.*	*Reference(s)*
	P. penetrans	*F. o.* f. sp. *pisi*	Oyekan & Mitchell, 1971
	Rotylenchulus reniformis	*F. o.* f. sp. *pisi*	Vats & Dalal, 1997
Bean	*M. incognita*	*F. o.* f. sp. *phaseoli*	Singh *et al.*, 1981
	M. javanica		Ribeiro & Ferraz, 1983
Cowpea	*M. javanica*	*F. oxysporum*	Thomason *et al.*, 1959
Watermelon	*M. arenaria* & *M. incognita*	*F. oxysporum*	Sumner & Johnson, 1973
Muskmelon	*M. incognita*	*F. oxysporum*	Bergeson, 1975
Cucumber	*M. incognita*	*F. o.* f. sp. *cucumerinum*	Fritzsche *et al.*, 1983
Squash	*M. incognita*	*F. oxysporum*	Caperton *et al.*, 1986
Cabbage	*M. incognita acrita*	*F. oxysporum*	Fassuliotis & Rau, 1969
Cauliflower	*M. incognita*	*F. o.* f. sp. *conglutinans*	Pathak & Keshar, 2004
Crossandra	*M. incognita*	*F. oxysporum, F. solani*	Khan & Reddy, 1992
	P. delattrei	*F. oxysporum, F. solani*	Srinivasan, 1974
Chrysanthemum	*Meloidogyne* spp.	*F. oxysporum*	Johnson & Littrell, 1969
Tuberose	*M. incognita*	*F. o.* f. sp. *dianthi*	Shylaja, 2005
Carnation	*Meloidogyne* spp.	*F. o.* f. sp. *dianthi*	Schindler *et al.*, 1961; Shylaja, 2004
	Criconemella curvata, M. hapla, Paratylenchus dianthus	*F. o.* f. sp. *dianthi*	Ferraz & Lear, 1976
Coffee	*M. incognita*	*F. o.* f. sp. *coffeae*	Negron & Acosta, 1989
	Meloidogyne arabicida	*Fusarium oxysporum*	Bertrand *et al.*, 2000
Black pepper	*M. incognita*	*F. solani*	Hamada *et al.*, 1985
Ginger	*M. incognita*	*F. solani*	Doshi & Mathur, 1987
Yam	*Scutellonema bradys*	*F. oxysporum*	Goodey, 1935

2.1.6. Mechanisms of Interactions

2.1.6.1. *Nematodes as wounding agents*

Nematode parasitism of plants is associated with wounding of their hosts. The feeding process of all plant parasitic nematodes produces a wound of some kind in the host plant, either by simple micropuncture or by rupturing or separating cells (Taylor, 1980). Ectoparasitic nematodes cause micropunctures on the plant root surface. Migratory endoparasites produce lesions on roots. Second-stage juveniles of sedentary endoparasites after root invasion migrate intercellularly through the cortex and establish contact with vascular tissue to induce giant cells or syncytia.

Powell (1979) remarked that wounding by nematodes in increasing the susceptibility of plants to invasion by *Fusarium oxysporum* appears to be logical in some cases, particularly in diseases involving migratory nematodes. It is most probable that wounding coupled with physiological changes in the host are important in interactions.

2.1.6.2. *Increased host susceptibility*

A vast majority of nematode interactions involve fungal pathogens, especially wilt fungi. Plant-parasitic nematodes play an assisting role by way of enhancing host susceptibility, leading to increased rate of development and severity of the disease caused by the fungal pathogens.

The bulk of studies on interaction between fungi and nematodes involve species of *Fusarium* and *Meloidogyne*. *Fusarium* spp. are highly destructive pathogens of a number of vegetable, fruit and other horticultural crops. All the wilt causing formae speciales and races belong to the species of *F. oxysporum*. Most *Fusarium* spp. penetrate the root tissue directly, often near the root tip (Nelson, 1981).

Mechanical plugging, toxins, enzymes and growth regulators singly or jointly have been implicated as causes of wilting (Mai and Abawi, 1987). Attack of root-knot nematodes of plants infected by *Fusarium* spp. increases the wilt symptom expression and death rate of plants. Evidence suggests that the initial phase of the root-knot nematode-*Fusarium* interaction occurs in rhizosphere, where the root exudates from root-knot infected plants stimulate fungal pathogen. The exudates also suppress actinomycetes, the antagonists of the wilt fungus (Cook and Baker, 1983; Walter, 1965; Bergeson, 1972).

The next phase in these interactions involves the effect of root-knot infection on penetration of wilt fungi. Initially it was thought that micropunctures caused by nematodes on plant root facilitate entry of the fungal plant pathogens. Later, it was demonstrated that severity of fungal-induced wilt diseases increased greatly when root-knot nematodes were added 3 to 4 weeks prior to fungus inoculation of the host. This led to the hypothesis that the nature of interactions between root-knot nematodes and *Fusarium* wilt fungi are physiological rather than physical. The ultimate phase of the interaction between the two pathogens occurs during the pathogenesis of the wilt fungus. Modifications in the host plant by root-knot nematodes is the key factor to this phase of interaction leading to increased wilt severity.

The sedentary females of *Meloidogyne* spp. established their feeding sites in the xylem parenchyma cells, bringing about significant changes in the morphology, anatomy and biochemistry of the host plant. Thus it is probably the major site of interaction between these pathogens (Mai and Abawi, 1987). Giant cells induced by *Meloidogyne* remain in a state of high metabolic activity through the continuous stimulation by the nematode (Webster, 1975). They contain maximum DNA, RNA and photosynthates about 3 to 4 weeks after infection (Bird, 1972). Concentration of sugars is also greater during the late stages of infection (Wang, 1973). The concentration of hemicellulose, organic acids, free amino acids, proteins and lipids are much higher in giant cells. This enriched medium benefits the fungal pathogens in their interactions with root-knot nematodes. The giant cells remain in a perpetual juvenile state which delays maturation and suberization of other vascular tissues, and thus *Fusarium* successfully penetrates and establishes in the xylem elements.

Noguera (1982) observed that purified extracts of tomato roots infected with *M. incognita* did not show rishitin, an antifungal substance normally present in healthy plants. Loss of resistance of the tomato variety to wilt fungus was attributed to inhibition of rishitin production. Inhibition of tylose formation by root-knot nematodes is also offered as a possible mechanism for increased wilt severity. Tyloses formed in the xylem vessels by expansion of xylem parenchyma through the pits do not develop from xylem parenchyma cells which are transformed into giant cells or physiologically altered adjacent cells (Webster, 1985). Nematode-free plants of tomato showed the formation of well-developed tyloses in the xylem vessels of stem, while in infected plants these were absent or much reduced.

2.1.6.3. *Modifications in host substrate*

Modification of host substrates was found to be primarily responsible for *Fusarium* and *Meloidogyne* interactions (Pitcher, 1965). Split-root (Bowman and Bloom, 1966), grafting (Sidhu and Webster, 1977) and stem layering techniques further demonstrated that predisposition to infection by *Fusarium* wilt fungi occurs through systematically translocatable metabolites produced at the site of nematode infection.

Negrón and Acosta (1989) found that *F. oxysporum* f. sp. *coffeae* caused increased root necrosis and chlorosis on the foliage of coffee plants (cv. Bourbón) if the plants had been inoculated with *M. incognita* 2 or 4 weeks previously. Sections taken from the roots of plants preinoculated with *M. incognita* were found to be colonized by *F. oxysporum* f. sp. *coffeae* in a uniquely different way from those where the fungus and nematode were either inoculated simultaneously, or where the fungus was inoculated alone. In the former case, the hyphae of *F. oxysporum* f. sp. *coffeae* were found to be abundant in the xylem vessels, giant cells and female nematodes. Giant cells, colonized by the fungus were in varying states of disrepair, with depleted or partially depleted contents. In comparison, plants that had been simultaneously inoculated with *M. incognita* and *F. oxysporum* f. sp. *coffeae* had fewer giant cells colonized by the fungus and no hyphae within the xylem. Taylor (1990) suggested that the 3-4-week nematode preinoculation, found to be critical in investigations of nematode–fungus disease complexes, could be linked to syncytial development which, under optimal conditions in a susceptible host, will take 3-4 weeks to reach peak activity.

Translocatable substances: The nematode-induced physiological changes can be systemic. In such cases, the nematode-induced factors or substances beneficial to fungi are hypothesized to be translocatable within the plant. Evidence of the involvement of translocatable substances or factors in the predisposition of plants to some fungal pathogens was provided by using split-root, double root and layered plant techniques. Adopting split-root techniques, one part of the root system of a wilt-resistant tomato plant was inoculated with *F. oxysporum* f. sp. *lycopersici* and the other with *M. incognita* (Bowman and Bloom, 1966; Hillocks, 1986). The plants showed wilting even though the nematode and fungus were added separate to root systems. Their results revealed that disease development on plants was dependent on being exposed to both *M. incognita* and the fungus. Nematode infestations may reduce the levels of fungitoxic compounds. Perhaps the sequence of events that leads to systemic induced changes is similar to, or the reverse of, that described in plants with systemic acquired resistance (SAR). Sidhu and Webster (1977) using layered tomato plants demonstrated a mobility of factors that predisposed plants to wilt fungus *F. oxysporum* f. sp. *lycopersici*. Powell (1979) described it as a message of the nematode-infected host sent to the fungus through a messenger, perhaps a growth-regulating substance. Substances introduced by the nematode or produced in the roots in response to the nematode are translocated throughout the plant.

Nematode-induced physiological changes to the host plant: The feeding sites of sedentary endoparasitic nematodes (giant cells or syncytia) are zones of high metabolic activity, having a large number of Golgi apparatus and mitochondria, while the cytoplasm is dense and contains many ribosomes. It is therefore no surprise that these nutrient-rich cells should become the substrate for fungal colonization.

Modifications within the rhizosphere: Bergeson *et al.* (1970) recorded a marked increase in the number of propagules of *F. oxysporum* f. sp. *lycopersici* found around the galls of *M. javanica*. Galls produced by *M. javanica* on tomato roots stimulated the number of spores of *F. oxysporum* f. sp. *lycopersici* while reducing the number of cells of actinomycetes antagonistic to *Fusarium*.

2.1.6.4. *Breaking of disease resistance*

Breaking of disease resistance by plant-parasitic nematodes is used in plant nematology literature to denote predisposition of plants resistant to a given pathogen. The gene(s) for resistance for a pathogen is rendered ineffective through physiological alterations in the plant caused by the nematode even though the gene(s) remains operational.

Bowman and Bloom (1966) found that the tomato cvs Rutgers and Homestead, previously resistant to *F. oxysporum* f. sp. *lycopersici*, developed symptoms of wilt during split-root experiments with *M. incognita*. Further studies (Sidhu and Webster, 1977) using root layering and grafting techniques confirmed that a nematode-induced factor could be passed through a resistant scion (a graft from a resistant tomato cultivar) and render it susceptible to *F. oxysporum* f. sp. *lycopersici*. Marley and Hillocks (1994) demonstrated that nematode-induced loss of resistance to *Fusarium udum* in pigeonpea (*Cajanus cajan*) was associated with reduced levels of the isoflavanoid phytoalexin cajanol. Previously, Marley and Hillocks (1993) determined that the rapid accumulation of cajanol in some pigeonpea cultivars was responsible for conferring resistance to the pathogen. However, cajanol content was 62% lower and resistance was lost during combined infections of *F. udum*, *M. incognita* and *M. javanica* where wilt disease incidence and severity were significantly higher than in plants inoculated with *F. udum* alone. This study clearly shows that nematode infestation reduced a chemical defense mechanism to *Fusarium* wilt in pigeonpea.

Temperature has been found to be critical in some nematode–fungus interactions. Interestingly, France and Abawi (1994) observed that a bean genotype with dual resistance against *F. oxysporum* f. sp. *phaseoli* and *M. incognita* exhibited visible wilt symptoms if both organisms were inoculated together at 27°C. It was suggested that the high temperature caused a breakdown of resistance to *M. incognita* and, in turn, the nematodes broke the resistance to the wilt-inducing pathogen.

The plants with polygenic resistance to fungal pathogens are frequently found to become susceptible to fungal attack during nematode infestations, whereas plants with a single dominant gene for resistance are rarely affected. This was observed by Abawi and Barker (1984) on tomatoes, where resistance to *F. oxysporum* f. sp. *lycopersici* was disrupted by infestations of *M. incognita* on cultivars with polygenic resistance, but not on those where resistance was expressed by a dominant single I-gene. Transgenic plants involving quantitative trait loci may have a greater capacity for providing durable resistance in the presence of interacting fungal pathogens and plant-parasitic nematodes.

There are reports which indicate that *Fusarium* wilt resistance in cultivars of cabbage, peas and other crops, possessing a single dominant gene for resistance against wilt fungi, is not altered by root-knot nematode infection (Fassuliotis and Rau, 1969; Walter, 1965). In tomato as well, some reports demonstrate that monogenic resistance to the wilt fungus is not broken down by root-knot nematodes (Binder and Hutchinson, 1959; Jones *et al.*, 1976; Hirano *et al.*, 1979; Abawi and Barker, 1984). In overall analysis, most apparently histophysiological changes in the host caused by nematode infection are responsible for rendering the gene(s) for resistance ineffective and a result the host is not able to express the resistant reaction.

In contrast, root-knot nematodes are reported to break the monogenic *Fusarium* wilt resistance in tomato cultivars (Jenkins and Coursen, 1957; Sidhu and Webster, 1974, 1977; Liburd, 1977; Webster, 1985). Sidhu and Webster (1974, 1977) suggested that this breakdown of resistance is genetically controlled and one pathogen induced susceptibility to another pathogen.

2.2. NEMATODE - VERTICILLIUM WILT DISEASE INTERACTIONS

Synergistic and additive responses between *Verticillium* spp. and several species of plant parasitic nematodes have been reported over the years (Table 2.2). The majority of these interactions occur with root-lesion nematodes, *Pratylenchus* spp. and potato cyst nematodes, *Globodera rostochiensis* and *G. pallida*. Synergistic interactions have been defined in terms of earlier symptoms (Bergeson, 1963; Burpee and Bloom, 1978; Faulkner and Skotland, 1965), higher disease incidence and/or severity (Conroy *et al.*, 1972; Faulkner and Skotland, 1965; Harrison, 1971; McKeen and mountain, 1960) and lower yields (Francl *et al.*, 1987; MacGuidwin and Rouse, 1990; Rowe *et al.*, 1985), as compared with the sum effects of the two pathogens separately.

2.2.1. Lesion Nematode - Verticillium Wilt Interactions

2.2.1.1. *Vegetable crops*

(i) *Potato*: An important interrelationship exists between *Pratylenchus* spp. and *Verticillium* spp. in potato (Morsink and Rich, 1968). In this situation, either pathogen alone may cause disease but damage is greater when nematodes and fungus occur together. Potatoes infected by *V. albo-atrum* also support greater reproduction of *P. penetrans* than fungus-free plants. Most evidence indicates strongly that *Verticillium* wilt generally promotes increases in reproduction of *Pratylenchus* spp.

(ii) *Tomato and eggplant*: *Verticillium* wilt seems to render plants more suitable for colonization by lesion nematodes, *Pratylenchus* spp. When the fungus is added to soil already infested with the nematodes, reproduction of the latter is increased in roots of both eggplant and tomato (Mountain and McKeen, 1962).

2.2.1.2. *Aromatic crops*

(i) *Mint*: Bergeson (1963) reported that when *Pratylenchus penetrans* infected *Mentha piperita* plants were inoculated with *Verticillium albo-atrum,* diagnostic symptoms of *Verticillium* wilt appeared approximately two weeks earlier than the plants inoculated with the *Verticillium* fungus in the absence of nematode. Physiological changes evidently occur in fungus infected plants, making them more susceptible or attractive to the nematode pathogen (Mountain and McKeen, 1962).

Presence of *Pratylenchus miniyus* increased both incidence and severity of *Verticillium* wilt (*Verticillium dahliae* f. sp. *menthae*) disease, reduced the period for the development of wilt by 2 to 3 weeks, increased the reproduction of nematode and reduced the dry weight of peppermint plants up to 68% when both the pathogens were present (Faulkner and Skotland, 1965). The optimum soil temperature for disease development when the fungus is present alone is 24°C, as compared to 27°C in plants exposed to both pathogens (Faulkner and Bolander, 1969).

2.2.2. Stunt Nematode – Verticillium Wilt Interactions

2.2.2.1. *Vegetable crops*

(i) *Tomato*: Reproduction of *Tylenchorhynchus capitatus* also increases in tomato roots when the *Verticillium* fungus is present (Mountain and McKeen, 1962).

Table 2.2: Nematode – Verticillium wilt disease interactions

Crop	*Nematode*	*Verticillium spp.*	*Reference(s)*
Strawberry	*Pratylenchus penetrans*	*Verticillium dahliae*	McKinley & Talboys, 1979
Cherry	*Meloidogyne* spp.	*V. dahliae*	Ndubizu, 1977
Tomato	*P. penetrans*	*V. albo-atrum*	Conroy *et al.*, 1972
	Trichodorus christiei	*V. albo-atrum*	Conroy & Green, 1974
	Globodera tabacum	*V. albo-atrum*	Miller, 1975
	Tylenchorhynchus capitatus	*V. albo-atrum*	Overman & Jones, 1970
	M. incognita	*V. albo-atrum*	Overman & Jones, 1970
	M. javanica	*V. dahliae*	Orion & Krikun, 1976
Potato	*P. penetrans*	*V. dahliae*	Martin *et al.*, 1982
	P. thornei	*V. dahliae*	Krikun & Orion, 1977
	P. neglectus / *Meloidogyne* spp.	*V. dahliae*	Scholte & s' Jacob, 1989 Hafez *et al.*, 1999
	M. hapla	*V. albo-atrum*	Jacobson *et al.*, 1979
	G. rostochiensis	*V. dahliae*	Corbett & Hide, 1971
Eggplant	*P. penetrans*	*V. albo-atrum*	McKeen & Mountain, 1960
Gerbera	*M. arenaria*	*V. dahliae*	Schlang & Sikora, 1978
Balsam	*P. penetrans*	*Verticillium* sp.	Muller, 1973
Peppermint	*P. penetrans*	*V. albo-atrum*	Bergeson, 1963
	P. neglectus	*V. dahliae*	Faulkner & Skotland, 1965
Mint	*P. penetrans*	*V. dahliae*	Johnson & Santo, 2001

2.2.3. Mechanisms of Interaction

2.2.3.1. *Utilization of nematode-induced wounds by soil-borne pathogens*

On potatoes, Storey and Evans (1987) have found that the pathology of the wilt fungus *V. dahliae* is dependent on the timing of invasion by the cyst nematode *G. pallida*, and also on the potato cultivar. The fungus was found to enter and use the invasion channels of juvenile *G. pallida* if it was introduced at the same time as the nematode on cvs Pentland Javelin, Maris Peer and Maris Anchor. On both Maris Peer and Maris Anchor, lignified tissue developed following the invasion of the nematodes. If the fungus was introduced 8 days after the nematode, wilt symptoms caused by *V. dahliae* were less severe than on plants treated with the fungus alone. When sections of root tissue were examined, it could be seen that the lignified tissue had sealed the nematode penetration sites. In fact, the living tissue was partially protected by areas of woody tissue where this response had taken place. This response was not produced by cv. Pentland Javelin after nematode invasion, and *V. dahliae* was still able to colonize roots to a larger extent than on plants where it was introduced alone.

2.2.3.2. *Increased host susceptibility*

Pratylenchus spp. have been implicated in synergistic interactions with other plant pathogens, particularly with *Verticillium* spp., on a number of crops. The mechanisms of these interactions are related to the development of necrotic infection courts and biochemical changes in the attacked plants. These alterations enhance the pathogenic capabilities of *Verticillium* spp. The necrotic lesions on roots serve as infection courts which facilitate the establishment of the fungus in the court and subsequent invasion. *V. dahliae* showed a distinct liking for the lesions on eggplant root caused by *P. penetrans*. According to Conroy *et al.* (1972) lesions in tomato roots were more important than any general physiological changes. Internal predisoposing factors in *Pratylenchus*-infected plants which are also claimed to be translocatable from the site of infection to the other parts of the plant are also involved (Faulkner *et al.*, 1970; Powell, 1979). The incubation period of *Verticillium* is shortened in *Pratylenchus*-infected plants (Bergeson, 1963; Faulkner *et al.*, 1970). *P. miniyus* was shown to shift susceptibility of peppermint to *Verticillium* wilt towards its own optimum temperature (Faulkner and Bolander, 1969). Therefore, physical and physiological changes in the plants infected with *Pratylenchus* are called together to enhance the susceptibility of plants to *Verticillium* wilt.

2.2.3.3. *Modifications in host substrate*

Translocatable substances: Evidence of the involvement of trans-locatable substances or factors in the predisposition of plants to some fungal pathogens was provided by using split-root, double root and layered plant techniques. Faulkner *et al.* (1970) used peppermint stem rooted at two places and inoculated one root system with *Verticillium dahliae* and the other with *Pratylenchus miniyus*. *Verticillium* wilt showed an increase even when the nematode infected the other root system. Substances introduced by the nematode or produced in the roots in response to the nematode are translocated throughout the plant.

2.2.3.4. *Pathogen-induced changes to the host plant*

Faulkner and Skotland (1965) observed that *Pratylenchus minyus* reached its reproductive peak at the same time as the maximum expression of wilt disease (*V. dahliae* f. sp. *menthae*) on peppermint plants (*Mentha piperita*). The authors suggested that *V. dahliae* may produce root growth-promoting substances such as indole-3-acetic acid (as previously recorded for *V. albo-atrum*), resulting in an enlarged root system, releasing greater volumes of root exudate and thereby attracting more plant-parasitic nematodes.

2.2.3.5. *Factors affecting synergistic interactions*

There are findings that highlight the specificity of certain disease complexes and the influence of biotic and abiotic factors on them. This is exemplified by studies on the *V. dahliae–Pratylenchus* complex of potato, where it has been found that the interaction between these organisms varies among different nematode species and populations, as well as fungal genotypes. For example, Bowers *et al.* (1996) observed that potato early dying disease (*V. dahliae*) was enhanced by populations of *P. penetrans* but not by *P. crenatus* or *P. scribneri*. Furthermore, greenhouse experiments undertaken by Hafez *et al.* (1999) demonstrated that populations of *P. neglectus* collected from Ontario, Canada would interact synergistically with *V. dahliae*, while populations of *P. neglectus* from Parma, Idaho did not increase disease or yield loss any more than treatment with *V. dahliae* alone. Restriction analysis of the ITS1 region on rDNA gene from nematodes from the Canada and Idaho populations revealed unique fragments for each population, implying variation within this species. The differences between nematode species and populations/pathotypes in their ability to accentuate *V. dahliae* wilt are likely to be related to their proficiency as parasites on potato. This is certainly evident from the work of Hafez *et al.* (1999), where the fecundity of the Canadian population was approximately 50% higher than that of the Idaho population, in which initial populations (P_i) were 5000 and 10000 nematodes per 5000 cm^3 soil. It is also possible that species and pathotypes of *Pratylenchus* spp. may favour different physical environmental conditions. Indeed, differential effects of soil pH and temperature between populations of *P. penetrans* and *P. crenatus* have been recorded.

Fungal genotype can also affect potato early dying complex, as shown by Botseas and Rowe (1994), who found that two pathotypes of *V. dahliae* vegetative compatibility group 4 (VCG 4) formed different relationships with *P. penetrans*. In greenhouse and field microplot experiments, the two pathotypes of *V. dahliae* (VCG 4A and B) exhibited no differences in aggressiveness when inoculated alone. However, plants inoculated with *V. dahliae* VCG 4A and grown in soil infested with *P. penetrans* had higher levels of disease severity, lower tuber yields and earlier senescence than plants inoculated with *V. dahliae* VCG 4B in the presence of *P. penetrans*. This type of specificity has also been reported by Johnson and

Santo (2001), who found that *P. penetrans* would interact synergistically with *V. dahliae* VCG 2B but not with VCG 4A on cultivated peppermint and Scotch spearmint. In these findings, the governing factor was host-specific fungal aggressiveness. Although Botseas and Rowe (1994) were unable to detect differences in aggressiveness between *V. dahliae* VCG 4A and B, a number of investigations have reported that VCG 4A is the more aggressive of the two pathotypes on potato. This is also relevant to the work on mint, where VCG 2B is described as the most aggressive pathotype.

Chapter 3

INTERACTION OF NEMATODES WITH ROOT-ROT FUNGI

Nematodes are important and, at times, vital to the development of root rots caused by fungi. Associations of this type may have greater overall importance than disease complexes involving nematodes with wilt-inducing fungi. Sedentary endoparasitic nematode pathogens (species of *Meloidogyne* and *Heterodera*), migratory endoparasitic nematodes (species of *Pratylenchus* and *Radopholus*), semiendoparasitic nematodes (*Tylenchulus semipenetrans* and *Rotylenchulus reniformis*) and ectoparasitic nematodes (*Tylenchorhynchus brassicae, Belonolaimus longicaudatus* and *Mesocriconemella xenoplax*) are involved in several root rot disease complexes (with species of *Pythium, Phytophthora, Rhizoctonia, Sclerotium* and *Macrophomina*). It is evident that root-rot complexes involving nematodes and fungi are among the most widespread and important interactions.

3.1. ROOT-KNOT NEMATODE – ROOT ROT DISEASE INTERACTIONS

3.1.1. Vegetable Crops

(i) *Tomato*: *M. javanica* increased the extent of damage by pre and post-emergence phases of damping off caused initially by *R. solani* and *Pythium debaryanum* in tomato (Nath *et al*., 1984). *M. javanica* also interacted with *Rhizopus nigricans* and *Pencillium digitatum* (which are thought of usually not pathogenic) and caused significant root necrosis (Nath and Kamalwanshi, 1989).

Soil fumigation usually decreased disease severity due to *R. solani* and *M. incognita* disease complex in tomato (Golden and Van Gundy, 1975).

(ii) *Brinjal*: *M. incognita* and *Ozonium texanum* var. *parasiticum* together acted synergistically by reducing the germination of brinjal to 28 per cent (Nath *et al*., 1976).

(iii) *Chilli*: *Pythium aphanidermatum* and *Rhizoctonia solani* were both found to interact with *M. incognita* on chilli, causing some loss of nematode resistance in two cultivars tested (Hasan, 1985), whereas the interaction of *P. debaryanum* with this nematode appeared to be due to physiological response of the plants to nematode infection, making the roots more susceptible to invasion by the fungus (Brodie and Cooper, 1964).

(iv) *French Bean*: *M. incognita* in association with soil-borne fungus such as *R. solani* was responsible for root-rot disease complex in French bean. In the presence of nematode, the root rot due to *R. solani* increased considerably (Reddy *et al.*, 1979).

(v) *Cowpea*: Inoculation of cowpea with *M. incognita* and *R. solani* led to breakdown of resistance to both organisms (Khan and Husain, 1989), and the greatest decrease in plant dry weight occurred if the nematode was inoculated two weeks before the fungus (Varshney *et al.*, 1987).

(vi) *Peas*: Peas were damaged by *M. incognita* or *R. solani* but plant growth was depressed even further when plants were inoculated with both organisms, with the maximum effect occurring when the two were inoculated simultaneously (Husain *et al.*, 1985).

(vii) *Okra*: Soil fumigation usually decreased disease severity due to *R. solani* and *M. incognita* disease complex in okra (Golden and Van Gundy, 1975).

(viii) *Radish*: Khan and Muller (1982) used gnotobiotic culture to investigate interactions between *R. solani* and *M. hapla* on radish and showed that prior infection of roots by the nematode helped the fungus colonization of roots.

3.1.2. Medicinal Crops

(i) *Coleus*: Simultaneous inoculation of *M. incognita* and *Macrophomina phaseolina* as well as nematode followed by fungus 15 days later, caused 100% root rot disease and significant reduction in plant growth parameters in *Coleus forskohlii* (Senthamarai *et al.*, 2006a).

3.1.3. Spice Crops

(i) *Ginger*: *M. incognita* had synergistic effect in rhizome rot/soft rot (*Pythium aphanidermatum)* of ginger, particularly when nematodes were inoculated to the plants prior to fungal inoculation. The damage to the crop was more and the onset of the disease was early when *M. incognita* was inoculated 40 days prior to fungal inoculation (Ramana *et al.*, 1999).

(ii) *Cardamom*: In cardamom, *M. incognita* is the predisposing factor for *Rhizoctonia solani* infection resulting in damping off and rhizome rot in nurseries (Ali and Venugopal, 1992, 1993).

3.2. CYST NEMATODE – ROOT ROT DISEASE INTERACTIONS

3.2.1. Vegetable Crops

(i) *Potato*: The potato cyst nematode, *Globodera rostochiensis* in association with the fungus *Rhizoctonia solani* caused root rot complex in potatoes.

(ii) *Tomato*: *Globodera rostochiensis* has been reported to interact with both *R. solani* and *Colletotrichum atramentarium* (the causal fungus of brown root rot) in tomato (Dunn, 1968; Dunn and Hughes, 1964).

3.3. BURROWING NEMATODE - ROOT ROT DISEASE INTERACTIONS

3.3.1. Plantation Crops

(i) *Areca nut*: *Cylindrocarpon obtusisporum* when introduced three weeks after *Radopholus similis* was responsible for extensive root lesioning (Sundararaju and Koshy, 1987).

3.4. STUNT NEMATODE - ROOT ROT DISEASE INTERACTIONS

3.4.1. Vegetable Crops

(i) *Cauliflower*: *Tylenchorhynchus brassicae* alone did not affect the percentage emergence of cauliflower seedlings. *R. solani* was highly destructive by itself, but the adverse effect on the emergence of seedlings was greater when nematode infestation occurred along with fungal infection (Khan *et al.*, 1971).

3.5. RENIFORM NEMATODE - ROOT ROT DISEASE INTERACTIONS

3.5.1. Vegetable Crops

(i) *Okra*: Okra plants succumbed to wilt at an early stage when the reniform nematode, *R. reniformis* and *R. solani* were present (Kumar and Sivakumar, 1981).

3.6. LESION NEMATODE - ROOT ROT DISEASE INTERACTIONS

3.6.1. Fruit Crops

(i) *Strawberry*: Black root-rot of strawberries, caused by *Rhizoctonia* spp. was exacerbated by *P. penetrans*, more so at higher nematode inoculum levels (LaMondia and Martin, 1989).

(ii) *Peach*: Nematodes of the genus *Pratylenchus* along with *Pythium* fungus has been associated with a large percentage of peach trees suffering from peach decline. *Criconemoides* spp. and *Tylenchorhynchus* spp. have also been associated with this condition (Hendrix *et al.*, 1965).

3.6.2. Vegetable Crops

(i) *Potato*: *Pratylenchus penetrans* increased the infection levels of *Colletotrichum coccodes* and *R. solani* on Russet Burbank potato roots (Kotcon *et al.*, 1985).

(ii) *Pea*: The minor pathogen, *Thielaviopsis basicola*, was assisted by *P. crenatus* in penetrating pea roots (Green *et al.*, 1983).

(iii) *Celery*: *P. penetrans* and *Trichoderma viride* caused more reduction in root and shoot growth in celery than either organisms alone (Edmunds and Mai, 1966).

3.6.3. Ornamental Crops

(i) *Chrysanthemum*: The presence of *P. coffeae* has been found to aggravate the disease severity of *Pythium aphanidermatum* and *Rhizoctonia solani* in chrysanthemum (Hasan, 1988).

3.7. RICE ROOT NEMATODE - ROOT ROT DISEASE INTERACTIONS

3.7.1. Tuber Crops

(i) *Taro*: A possible interaction occurred with *Hirschmanniella miticausa* and *Corticium solani* causing 'mitimiti' disease of taro, reported by Bridge *et al*. (1983).

3.8. STING NEMATODE - ROOT ROT DISEASE INTERACTIONS

3.8.1. Ornamental Crops

(i) *Chrysanthemum*: Chrysanthemum roots inoculated with *Belonolaimus longicaudatus* and *Pythium aphanidermatum* developed symptoms of *Pythium* root rot earlier and more extensively than those inoculated with the fungus alone (Littrell and Johnson, 1969).

3.9. RING NEMATODE - ROOT ROT DISEASE INTERACTIONS

3.9.1. Fruit Crops

(i) *Peach*: *Mesocriconemella xenoplax* was associated with *Fusarium* spp. and *Pythium* spp. on roots of peach trees in orchards with a history of peach tree short life, and may have been involved in an interaction (Nyczepir and Lewis, 1984).

3.10. FUNGIVOROUS NEMATODE - ROOT ROT DISEASE INTERACTIONS

3.10.1. Vegetable Crops

(i) *Cucumber*: Choi *et al*. (1988) exploited the capacity of mycophagous nematodes when they applied large numbers of *Aphelenchus avenae* to the soil in which cucumber seedlings were growing and effectively suppressed pre-emergence damping-off due to *Rhizoctonia solani*.

3.11. MECHANISMS OF INTERACTION

3.11.1. Nematodes Facilitate Access to the Root Surface

The nematodes interact with fungi in complex diseases by providing access to the root tissues through the wounds they cause. There are many reports of fungal pathogens growing in and extending the lesions caused by burrowing nematode or root lesion nematodes, and growing along the invasion tracks left by invading

Table 3.1: Nematode – Root rot disease interactions

Crop	*Nematode*	*Fungi*	*Reference(s)*
Banana	*Meloidogyne* spp.	*Rhizoctonia solani*	Sikora & Schlosser, 1973
	Radopholus similis	*Cylindrocarpon musae*	Booth & Stover, 1974
	R. similis	*Fusarium solani* *Rhizoctonia* spp.	Stover, 1966
	Helicotylenchus multicinctus	Unidentified	Orion *et al.*, 1999
Pineapple	*Meloidogyne* spp.	Various	Godfrey, 1936
Peach	*Mesocriconemella xenoplax*	Various	Nyczepir & Lewis, 1984
Strawberry	*Pratylenchus penetrans*	*R. fragariae*	LaMondia & Martin, 1989
Potato	*M. incognita*	*R. solani*	Sharma & Gill, 1979
	Globodera sp.	*R. solani*	Grainger & Clark, 1963 Back *et al.*, 2000
	Globodera sp.	*R. solani* *Polyscytalum pustulans*	Dunn & Hughes, 1967
	P. penetrans	*R. solani* *Colletotrichum coccodes*	Kotcon *et al.*, 1985
Tomato	*M. incognita*	*R. solani*	Chahal & Chhabra, 1984 Arya & Saxena, 1999
	M. incognita	*Sclerotium rolfsii*	Shukla & Swarup, 1970
	M. javanica	*Aspergillus niger* *Curvularia trifolii* *R. bataticola* *Rhizopus nigricans*	Nath & Kamalwanshi, 1989
Brinjal	*M. incognita*	*R. bataticola*	Chhabra & Sharma, 1981
	M. incognita	*R. solani*	Hazarika & Roy, 1974
Chilli	*M. incognita*	*Pythium aphanidermatum* /*R. solani*	Hasan, 1985
Okra	*M. incognita*	*R. solani*	Golden & Van Gundy, 1975 Chhabra *et al.*, 1977
	M. incognita	*R. bataticola*	Chhabra & Sharma, 1981 Sharma *et al.*, 1980
	R. reniformis	*R. solani*	Kumar & Sivakumar, 1981
Cauliflower	*Tylenchorhynchus brassicae*	*R. solani*	Khan *et al.*, 1971
French bean	*M. incognita*	*Macrophomina phaseolina*	Al-Hazmi, 1985
	M. incognita	*Rhizoctonia solani*	Parvatha Reddy *et al.*, 1979
Cowpea	*M. incognita*	*R. solani*	Varshney *et al.*, 1987
	M. incognita *R. reniformis*	*R. solani*	Khan & Husain, 1989

Table 3.1: (*Contd...*)

Table 3.1: (*Contd...*)

Crop	*Nematode*	*Fungi*	*Reference(s)*
Pea	*M. incognita*	*R. solani*	Husain *et al.*, 1985
	P. crenatus	*Thielaviopsis basicola*	Green *et al.*, 1983
Celery	*M. hapla*	*Pythium polymorphon*	Starr & Mai, 1976
Radish	*M. hapla*	*R. solani*	Khan & Muller, 1982
Chrysanth-emum	*M. incognita / B. longicaudatus*	*P. aphanidermatum*	Littrel & Johnson, 1969
	P. coffeae	*R. solani, P. aphanidermatum*	Hasan, 1988
Gerbera	*M. arenaria*	*Phytophthora cryptoge*	Schlang & Sikora, 1978
Tea	*P. loosi*	Soft root rot	Arulpragasam & Addaickan, 1983
Betel vine	*M. incognita*	*S. rolfsii*	Acharya *et al.*, 1987
	Meloidogyne spp.	*Phytophthora* spp.	Venkata Rao *et al.*, 1973
	R. reniformis	*Phytophthora palmivora*	Jonathan *et al.*, 1997
Solanum khasianum	*M. incognita*	*Corticium rolfsii*	Prasad *et al.*, 1980
Coleus forskohlii	*M. incognita*	*M. phaseolina*	Senthamarai *et al.*, 2006a
Areca nut	*R. similis*	*Cylindrocarpon obtussisporum*	Sundararaju & Koshy, 1982, 1987
Cardamom	*M. incognita*	*R. solani*	Ali & Venugopal, 1992
Ginger	*M. incognita*	*Pythium myriotylum*	Lanjewar & Shukla, 1985
		P. aphanidermaum	Ramana *et al.*, 1998
Yam	*Scutellonema bradys*	*Rhizoctonia solani*	Goodey, 1935
Taro	*Hirschmanniella miticausa*	*Corticium solani*	Bridge *et al.*, 1993

juveniles of cyst and root-knot nematodes. For instance, *Cylidrocarpon musac* is only common in the lesions formed in banana roots by *Radopholus similis* (Booth and Stover, 1974). The increased lateral root production following nematode invasion also provides entry sites for fungi. Nematodes with a sedentary life style, such as root-knot and cyst nematodes, have enlarged females which usually rupture the root cortex and thereby provide entry sites for fungi. Since it takes about three to four weeks from invasion for this to occur, this may at least partly explain why inoculation of nematodes some time before the fungus leads to maximum synergistic effect. In the case of root-knot nematodes, penetration through ruptures formed in the root by females may be facilitated by the fungus growing more extensively on galled rather than healthy root tissue, as occurred with *M. incognita* and *R. solani* on tomato roots (Golden and Van Gundy, 1975). As galled roots degenerate, they will also develop cracks around the knots which will provide invasion access for fungi.

3.11.2. Nematode or Fungus Infection Causes Physiological Changes in the Host

The nematode made the roots a better environment for fungal development, perhaps most simply by increasing the nutrient supply available. Such a mechanism seemed to operate in tomato roots attacked by *M. incognita*, where *R. solani* colonized the nutrient rich nematode feeding cells very heavily (Golden and Van Gundy, 1975). Probably the same mechanism may operate with *R. solani* and *M. hapla* on radish (Khan and Muller, 1982). Further evidence for physiological predisposition of plants to fungus attack by nematodes comes from Nath *et al.* (1984), who were unable to reproduce the effects of *M. javanica* on invasion of tomatoes by a range of fungi by mechanical wounding. The existence of a general, a systemic effect on susceptibility to this fungus was confirmed by El-Sherif and Elwakii (1991), who used split tomato root systems and inoculated one-half with *M. incognita* and the other with the fungus. An increase in susceptibility to fungal invasion was evident in the half of the root system free of nematodes. Thus, there is much evidence to indicate that physiological changes caused by nematodes increase plant susceptibility to fungi.

In the same way that nematodes can cause changes in the host to predispose it to fungal attack, fungi can predispose plants to nematode attack. Again there is much circumstantial evidence of increased nematode population densities compared with fungus-free controls (Varshney *et al.*, 1987).

3.11.3. Increased Host Susceptibility

Some species off *Pythium* (*P. aphanidermatum, P. ultimum*), *Rhizoctonia* (*R. solani, R. bataticola*), *Fusarium* (*F. solani*) and *Phytophthora parasitica*, *Sclerotium rolfsii* and *Colletotrichum coccodes* are prominent amongst the root-rot fungi that are known to interact with different plant parasitic nematodes. Like wilt-fungi, root-rot fungi are capable of causing root diseases on their own and have their own inherent mechanism of root penetration. The role of the nematode in root-rot diseases in general is related to assisting the fungal pathogen in its pathogenesis and increasing host susceptibility. Nematodes, by wounding, provide the fungal pathogen access to root tissue. The lesions caused by lesion or burrowing nematodes or invasion tracks formed by penetrating juveniles of root-knot or cyst nematodes provide a better substratum for establishment and colonization of the fungal pathogens (Booth and Stover, 1974). Further physiological alterations by the nematode improve the nutrient status of the host for the fungal pathogen. Golden and Van Gundy (1975) showed that galled regions of tomato roots infected with *M. incognita* were heavily colonized by *R. solani*, indicating its preference for the region due to the presence of rich nutrient medium in the galls. A similar pattern of colonization by *R. solani* was observed by Khan and Muller (1982) on radish roots infected with *M. hapla*. Therefore physiological alterations which ensure better nutrient availability to the penetrated fungal pathogen are the key factors of synergistic damage caused to the host.

3.11.4. Modifications in Host Substrate

Localized modifications: In interaction of root-knot nematode and *R. solani*, the localized substrate modification by root-knot nematode provided enriched medium for growth, establishment and further invasion of *R. solani*. The host substrate was found to be most favourable 3 to 4 weeks after nematode inoculation. Nutrient rich galls caused by root-knot nematodes favour luxuriant growth of *R. solani*. *R. solani* preferred *M. hapla*–induced galls on radish, the mycelium accumulated over them showing vigorous growth and abundant sclerotial formation. Extensive necrosis of the galls occurred and roots became obliterated (Khan and Muller, 1982). Galled roots of okra and tomato infected with *M. incognita* were highly susceptible to *R. solani*. Fungal sclerotia were formed only on gall tissues. Penetration studies indicated that the fungus was specifically attracted to nematode gall tissue and the sclerotia were selectively formed on the galls. *R. solani* responded to stimuli which originated from the nematode-infected roots and passed through semi-permeable cellophane membranes, by producing sclerotia on the membranes just opposite to the galls. The leakage of nutrients from the roots attracted the fungus to the galls and initiated sclerotial development (Golden and Van Gundy, 1975).

3.11.5. Nematode or Fungus Infection Causes Breakdown of Resistance in the Host

If a cultivar has been specifically produced to be resistant to a particular pathogen, it is important that the resistance be stable and not broken by another pathogen. Unfortunately, there are many examples in the literature of a fungus or nematode breaking crop cultivar resistance to the other. There are probably more examples of nematodes breaking resistance to fungal diseases probably because the cultivars with resistance to fungal disease were produced before nematode-resistant cultivars became available.

Vargas *et al.* (1996) observed a similar effect on chilli (*Capsicum annuum*), where the nematode *Nacobbus aberrans* caused a loss of resistance to *Phytophthora capsici* even if the nematode and oomycete were physically separated on split roots.

Hasan (1985) reported loss of resistance to *M. incognita* in two chilli cultivars infected with *R. solani* or *Pythium aphanidermatum*. A greenhouse experiment was devised to test the effect of *R. solani* and *P. aphanidermatum* on the resistance of chilli cvs Jwala (resistant) and Longthin Faizibadi (moderately resistant) to *M. incognita*. On both cultivars, the presence of *R. solani* or *P. aphanidermatum* caused a significant increase in the reproductive capacity (number of egg masses and eggs produced) of *M. incognita*. In terms of resistance, the ratings of Jwala and Longthin

Faizibadi were demoted to moderately resistant and susceptible, respectively (Hasan, 1985).

Hasan and Khan (1985) found that tomato cultivars could lose resistance to *M. incognita* in the presence of *R. solani* and *Sclerotium rolfsii*. Khan and Husain (1989) reported experiments with cultivars of cowpeas in which resistance to both nematodes and fungi was broken by the other pathogen.

Chapter 4

NEMATODE - MYCORRHIZAE INTERACTIONS

Plant-parasitic nematodes often encounter roots that have been transformed structurally and physiologically by mycorrhizal fungi. Obligate dependence on a host for reproduction and concomitant infections in feeder roots has resulted in an association between nematodes and Arbuscular mycorrhizal fungi (AMF). Thus, it is logical to test for interaction between these two groups of organisms in terms of their effects on plant growth or yield and in terms of each organism's effect on the other. The interaction of nematodes and these beneficial symbionts is unlike those between parasitic nematodes and pathogenic organisms.

4.1. EFFECT OF FUNGAL FEEDING NEMATODES ON MYCORRHIZAL FUNGI

Species in the genera *Aphelenchus, Aphelenchoides, Bursaphelenchus* and *Ditylenchus* are mycetophagous and potentially can reduce the growth of AM fungi. Grazing of hyphae external to the root may reduce the potential yield of mycorrhizal plants and such reductions may be induced by naturally occurring population densities of nematodes (Finlay, 1985; Ingham, 1988). Conceivably, even limited feeding can break the chain of nutrient translocation in linear stretches of hyphae.

Riffle (1971) found that *Aphelenchus cibolensis* fed and reproduced on 53 of 58 ectomycorrhizal species tested under culture conditions. Thirty-two fungal species had their radial growth reduced by more than 20% and aerial hyphal growth was reduced from 21% to 100% in 15 species. Sutherland and Fortin (1968) found that *A. avenae* reduced the radial growth of seven ectomycorrhizal species *in vitro*. In another *in vitro* study, *Aphelenchoides bicaudatus* severely decreased the growth of all five ericoid endomycorrhizal fungi tested (Shafer *et al.*, 1981).

Reproduction of *Aphelenchoides composticola* on *Agaricus bisporus* and *Glomus clarum* was associated with reduced yield.

4.2. EFFECT OF PLANT-PARASITIC NEMATODES ON MYCORRHIZAL FUNGI

Reduced root infection and sporulation of *Glomus fasciculatum* and *G. epigaeum* resulted from inoculation of *Heterodera cajani* on cowpea (Jain and Sethi, 1987). Mycorrhizal structures of *G. mosseae* developed normally in feeder roots of *Citrus limon* in the absence of *Tylenchulus semipenetrans*, but when the nematode was present vesicles were not formed (O'Bannon *et al.*, 1979). Vesicle formation and mycelial growth of *G. etunicatum* were reduced in mycorrhizal roots of *C. limon* infected by *Radopholus similis* when compared to AM fungi without nematodes (O'Bannon and Nemec, 1979). Cortical tissue destruction by *R. similis* presumably left fewer habitable sites for *G. etunicatum*. Similarly, Umesh *et al.* (1988) noted decreased colonization of banana by *G. fasciculatum* in roots infected by *R. similis* but there was no effect on sporulation.

An effect of nematodes on AM fungi or possibly a direct effect on mycorrhizae was suggested when increased AM fungi root infection followed treatment of soil with nematicides (Menge, 1982).

Apparently, AM fungi are sensitive to transformations of roots induced by sedentary nematodes (galls, syncytia and nurse cells) and, without doubt, are deterred from colonizing roots severely crippled by migratory endoparasites. Ectoparasitic nematodes might be expected to interact with AM fungi in a manner similar to migratory endoparasites. The first organism to arrive at a habitable site on the root would appear to have some advantage; moreover, population density will play a key role in the relationship.

4.3. EFFECT OF MYCORRHIZAL FUNGI ON PLANT-PARASITIC NEMATODES

Numbers of *R. similis* in banana roots were significantly lower when the endomycorrhizal fungus *G. fasciculatum* was applied simultaneously or seven days prior to nematode inoculation (Umesh *et al.*, 1988). No significant differences were observed in juvenile penetration of *M. incognita* between tomato seedlings infected *G. fasciculatum* and non-mycorrhizal seedlings (Suresh *et al.*, 1985). The time required for inoculated second-stage juveniles of *M. hapla* to mature to adults was greater by 1000 degree hours (base = 9°C) in AMF than in non-mycorrhizal onion plants supplemented with phosphorus (MacGuidwin *et al.*, 1985).

Sitaramaiah and Sikora (1982) found that maturation of females of *Rotylenchulus reniformis* was delayed and the nematode reproduced less on endomycorrhizal tomato roots (symbiont: *G. fasciculatum*) compared to non-colonized roots. Significantly reduced galling and populations of *M. incognita* were found when *Piper nigrum* plants were inoculated with *G. fasciculatum* or *G. etunicatum* (Sivaprasad *et al.*, 1990). Mycorrhizae induced by *G. fasciculatum* on cowpea decreased cyst production and reproduction of *Heterodera cajani*, but *G. epigaeum* tended to exhibit a reverse trend (Jain and Sethi, 1987).

Zambolin and Oliveira (1986) studied the interaction between *G. etunicatum* and *M. javanica* on bean grown at four soil P levels. At added P rates of 0, 50, 100 and 200 mg/kg soil, *M. javanica* eggs/plant were 218, 129, 36 and 55 × 10^3, respectively, on mycorrhizal plants and 118, 204, 235 and 51 × 10^3, respectively, on non-mycorrhizal plants. With no added P, the approximately two-fold increase in eggs/plant on mycorrhizal plants compared with non-mycorrhizal plants.

Biochemical alteration of the plant root has been hypothesized as one explanation of reduced nematode infection (Smith, 1987). Phenolic compounds have long been thought to play a role in disease resistance (Goodman *et al.*, 1967) and they have been shown to be formed after infection by AM fungi (Sylvia and Sinclair, 1983). Morandi (1987) found that populations of *M. javanica* were significantly reduced in roots of tomato by pre-inoculation with *G. fasciculatum*. In addition, increases in lignin and phenols in mycorrhizal roots were associated with reduced reproduction of *M. javanica* on tomato (Singh *et al.*, 1990) and *Radopholus similis* on banana (Umesh *et al.*, 1988).

4.4. EFFECTS OF NEMATODE-MYCORRHIZAL INTERACTIONS ON PLANT GROWTH

Generally speaking, plants with AMF and nematode parasites yield less than mycorrhizal plants without nematodes and more than non-mycorrhizal plants with nematodes. The interaction between *G. intraradices, M. incognita* and cantaloupe (*Cucumis melo*) was studied at three P levels (Heald *et al.*, 1989). In soil amended with 50 µg/g soil, *M. incognita* suppressed the growth of non-mycorrhizal plants by 84% compared to the control but growth of mycorrhizal plants inoculated with *M. incognita* was retarded by only 21%. A similar trend occurred in plants grown in soil with 100 µg/g soil and plant growth was reduced if no P was added to soil. Smith and Kaplan (1988), working with *Radopholus citrophilus, G. intraradices* and *C. limon*, found increased plant growth that appeared to have resulted from improved P nutrition rather than antagonism between fungus and nematodes. Kotcon *et al.* (1985) reported that growth of onion 'Downing Yellow Globe', inoculated with *G. fasciculatum* in soil infested with 500 or 3,000 *M. hapla* eggs/100 cm^3, was not suppressed even though final population densities of *M. hapla* were greater on mycorrhizal plants. On the other hand, MacGuidwin *et al.* (1985) reported that growth of onion 'Krummery Special' in soil infested with 667 *M. hapla* eggs/100 cm^3 caused some plant death, and inoculation with *G. fasciculatum* did not increase host tolerance or affect nematode infection, development, or reproduction.

4.5. MECHANISMS OF INTERACTION

Root colonization by mycorrhizal fungi and their sporulation are affected by plant-parasitic nematodes. At the same time, development and reproduction of the nematodes are influenced by AM fungi. Direct interaction also occurs between sedentary females of endoparasitic nematodes and AM fungi. Some nematodes

Table 4.1: Interaction of AM fungi and phosphorus with plant-parasitic nematodes and their effect on host response and nematode infection, development and reproduction

Host	*Fungus*	*Nematode*	*AMF/ P effect*	*Interaction component*				*Reference(s)*
				Host	*Infection*	*Devp*	*Reprdn*	
Bean	*G. etunicatum*	*M. javanica*	AMF	+	nd	nd	–	Zambolin &
			P	+	nd	nd	+, –	Oliveira, 1986
Onion	*G. fasciculatum*	*M. hapla*	AMF	+	nd	nd	+	Kotcon *et al.*, 1985
			P	nd	nd	nd	nd	
	G. fasciculatum	*M. hapla*	AMF	0	0	0	0	MacGuidwin *et al.*,
			P	nd	0	0	0	1985
Potato	*G. etunicatum*	*G. rostochiensis*	AMF	+	+	nd	–	Welsh & Sikora,
			P	nd	nd	nd	nd	1987
Tomato	*G. margarita, G. mosseae*	*M. incognita*	AMF	0	0	0	0	Thomson *et al.*,
			P	+	+	0	+,0	1983
	G. fasciculatum	*R. reniformis*	AMF	+,0	-	-	–,0	Sitaramaiah &
			P	nd	nd	nd	nd	Sikora, 1982
	G. margarita, G. mosseae, G. fasciculatum, G. tenue	*M. hapla*	AMF	+	+,–	–	nd	Cooper &
			P	+	+	+	nd	Grandison, 1986

+ = Increase, 0 = No change, – = Decrease, nd = Not determined

feed upon mycorrhizal fungi. As a result of such interactions, plant growth may be increased or decreased, or there may not be any discernible effect on it.

4.5.1. Inhibition of Mycorrhizal Root Colonization by Nematodes and Decrease in Plant Growth

Destruction of root tissue by migratory endoparasitic nematodes and various transformations induced by sedentary endoparasitic nematodes greatly reduce the root colonization by AM fungi. Mechanisms of inhibition of root colonization by nematodes are apparently physical alterations of root tissues and biochemical changes within the roots.

4.5.2. Inhibition of Nematode Parasitism and Increase in Plant Growth

AM fungi increase host tolerance to plant parasitic nematodes. Mechanisms of adverse effects of AM fungi on nematode parasitism are also physical and biochemical. AM fungi indirectly affect host-parasite relationship of nematodes by influencing root growth, root exudation, nutrient absorption and physiological response of the hosts (Gerdemann, 1968; Hayman, 1982; Harley and Smith, 1983). Some mycorrhizal fungi like *Glomus* sp. occur within the cysts of *Heterodera* and *Globodera* species (Tribe, 1977). The invaded cysts contain chlamydospores of the fungus instead of eggs (Tribe, 1977). These capabilities of the AM fungi may be additionally responsible for inhibition of such nematodes.

Mechanisms that could also account for the protective activity ascribed to AM fungi include enhanced plant nutrition, root damage compensation, competition for nutrients and infection site, competition for host photosynthates, anatomical or morphological changes in the root system, microbial changes in the mycorhizosphere, and inactivation of plant defense mechanisms.

4.5.3. Enhanced Plant Nutritional Status

The obvious AMF contribution to reduction of root nematodes is to increase nutrient uptake, particularly P and other minerals, because AM symbiosis results in more vigorous plants; the plant itself may thus be more resistant or tolerant to nematode attacks (Linderman, 1994). In addition to P, AMF have been shown to enhance the uptake of K, Ca, Cu, Zn and Mn; increase the sugar, amino acids (phenylalanine, serine) and chlorophyll synthesis and water uptake. The evidence to support the improved nutrition idea comes from experiments where effects comparable to AM effects were observed when more P fertilizer was added.

4.5.4 Production of Nematostatic Compounds

AMF may inhibit nematode activities by altering the root growth, root exudation, nutrient absorption, root physiology in ways unrelated to P nutrition and enhanced root growth (Gerdemann, 1968; Hayman, 1982; Harley and Smith, 1983). These changes could alter chemotactic attraction of nematodes to roots, affecting those species which require a hatching stimulus, or directly retard nematode development within root tissues.

4.5.5. Competition for Host Photosynthates

Both organisms utilize host photosynthates and in their combined demands nematodes may be inhibited because nematode activities are negatively affected by overcrowding or lack of infection sites on jointly infected plants (Smith, 1987). AMF develop intensively inside roots and within the soil by forming an extensive extra radical network, and this helps plants noticeably in exploiting mineral nutrients and water from the soil. P is the key element obtained by plants through the symbiosis. In exchange, mycorrhizal plants provide the AM fungus with photosynthetic C which is recycled and delivered to the soil via AM fungal hyphae. The extra radical hyphae of AMF, therefore, act as a direct conduit for host C into soil and contribute directly to its C pools, bypassing the decomposition process (Jeffries *et al.*, 2003). As a consequence of this, the amount and activity of other soil biota are stimulated; particularly the microorganisms having antagonistic activity against soil-borne nematodes (Linderman, 2000). AMF almost totally depend on soluble carbohydrates produced by the host for their carbon source. One of the mechanisms affecting the nematode activities in a mycorrhizal system is the conversion of carbohydrates received from the host into forms that can not be used by the nematodes. P in the fungus may restrict the availability of P for nematode development and reproduction.

4.5.6. Competition for Space or Infection Site

Direct competition between AM fungi and endoparasitic nematodes for space may result in inhibition of the nematode. Both localized and nonlocalized mechanisms could be responsible for competition for infection sites. Because AMF are soil-borne fungal pathogens and plant parasitic nematodes occupy similar root tissues, direct competition for space has been postulated as a mechanism of pathogen inhibition by AMF. When AMF are reported to suppress nematode disease, they generally must be established and functioning before invasion by nematodes.

4.5.7. Root Damage Compensation

AMF through an increase in P nutrition, enhance root growth and expand the absorptive capacity of the root system for nutrients and water. This compensatory process induced by AMF may explain the increased tolerance of mycorrhizal and P fertilized plant. It has been suggested that AM fungi increase host tolerance to nematode attack by compensating for the loss of root biomass or function caused by soil-borne nematode pathogens (Linderman, 1994). This represents an indirect contribution to biocontrol through the conservation of root system function both by AM fungal hyphae growing out into soil and increasing the absorbing surface of the roots, and by the maintenance of root-cell activity through arbuscle formation.

4.5.8. Anatomical and Morphological Changes in Root Systems

It has been demonstrated that AM colonization induces remarkable changes in root system morphology as well as the meristematic and nuclear activities of root cells. Lignification of roots prevents penetration of mycorrhizal plants from nematodes. Increased lignification of cells in the endodermis of mycorrhizal tomato and cucumber plants has been observed. This might affect rhizosphere interactions, particularly nematode infection and development. The most frequent consequence of AM colonization is an increase in branching, resulting in a relatively larger proportion of higher-order roots in the root system.

4.5.9. Changes in Chemical Constituents of Plant Tissues

Structural and biochemical changes have been observed in the cell walls of the plants colonized by AM fungi. Enhanced lignification of endosperm cell walls and vascular tissues can be observed. Increased concentrations of chitinase and arginine accumulation have been reported in AM roots. Synthesis of β-1, 3-glucans has been reported in mycorrhizal pea-root, and also been detected within the structural host-wall material around the point of penetration of AM hyphae into plant cell.

It is most likely that more than one mechanism is responsible for the role of AM fungi as biocontrol agents (Mukherji, 1999). Triggering of a specific plant defense response due to AM colonization can be a most obvious basis for the protective capacity of AM fungi. Changes in hormones, amino acids and permeability of cell membrane occur in roots due to mycorrhizal symbiosis (Hayman, 1982). According to Smith (1983) possibly a factor(s) inherent to mycorrhizal symbiosis transforms mycorrhizal roots unfavourable to nematodes. He speculated about the possibility of production or increase of compounds with nematostatic properties in mycorrhizal roots. Some reports indicate that formation of phenolic compounds and phytoalexins in mycorrhizal roots adversely affect nematode pathogenesis (Morandi, 1987; Smith, 1987).

Among the compounds involved in plant defense (Bowles, 1990) in relation to AM formation are phytoalexins, enzymes of the phenylpropanoid pathway, chitinases, β-1, 3-glucanases, peroxidases, pathogenesis-related (PR) proteins, callose, hydroxyproline-rich glycoproteins and phenolics.

Chapter 5

INTERACTION OF NEMATODES WITH VIRUSES

The demonstration by Hewitt *et al.* (1958) that certain nematodes are vectors of plant viruses initiated research in Nematology and Virology that resulted in understanding of the transmission and etiology of an important group of soil-borne plant virus diseases. The first reliable proof of transmission of a plant virus by a plant-parasitic nematode was given when Hewitt *et al.* (1958) succeeded in transmitting grapevine fanleaf virus by *Xiphinema index*. This discovery incited an intensive search for nematode vectors of other soil-borne viruses.

Species of *Xiphinema, Longidorus, Paralongidorus, Trichodorus* and *Paratrichodorus* transmit a number of plant viruses. Nematode transmitted viruses belong to two groups, NEPO- and TOBRA viruses. NEPO is an acronym of Nematode-transmitted Polyhedral, and TOBRA of TOBacco RAttle. *Xiphinema, Longidorus* and *Paralongidorus* transmit NEPO viruses, while *Trichodorus* and *Paratrichodorus* transmit TOBRA viruses. Nematode transmission of viruses is non-circulative. 7% of the nearly 550 vector transmitted virus species recorded so far are disseminated by nematodes (Astier *et al.*, 2001).

Viruses are categorized into over 30 distinct groups which are classified as families and genera. They include viruses which consist of double-stranded DNA, single-stranded DNA, double-stranded RNA and single-stranded RNA. Two single-stranded RNA virus genera, Nepovirus (NEPO) and Tobravirus (TOBRA), have nematode vectors. Nepoviruses are in the Comoviridae family while Tobraviruses are not yet assigned to a family. Nepoviruses and Tobraviruses have nematode vectors (Lamberti and Roca, 1987). Nepoviruses are isometric (polyhedral) particles around 30 nm in diameter (1 nm=10^{-9} m, 1 μm=10^{-6} m). The only known nematode vectors are in the genera *Xiphinema, Longidorus* and *Paralongidorus*. Tobraviruses are straight tubular particles with two size ranges, 180-210 nm and 45-115 nm. *Trichodorus* and *Paratrichodorus* are vectors.

5.1. NEMATODES AS VECTORS OF PLANT VIRUSES

A biologically unequivocal role of nematodes as vectors has been ascertained only for nepoviruses/longidorids and tobraviruses/trichodorids. Longidorid nematodes are characterized by an axial stylet measuring up to 200 µm long consisting of an elongated spear (odonostyle) plus an extension (odontophore) usually half as long as the odonostyle; by a typical dorylaimoid esophagus with the stylet connected to the cylindrical, muscular and glandular bulb by a slender food canal; and by their 1.5 – 13.0 mm long slender body. The tail is usually short, hemispheroid or conoid. Only some members of the genera *Xiphinema, Longidorus* and *Paralongidorus* are virus vectors (Fig. 5.1). Species of *Longidorus* are characterized by a single guiding ring around the anterior part of the odontostyle; by a simple junction between odontostyle and odontophore; and by an odontophore usually without basal flanges. In *Xiphinema* the guiding sheath is around the posterior part of the odontostyle. The base of the odontostyle is forked at the junction with the odontophore and the odontophore has prominent basal flanges.

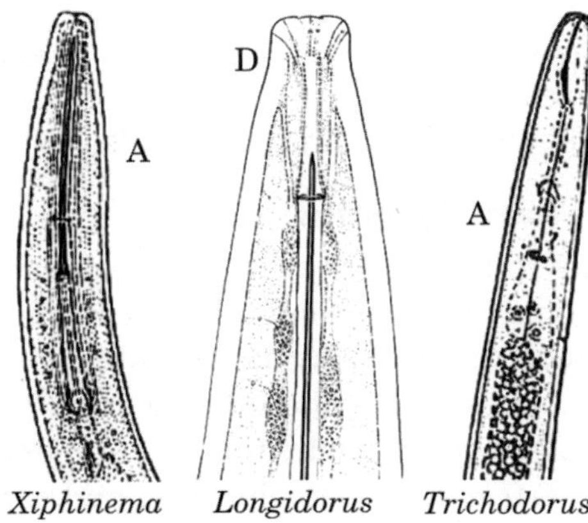

Fig. 5.1: Genera of nematode vectors of plant viruses

Trichodorid nematodes are readily recognized by their ventrally curved mouth spear (onchiostyle), which is an elongated dorsal tooth without a lumen and without basal knobs or flanges. The esophagus consists of a narrow anterior part swelling posteriorly to form an elongate or pear-shaped basal bulb. Virus vectors occur in the genera *Trichodorus* and *Paratrichodorus*. Both are similar in appearance, the most evident difference being the abnormally strong swelling of the cuticle in *Paratrichodorus* when heat-relaxed or fixed in acid fixatives. Furthermore, males of *Paratrichodorus* have a bursa, which is absent in *Trichodorus*.

5.1.1. Viruses Transmitted by Nematodes

The plant viruses transmitted by nematodes can be differentiated on the basis of particle morphology and relationship with vectors into two groups, nepo- and tobraviruses (Harrison *et al.*, 1971). All nepoviruses have isodiametric particles with icosahedral symmetry and are 20-30 nm in diameter. The genome is bipartite, with two functional ribonucleic acids (RNA-1 and RNA-2), separately encapsidated. Common biological properties are host responses and seed/pollen transmission.

Only three viruses, namely tobacco rattle, pea early-browning and pepper ringspot tobraviruses are transmitted by trichodorid nematodes. They have rigid, rod-shaped particles of three different lengths: very short, ± 45 nm; short, 50-100 nm; and long, 185-200 nm. The bipartite genome has two functional RNA species; a larger RNA-1 and a smaller RNA-2. The short particles encapsidate one molecule of RNA-2, the long particles one molecule of RNA-1. The three tobraviruses are distinguished genetically, *i.e.* by the sequence homology of RNA-1 (Robinson and Harrison, 1985).

5.1.2. Vector Specificity

- 11 spp of *Xiphinema* transmit 13 NEPO viruses.
- 11 spp of *Longidorus* transmit 10 NEPO viruses.
- 14 spp of *Trichodorus* transmit various strains of two TOBRA viruses: tobacco rattle and pea early browning.

Lamberti (1987) suggests that since several trichodorid spp transmit the same virus, and that both viruses are transmitted by the same nematode, vector specificity is less developed in trichodorids than in longidorids.

5.1.3. Transmission Characteristics

- Trichodorids may retain the virus for up to a year.
- Acquisition time may be less than an hour to several days, depending on the feeding characteristics of the nematode.
- Retention sites: *Longidorus* - odontostyle area
Xiphinema - odontophore and esophagus region
Trichodorus - onchiostyle and esophagus.
- Selectively and specifically adsorbed at retention site - indicating a specific association between protein coat of virus and cuticular surface.
- Dissociation with cuticle probably occurs as glandular secretions pass forward into the plant cell. Virus does not pass through the egg stage.
- Virus does not replicate in the nematode.
- Virus is not retained through a molt.

In *Xiphinema* and *Trichodorus*, although the lining of the esophagus is not shed at a molt, it undergoes structural changes and virus particles may pass into the intestine. In *Longidorus* the stoma, odontostyle and guiding sheath are shed.

5.1.4. Nepo Viruses

- *X. americanum* (*sensu lato*) transmits tobacco ringspot and tomato ringspot.
- The virus is acquired within 24 hours of the initiation of feeding.
- Virus particles are transmitted by both adults and juveniles.
- *X. americanum* transmits the tomato ringspot virus strains: peach rosette mosaic virus, peach yellow bud mosaic virus, cherry rasp leaf virus, and grapevine yellow vein virus.
- Particles are carried in the lumen of esophagus.

Lamberti *et al*. consider that *X. americanum* is not a vector of tomato ringspot virus, although it is widely reported to be. They contend that the reports represent a mis-identification especially in California where they feel that the vector is *X. californicum*. However, Griesbach and Maggenti (1989) found a wide range of vectoring capabilities of tomato ringspot viruses with California populations of *X. americanum* sensu lato. On the basis of considerable overlap of morphological and morphometric characters among putative species, they synonymized *X. americanum* and *X. californicum* (Griesbach and Maggenti, 1990). That synonymy was later rejected (Robbins and Brown, 1991). The situation is not yet clearly resolved; studies on developmental biology and the application of molecular techniques are providing further information on the diversity within the whole *X. americanum* group (Halbrendt and Brown, 1993; Vrain, 1993).

Lamberti suggests that the same mis-identification occurs with the vector of Cherry Rasp Leaf virus in California. There is a need for the application of molecular biology techniques in this problem of diagnostics.

- *X. californicum* transmits prune brown line and *Prunus* stem pitting strains of tomato ringspot virus readily and the cherry leaf mottle strain rarely (Hoy, PhD thesis, UC, Davis).
- *X. diversicaudatum* transmits *Arabis* mosaic virus to variety of crops. Adults retain the virus for at least 8 months.
- *X. index* - grapevine fanleaf virus can be acquired in 5 to 15 min, persists up to 9 months when nematode not feeding.
- *Longidorus elongatus* transmits raspberry ringspot and tomato blackring viruses.

5.1.5. Tobra Viruses

Trichodorus viruliferus - transmits tobacco rattle and pea early browning virus. It is sometimes involved in Docking disorder in sugar beets in England. *Trichodorus similis* - transmits tobacco rattle virus. At least 14 species of Trichodoridae have been demonstrated to vector Tobra viruses.

Tobacco rattle virus (genus *Tobravirus*) is a multi-component virus with rod-shaped long (RNA-1) and short (RNA-2) particles (Mojtahedi *et al*., 2000).

- M-type (normal) viral isolates contain both RNA-1 and RNA-2; they code for coat protein synthesis and are readily transmitted mechanically and by trichodorid nematodes (Stevenson *et al*., 2001).
- NM-type viral isolates have only contained RNA-1. They can be sap-transmitted and move through the plant systemically. Since they do not code for coat protein synthesis are not transmitted by trichodorid nematodes (Dale *et al*., 2004).
- The multi-component nature of the particles results in considerable variation among isolates (Brown *et al*., 2000; Mojtahedi *et al*., 2001).

Corky ringspot is caused by tobacco rattle virus (Tobravirus), which is vectored by stubby-root nematodes (*Paratrichodorus* spp. and *Trichodorus* spp.). The condition is sometimes referred to as Spraing, a Scottish word meaning a bright streak or stripe (de Bokx, 1972).

Corky ringspot symptoms vary depending on virus strain, potato cultivar, and time of infection. Symptoms often include of brown necrotic rings, arcs, and diffuse spots which are considered quality defects and may result in after-harvest devaluation or rejection of either table or processed potatoes.

Table 5.1: Virus species transmitted by nematodes

Name	*Acronym*	*Genus*	*Family*
Arabis mosaic virus	ArMV	*Nepovirus*	*Comoviridae*
Grapevine fanleaf virus	GFLV	*Nepovirus*	*Comoviridae*
Strawberry latent ringspot virus	SLRV	*Sadwavirus*	Unassigned
Tobacco ringspot virus	TRSV	*Nepovirus*	*Comoviridae*
Tomato black ring virus	TBRV	*Nepovirus*	*Comoviridae*
Pepper ringspot virus	PepRSV	*Tobravirus*	Unassigned
Tobacco rattle virus	TRV	*Tobravirus*	Unassigned
Pea early browning virus	PEBV	*Tobravirus*	Unassigned

5.1.6. Nematode Feeding and Virus Acquisition

Both nematode groups acquire viruses when ingesting cytoplasm of virus-infected plants. Ingested virus particles are selectively and specifically adsorbed at the retention sites. Viruses not transmitted by the nematode, if present in the cell sap and ingested, are not retained at the sites. The time needed for successful acquisition is short, *e.g. Xiphinema index* has been shown to acquire grapevine fanleaf nepovirus in less than 5 min (Alfaro and Goheen, 1974). However, the number of successful transmissions increases with increase in access periods.

5.1.7. Virus Retention and Transmission

After the ingestion of virus particles during the feeding process, nematode vectors retain their corresponding viruses for some time. Generally *Xiphinema* and *Trichodorus* spp. are said to retain the viruses they transmit for longer periods (10-12 months) and most *Longidorus* vectors for shorter periods, usually shorter than three months. In general, one virus is transmitted by one nematode species.

Nepo- and tobraviruses are retained in their appropriate vectors at distinct sites. In *Longidorus* vectors the virus particles associate with the interior surface of the lumen of the odontostyle and with the guiding sheath. In *Xiphinema* they are adsorbed to the cuticular lining of the lumen of the odontostyle and of the esophagus, and in *Trichodorus* and *Paratrichodorus* they are attached to the cuticle lining the esophagus but not to the onchiostyle (Fig. 5.2). All surfaces serving as retention sites are shed during molting, together with the outer cuticle of the body. Consequently, the adhering virus particles are lost and cannot be transferred from one developmental stage to the following. Nematode transmission of viruses is non-circulative.

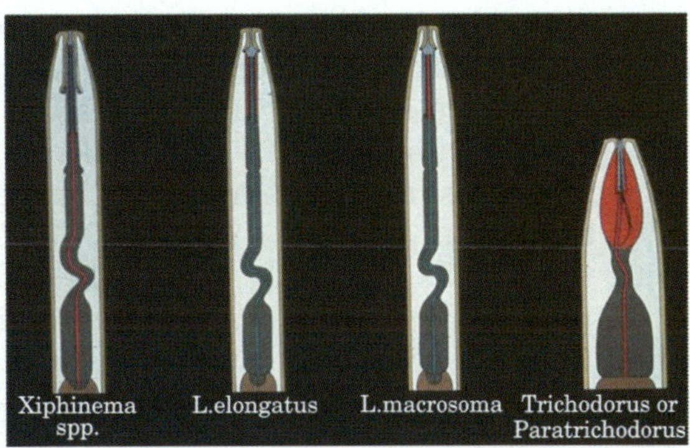

Fig. 5.2: Esophageal region of dorylaimoid nematodes where virus particles are retained (dark region)

After the specific retention, the dissociation of virus particles from their sites of retention is a prerequisite for successful transmission. It occurs when nematode saliva passes from the basal bulb anteriorly through the esophagus and the stylet into the plant cell during the feeding process. The saliva modifies the pH within the lumen and alters the surface charge of the virus particles (Martelli and Taylor, 1989). The modified lumen pH, by altering the surface charge of the virus particles, brings about their dissociation from the retention site (Taylor and Robertson, 1977; Taylor and Brown, 1981). This leads to a detachment of a number of particles. The long retention periods observed show that not all particles adsorbed to the retention site are released at the same time, *i.e.* during a single feeding act.

5.1.8. Determinants of Noncirculative Semipersistent Viruses

Previous work with pseudorecombinant isolates indicated that RNA2 carries the determinants for transmission of the nematode-borne nepoviruses (Harrison *et al.*, 1974; Harrison and Murant, 1978) and tobraviruses (Ploeg *et al.*, 1993). These results were confirmed by the use of full-length infectious chimeric cDNA clones of *Pea early browning virus* (PEBV) (Brown *et al.*, 1995a; MacFarlane and Brown, 1995). For nepoviruses, the CP seems to be the sole determinant of transmission specificity, as shown for GFLV (Belin *et al.*, 2001; Andret- Link *et al.*, 2004). The retention of virions at specific sites on the walls of the nematode's food canal could explain the specificity of transmission (Belin *et al.*, 2001). Homology modeling of the GFLV CP subunits and capsid structure with the 3.5 Å resolution crystal structure of *Tobacco ringspot virus* (TRSV) (Chandrasekar and Johnson, 1997), the type member of the genus *Nepovirus*, was further used to examine the surface topography of the capsid structure. Putative sites of interaction with a vector receptor were identified in a depression at the bottom of pronounced protrusions near the icosahedral fivefold axis (Andret *et al.*, 2003). Interestingly, the receptor-binding site for poliovirus is also located in a deep invagination of the virus surface (Hogle, 2002). Mutagenesis experiments are ongoing in laboratory to validate the GFLV 3D model and determine the functional significance of the putative CP sites responsible for transmission specificity. For tobraviruses, one or two nonstructural proteins, in addition to the CP, are involved in transmission (MacFarlane *et al.*, 1995). The nonstructural proteins 2b and/or 2c may facilitate retention of virions in the nematode feeding apparatus in a manner similar to the HC-bridge of poty- and caulimoviruses (MacFarlane *et al.*, 2002). Protein 2b of *Tobacco rattle virus* (TRV) can interact with the CP, as shown by yeast-2-hybrid binding data, and removal of the CP C-terminal domain abolishes this interaction (Visser and Bol, 1999). Furthermore, immunogold localization studies show *in vivo* binding of the TRV and PEBV 2b proteins to virions (Vellios *et al.*, 2002).

5.1.9. Vector Ligands Involved in the Specificity of Virus Transmission

Vector Ligands that Could Act as Receptor: In the case of nematode vectors, the site of virus retention is the inner surface of the odontostyle in *Longidorus* species whereas it is the oesophageal tract, extending from the anterior end of the odontophore posteriorly into the oesophageal bulb, in *Xiphinema*, *Paratrichodorus*, and *Trichodorus* species. For *Xiphinema* and trichodorid nematodes, a discontinuous layer of carbohydrate-staining material is identified lining the oesophageal tract. In *X. diversicaudatum*, particles of *Arabis mosaic virus* (Ar-MV) and *Strawberry latent ringspot virus* (SLRV) are absorbed only where this layer accumulates. Consequently, virus retention in *Xiphinema* and trichodorids could involve interactions between carbohydrate moieties in the nematode's oesophageal tract and surface structures of the virus capsid. An alternative hypothesis is that surface charges on virions interact with oppositely charged areas associated with the cuticle lining the nematodes' feeding apparatus. Fundamental differences occur in the morphology of the cuticle associated with the odontostyle region in nematodes, which is similar to the external cuticle, compared to that of the oesophageal tract, which represents the internal cuticle. These two cuticle types have differential isoelectric points and labeling of the odontostyle region in *L. elongatus* with cationized ferritin reveals a strong negative charge associated with the surface of the odontostyle and the wall of the lumen. Therefore, particularly with *Longidorus* species, surface charges may determine virus retention. Studies on the accumulation of *Tomato black ring virus* (TBRV) show that, late in infection or after purification, the CP in virions is reduced in size. This change results from the removal of nine amino acids at the C-terminal end, suggesting that this region is exposed at the surface of the protein where it could be available for interaction with nematode surfaces during vector transmission. However, not all nepovirus CPs possess a protruding C terminal peptide (MacFarlane *et al.*, 2002).

The CP subunits of tobraviruses form a tight helical array with their N- and C-termininal located on the external surface of virions. Nuclear magnetic resonance studies of *Pepper ringspot virus* (PePRSV) reveal that the C terminal region of the CP is unstructured and presumably extends away from the surface of virions (MacFarlane *et al.*, 2002). This protruding flexible domain could be involved in the specific attachment of virions to sites of retention within nematode vectors. Electron microscopy studies show a gap of 5-7 nm between the surface of TRV particles and the cuticle lining the oesophageal lumen (MacFarlane *et al.*, 2002). This space is too large to be bridged by a CP C-terminal peptide of only 22 (TRV-PpK20) to 38 (PepRSV) amino acids. However, the observed gap might be bridged by one or more of the additional nonstructural (2b and 2c) proteins encoded by the tobravirus RNA2, and these helper proteins might link the C-terminal peptide with the carbohydrate-staining material lining the nematodes' oesophageal tract.

Table 5.2: Nepo viruses transmitted by *Xiphinema* species

Crop	***Nematode***	***Virus***	***Reference(s)***
Fruits, vegetables, ornamentals	*Xiphinema americanum sensu lato*	Tomato ringspot	Breece & Hart, 1959
Fruits	*X. americanum*	Cherry rasp leaf	Nyland *et al.*, 1969
Fruits	*X. americanum sensu stricto*	Peach rosette mosaic	Klos *et al.*, 1967
Fruits, vegetables, ornamentals	*X. californicum*	Tomato ringspot	Hoy *et al.*, 1984
Fruits, vegetables, ornamentals	*X. diversicaudatum*	Arabis mosaic	Jha & Posnette, 1959
Fruits, vegetables, ornamentals	*X. diversicaudatum*	Strawberry latent ringspot	Lister, 1964
Cherry	*X. diversicaudatum*	Cherry rasp leaf	Flegg, 1969
Raspberry	*X. diversicaudatum*	Raspberry ringspot	Valdez, 1972
Carnation	*X. diversicaudatum*	Carnation ringspot	Fritzsche & Schmelzer, 1967
Grapevine	*X. index*	Grapevine fan leaf	Hewitt *et al.*, 1958
Grapevine	*X. index*	Grapevine chrome mosaic	Mali *et al.*, 1975
Fruits, vegetables, ornamentals	*X. index*	Arabis mosaic	Fritzsche & Thiele, 1979
Grapevine	*X. italiae*	Grapevine fan leaf	Cohn *et al.*, 1970
Fruits, vegetables, ornamentals	*X. rivesi*	Tomato ringspot	Forer *et al.*, 1981
Fruits, vegetables, ornamentals	*X. coxi*	Arabis mosaic Cherry leaf roll	Fritzsche & Kegler, 1964

Table 5.2: (*Contd...*)

Table 5.2: (*Contd...*)

Crop	*Nematode*	*Virus*	*Reference(s)*
Strawberry	*X. coxi*	Strawberry latent ringspot	Putz & Stocky, 1970
Tomato	*X. brevicolle*	Tomato ringspot	Fritzsche & Kegler, 1968
Cherry	*X. vuittenezi*	Cherry leaf roll	Flegg, 1969
Cowpea	*X. basiri*	Cowpea mosaic	Caveness *et al.*, 1975
Fruits, vegetables, ornamentals	*X. bakeri*	Arabis mosaic	Iwaki & Komura, 1974
Cowpea	*X. ifacolum*	Cowpea mosaic	Caveness *et al.*, 1975

Table 5.3: Nepo viruses transmitted by *Longidorus* and *Paralongidorus* species

Crop	*Nematode*	*Virus*	*Reference(s)*
Vegetables	*Longidorus apulus*	Artichoke Italian latent (Italian strain)	Rana & Roca, 1973
Vegetables	*L. apulus*	Artichoke Italian latent (Apulian starin)	Roca *et al.*, 1975
Fruits, vegetables, ornamentals	*L. attenuatus*	Tomato blacking (English strain)	Harrison *et al.*, 1961
Fruits	*L. diadecturus*	Peach rosette mosaic	Allen *et al.*, 1984
Fruits	*L. elongatus*	Raspberry ringspot (Scottish strain)	Taylor, 1962
Sugar-beet	*L. elongatus*	Tomato blacking(Beet ring spot strain)	Harrison *et al.*, 1961
Fruits	*L. elongatus*	Peach rosette mosaic	Fritzsche, 1968
Cherry	*L. elongatus*	Cherry leaf roll	Jones *et al.*, 1981
Carnation	*L. elongatus*	Carnation ring spot	Fritzsche *et al.*, 1979
Vegetables	*L. fasciatus*	Artichoke Italian latent (Greek strain)	Roca *et al.*, 1982
Fruits	*L. macrosoma*	Raspberry ringspot (English strain)	Harrison, 1962
Fruits	*L. macrosoma*	Raspberry ringspot (Scottish strain)	Trudgill & Brown, 1978
Fruits	*L. macrosoma*	Cherry leaf roll	Jones *et al.*, 1981
Fruits	*L. macrosoma*	Pear stony pit	Kegler *et al.*, 1976
Fruits	*L. macrosoma*	*Prunus* necrotic ring spot Carnation ring spot	Fritzsche, 1968
Mulberry	*L. martini*	Mulberry ring spot	Yagita & Komura, 1972
Raspberry	*L. profundorum*	Raspberry ringspot	Fritzsche & Kegler, 1968

Table 5.3: (*Contd...*)

Table 5.3: (*Contd...*)

Crop	*Nematode*	*Virus*	*Reference(s)*
Fruits, vegetables, ornamentals	*L. caespiticola*	Arabis mosaic Raspberry ringspot	Valdez, 1972
Raspberry	*L. leptocephalus*	Raspberry ringspot	Valdez, 1972
Cherry	*L. leptocephalus*	Cherry leaf roll	Jones *et al.*, 1981
Fruits	*Paralongidorus maximus*	Strawberry latent ring spot, Arabis mosaic, Raspberry ringspot	McElroy *et al.*, 1977
Cherry	*P. maximus*	Cherry leaf roll	Jones *et al.*, 1981

Table 5.4: Tobra viruses transmitted by *Trichodorus* and *Paratrichodorus* species

Nematode	*Virus*	*Host range of virus*	*Reference(s)*
Trichodorus cylindricus	Tobacco rattle (Dutch)	Wide	Van Hoof, 1968
T. hooperi	Tobacco rattle	Wide	Alphey, 1973
T. primitivus	Tobacco rattle (English)	Wide	Harrison, 1961
T. primitivus	Tobacco rattle (German)	Wide	Sanger, 1961
T. primitivus	Tobacco rattle (Scottish)	Wide	Mowat & Taylor, 1962
T. primitivus	Pea early browning (English)	Leguminosae	Harrison, 1967
T. similis	Tobacco rattle (Dutch gladiolus notch-leaf)	Wide	Cremer & Kooistra, 1964
T. viruliferus	Tobacco rattle (Dutch)	Wide	Van Hoof, 1968
T. viruliferus	Pea early browning (English)	Leguminosae	Gibbs & Harrison, 1964
T. viruliferus	Pea early browning (Italian No. 6)	Leguminosae	Van Hoof *et al.*, 1966
P. allius	Tobacco rattle (Oregon)	Wide	Jensen & Allen, 1964
P. allius	Tobacco rattle (California)	Wide	Ayala & Allen, 1966; Ayala & Allen, 1968
P. anemones	Tobacco rattle	Wide	Kurppa *et al.*, 1979
P. anemones	Pea early browning (English)	Leguminosae	Harrison, 1967
P. minor	Tobacco rattle (Wisconsin)	Wide	Walkinshaw *et al.*, 1961 Ayala & Allen, 1966
P. minor	Tobacco rattle (Japan)	Wide	Komuro *et al.*, 1970
P. nanus	Tobacco rattle (English)	Wide	Cooper & Thomas, 1970
P. nanus	Tobacco rattle (Dutch)	Wide	Van Hoof, 1968
P. pachydermis	Tobacco rattle (Dutch)	Wide	Sol & Seinhorst, 1961
	Tobacco rattle (English)	Wide	Gibbs & Harrison, 1964
	Pea early browning (English)	Leguminosae	Gibbs & Harrison, 1964
	Pea early browning (Dutch)	Leguminosae	Van Hoof, 1962
P. porosus	Tobacco rattle (California)	Wide	Ayala & Allen, 1966

Table 5.4: (*Contd...*)

Table 5.4: (*Contd...*)

Nematode	*Virus*	*Host range of virus*	*Reference(s)*
P. teres	Tobacco rattle (English)	Wide	Van Hoof, 1968
P. teres	Tobacco rattle (Oregon)	Wide	Jensen *et al.*, 1974
P. teres	Pea early browning (Dutch)	Leguminosae	Van Hoof, 1962
P. christiei	Tobacco rattle (Wisconsin)	Wide	Walkinshaw *et al.*, 1961
P. christiei	Tobacco rattle (California)	Wide	Ayala & Allen, 1966
P. tunisiensis	Tobacco rattle	Wide	Roca & Rana, 1981

5.2. NEMATODE-VIRUS INTERACTIONS

The plant viruses have no ascertainable direct influence on carrier nematodes, vectors or non-vectors. Indirect effect via the host plant is well documented. The indirect host-mediated effects are based on changes in host plant metabolism caused by viruses and nematodes, respectively. Development and multiplication of nematodes can be enhanced or inhibited depending on the virus, nematode or plant species.

Tomato blackring nepovirus enhanced *Ditylenchus dipsaci* but suppressed *Aphelenchoides ritzemabosi* (Weischer, 1975). Root-knot nematode *Meloidogyne incognita* produced five to ten times more individuals on cardamom plants infected with 'katte' mosaic virus than on healthy plants (Ali, 1988). Similarly plants of *Zinnia elegans* infected with zinnia mosaic virus were better hosts for *M. incognita* than healthy plants (Jabri *et al.*, 1985). The population build up of both the ectoparasite *Tylenchorhynchus brassicae* and the semi-endoparasite *Rotylenchulus reniformis* on eggplant was promoted when plants were infected with brinjal mosaic virus (Naqvi and Alam, 1975). In *Solanum khasianum* the root-knot index was higher on plants inoculated with tobacco mosaic tobamovirus than on healthy plants (Ismail *et al.*, 1979). Inhibitory effects of *M. javanica* were observed in *Cucurbita pepo* infected with watermelon mosaic potyvirus. Virus infection retarded the establishment of these nematodes in the roots as compared with healthy plants (Huang and Chu, 1984).

Alam *et al.* (1990) observed antagonistic effects between tomato mosaic tobamovirus and *M. incognita*. When virus infection preceded nematode inoculations nematodes were suppressed, and when nematodes were the first agent the virus was inhibited. The changes caused by one pathogen were detrimental to the other. In two out of three tomato varieties, the egg production of *M. incognita* was significantly increased by the presence of tobacco mosaic tobamovirus but in the third it was not. Similarly arabis mosaic nepovirus inhibited *D. dipsaci* in petunias.

The concomitant presence of nematode and virus in most cases results in a synergistic effect that aggravates the plant damage considerably. The growth of tomato plants was significantly reduced in simultaneous inoculations with *M. incognita* and tobacco mosaic virus or when nematode preceded the virus inoculation (Goswami and Chenulu, 1974). In case of brinjal and bottle gourd, root-knot development has been enhanced by the presence of mosaic virus (Mahmood *et al.*, 1974).

Combined effect of *M. incognita* and mycoplasma-like organisms (little leaf) on growth of brinjal was more than their individual effects (Dhawan and Sethi, 1977).

Chapter 6

INTERACTION OF NEMATODES WITH BACTERIA

Some of the most striking complexes between nematodes and bacteria include those bacteria which induce wilting in the host as one of the primary symptoms. Bacterial wilt of several host plants, caused by *Ralstonia solanacearum*, is one of the world's best-known diseases of this type. It has been known for a long time that nematodes, present along with the bacterial pathogen, greatly increase the incidence of this disease (Pitcher, 1963).

6.1. ROOT-KNOT NEMATODE-BACTERIA INTERACTIONS

6.1.1. Fruit Crops

(i) *Peach*: Crown gall of peach, caused by *Agrobacterium tumefaciens*, is increased by high populations of *M. javanica* (Nigh, 1966).

6.1.2. Vegetable Crops

(i) *Tomato*: Pani and Das (1972) have reported the association of root-knot nematode with bacterial wilt of tomato.

M. incognita contributes to an increase in bacterial wilt development (Libman *et al.*, 1964). The reason given for increased wilt incidence in the presence of nematodes was the micro punctures made by the nematodes.

Napiere (1980) and Napiere and Quinio (1980) found that wilt disease development occurred earlier and with a higher mortality rate in both wilt-resistant and susceptible tomato cultivars grown in *R. solanacearum* and *M. incognita* – infested soil.

Bacterial canker of tomato induced by *Clavibacter* (*Corynebacterium*) *michiganense* was increased when the roots were infected with *Meloidogyne* spp. (De Moura *et al.*, 1975).

Brinjal: The root-knot nematode, *M. incognita*, present along with *Ralstonia solanacearum*, greatly increases the incidence of bacterial wilt of brinjal. The root-knot nematode is responsible for breaking bacterial wilt resistance in "Pusa Purple Cluster" cultivar of brinjal (Reddy *et al.*, 1979). *M. incognita* interacts with *R. solanacearum* in increasing the speed of development and severity of wilt disease of eggplant. Maximum disease occurred when the bacterium and nematode were inoculated simultaneously. Inoculation with *R. solanacearum* alone resulted in less severe disease. It was concluded that root-knot nematodes modify the plant tissues facilitating bacterial colonization (Nayar *et al.*, 1988).

6.1.3. Ornamental Crops

(i) *Carnation*: Stewart and Schindler (1956) studied wilting of carnation cutting infected with *Pseudomonas caryophylli* in association with *Meloidogyne* spp. The results indicated that root-wounding by *Meloidogyne* spp. increased the rate of wilting in the presence of bacterial pathogens.

(ii) *Gladiolus*: Root infection with *M. javanica* greatly increased the severity of gladiolus scab caused by *Pseudomonas marginata* (El-Goorani *et al.*, 1974). This study show the systemic nature of physiological changes brought about by root-knot nematodes in favouring infection of foliar bacterial plant pathogens.

6.1.4. Spice Crops

(i) *Ginger*: Bacterial wilt of ginger, caused by *R. solanacearum* is influenced by *M. incognita* (Samuel and Mathew, 1983).

6.2. CYST NEMATODE – BACTERIA INTERACTIONS

6.2.1. Vegetable Crops

(i) *Potato*: The potato cyst nematode, *Globodera rostochiensis* in association with the wilt bacterium, *R. solanacearum* causes wilt complex in potatoes. Jatala *et al.* (1990) reported the interactions between *R. solanacearum* and nematodes, especially *Meloidogyne*, *Pratylenchus* and *Globodera* spp. in potato in relation to breeding for combined resistance.

6.3. RENIFORM NEMATODE – BACTERIA INTERACTIONS

6.3.1. Fruit Crops

(i) *Grapevine*: Lele *et al.* (1978) described an interaction between *Rotylenchulus reniformis* and *A. tumefaciens* in grapevine. Initial damage by the reniform nematode facilitated entry, establishment and disease development.

6.4. LESION NEMATODE – BACTERIA INTERACTIONS

6.4.1. Ornamental Crops

(i) *Rose*: Serious losses have been recorded in southern California apparently caused by a new strain of *Agrobacterium rhizogenes* (cause hairy root of roses) in association with infestations of the lesion nematode, *P. vulnus* (Munnecke *et al.*, 1963).

6.5. SPIRAL NEMATODE – BACTERIA INTERACTIONS

6.5.1. Vegetable Crops

(i) *Tomato*: The spiral nematode, *Helicotylenchus nannus* contribute to an increase in bacterial wilt development (Libman *et al.*, 1964). The reason given for increased wilt incidence in the presence of nematodes was the micro punctures made by the nematodes.

6.5.2. Ornamental Crops

(i) *Carnation*: Stewart and Schindler (1956) studied wilting of carnation cuttings infected with *Pseudomonas caryophylli* in association with *Helicotylenchus nannus*. The results indicated that root-wounding by *H. nannus* increased the rate of wilting in the presence of bacterial pathogens.

6.6. RING NEMATODE – BACTERIA INTERACTIONS

6.6.1. Fruit Crops

(i) *Plum*: The susceptibility of young fresh plum trees to *Pseudomonas syringae* increased when they were infected with *Mesocriconemella xenoplax*.

6.7. FOLIAR NEMATODE – BACTERIA INTERACTIONS

6.7.1. Fruit Crops

(i) *Strawberry*: The cauliflower disorder in strawberry is a complex requiring the nematode, *Aphelenchoides ritzema-bosi* and the bacterium, *Rhodococcus fascians* for expression of the complete disease syndrome (Crosse and Pitcher, 1952; Pitcher and Crosse, 1958).

6.8. MECHANISMS OF INTERACTION

6.8.1. Nematodes as Vectors of Plant-Pathogenic Bacteria

Bacteria can be transmitted by nematodes externally on their body surface or internally within the alimentary canal. In this type of relationship, the role of nematodes is mainly in carrying the pathogen from soil to plant or from the outer plant tissue to meristematic tissue. Crosse and Pitcher (1952) and Pitcher and Crosse (1956, 1958) found that *Rhodococcus fascians* (=*Corynebacterium fascians*) was unable to reach the meristem of the strawberry plant without the aid of *Aphelenchoides ritzemabosi* or *A. fragariae.* Without the active participation of the invading nematodes, the bacteria can not reach the meristem and the plants remain healthy. This complex disease requires the active contribution of both organisms for full disease syndrome. The bacterium is capable of stimulation of

Table 6.1: Nematode – bacterial disease interactions

Crop/Disease	***Nematode***	***Bacterium***	***Reference(s)***
Grapevine – Crown gall	*Rotylenchulus reniformis*	*Agrobacterium tumefaciens*	Lele *et al.*, 1978
Peach – Crown gall	*Meloidogyne javanica*	*A. tumefaciens*	Nigh, 1966
	M. incognita	*A. radiobacter* var. *tumefaciens*	Zutra & Orion, 1982
Strawberry – Cauliflower disease	*Aphelenchoides ritzemabosi/A. fragariae*	*Rhodococcus fascians*	Crosse & Pitcher, 1952; Pitcher & Crosse, 1958
Potato – Wilt	*M. incognita acrita*	*Ralstonia solanacearum*	Jatala & Martin, 1977a, 1977b
	G. rostochiensis	*R. solanacearum*	Pani & Das, 1972
	G. pallida	*R. solanacearum*	Jensen, 1978
Tomato – Wilt	*M. incognita*	*R. solanacearum*	Napiere & Quinio, 1980
	M. hapla	*R. solanacearum*	Libmann *et al.*, 1964
	Meloidogyne spp.	*Clavibacter* spp.	Hunt *et al.*, 1971
	Helicotylenchus nannus	*R. solanacearum*	Libmann *et al.*,1964
Tomato – Canker	*Meloidogyne* spp.	*C. michaganense*	De Moura *et al.*, 1975
Eggplant – Wilt	*M. incognita*	*R. solanacearum*	Reddy *et al.*, 1979
	M. javanica	*R. solanacearum* biotype 3	Sitaramaiah & Sinha, 1984a, 1984b
Bell pepper – Wilt	*M. incognita*	*R. solanacearum*	Naik, 2004
Beans – Wilt	*M. incognita*	*Curtobacter flaccumfascians*	Schuster, 1959
Garlic – 'Cafe au lait' bacteriosis	*D. dipsaci*	*P. fluorescens*	Caubel & Samson, 1984
Roses – Hairy root	*Pratylenchus vulnus*	*A. rhizogenes*	Munnecke *et al.*, 1963
Carnation – Wilt	*Ditylenchus* spp.	*Pseudomonas caryophylli*	Schuster, 1959
	H. nannus	*P. caryophylli*	Stewart & Schindler, 1956
	Meloidogyne spp.	*P. caryophylli*	Stewart & Schindler, 1956
Gladiolus – Scab	*M. javanica*	*P. marginata*	El-Goorani *et al.*, 1974
Ginger – Wilt	*M. incognita*	*R. solanacearum*	Samuel & Mathew, 1983

dormant meristems if transported at the site. The nematode acts as a vector and efficient inoculant. Since the bacterium eventually dominates over the nematode, it is likely that the nematode directly stimulates bacterial growth either by providing a useful metabolite or modifying the host substrate (Pitcher, 1963).

The role of nematodes in these associations was considered to be restricted to localized transport of the pathogens (Pitcher, 1965, 1978). Undoubtedly, the transport is the essential step of the interactions. Adhesion-receptor theory has been proposed to explain the mechanism of adhesion of bacterial cells to nematode cuticle and the specificity between the bacterium and the nematode involved (Riley and McKay, 1990). As the disease complex does not develop in the absence of the nematode, further interaction between the bacterium and nematode to produce the disease syndrome is logically expected. The nature of this interaction may be physiological, which might be helpful in establishment of the bacterium at their respective infection sites.

Some nematodes have been implicated as disseminators of plant pathogenic bacteria, carrying them internally within the alimentary tract by ingesting bacterial cells.

6.8.2. Nematodes as Wound Agents

The role of wounds was considered to be of great importance to explain the nematode-bacteria complex. Feeding of nematodes causes physical damage to the host plant and also provides avenues for the quick, direct entry and passage of pathogenic bacteria, especially when the pathogen is not strong enough to break the mechanical barriers of the host. Continuous feeding of the nematode also exposes a greater surface area for colonization. The availability of necrotic cells on the root also provides quick establishment and from there the bacterial pathogens are able to invade surrounding healthy tissue. The nematode may also carry the bacterial pathogens externally to deeper tissues. Wounding of roots may have some indirect effects as it may increase the root exudation pattern and positively influence the rhizosphere microflora in such a way that the bacterial pathogens are stimulated.

Nematodes provide wounds to facilitate bacterial entry and modify the host tissue to enrich the substrate nutritionally to the advantage of plant bacteria. Plant bacteria enter their host plants mostly through wounds. Wounds need to be fresh (before healing occurs) and while leakage of cellular contents provides a congenial atmosphere of high humidity and good nutritive substrate for infection and colonization. Wilt-inducing bacterial pathogens depend mainly on wound penetration to establish an infection court (Goodman *et al.*, 1967). Wounds caused by ecto- and endo-parasitic nematodes on underground plant parts favour bacterial pathogens of plants. Interactions where nematodes were claimed to serve as wounding agents have been observed in carnations, bean and tomato (Stewert and Schindler, 1956; Schuster, 1959; Libman and Leach, 1962). Root wounding with *Meloidogyne* spp. and *Helicotylenchus nannus* increased the rate of wilting in the presence of *Pseudomonas caryophylli* in carnation (Stewert and Schindler, 1956). Both *M. hapla* and *H. nannus* increased bacterial wilt (*Ralstonia solanacearum*) severity in tomato (Libman *et al.*, 1964). The reason given for increased wilt incidence in the presence of nematodes was the micropunctures made by nematodes. The interaction of *Meloidogyne* spp. with *R. solanacearum* has been clearly demonstrated in potato (Feldmesser and Goth, 1970; Nirula and Paharia, 1970; Jatala and Martin, 1977), tomato (Davide, 1972; Jenkins, 1972; Napiiere, 1980; Sellam *et al.*, 1982; Thakur, 1985) and eggplant (Jenkins, 1972; Sitaramaiah and Sinha, 1984a, 1984b). Interaction of *M. incognita* with wilt bacterium *Curtobacterium flaccumfascians* in beans (Schuster, 1959) has also been reported. In all these studies, mechanical root injury-root wounding of plant cells by the plant parasitic nematodes-is an important predisposing factor for direct introduction, establishment and multiplication of bacterial pathogens into the host tissue.

6.8.3. Nematodes as Host Modification Agents to Increase Susceptibility

Endoparasitic nematodes such as *Meloidogyne* spp. become sedentary in their host, and usually induce surrounding cells to undergo profound changes in structure and physiology. Results of histochemical tests on root-knot nematode galls demonstrated that giant cell walls contain cellulose and pectin but no lignin, suberin, starch or ninhydrin-positive substances. However, giant cell protoplasm contains carbohydrates, fat, RNA and a large amount of protein. Starch disappears in galled tissues while cellulose, sugars, phosphorylated intermediates, keto acids, free amino acids, protein, nucleic acids, phosphorus and nitrogen increases when compared to healthy tissues. These findings, as well as results from histochemical studies, indicated that *Meloidogyne*-infected roots are very active metabolically (Krusberg, 1963; Sitaramaiah, 1989).

The galled plant tissues were reported to contain more auxin or indole compounds than healthy tissues. During the feeding process *Meloidogyne* spp. secrete a proteolytic enzyme which releases IAA or tryptophan (Endo, 1971; Bird, 1974). The tryptophan is metabolized to IAA, resulting in a hormonal imbalance which is not confined to only to the site of action but also affects other tissues. These hormonal imbalances and other physiological changes in the plants infected with nematodes as primary pathogen can influence positively the pathogenesis of bacterial pathogens. Due to a change in host physiology, the mechanism of inhibiting production of toxins by bacterial wilt pathogen may be destroyed, thus enabling the wilt bacteria to break down host resistance.

In the interactions involving nematodes and bacteria, disease severity was always greater when the plants were infected with nematodes several days or weeks prior to exposure to bacterial pathogen than when plants were infected with both organisms simultaneously (Sitaramaiah and Sinha, 1984a, b). This shows that although physical injury plays an important role, it may not completely explain the predisposition of plants to bacterial pathogens.

Roots infected with *Meloidogyne* spp. are nutritionally richer than non-infected roots. Root galls have been found to contain more amino acids, auxins, growth-promoting substances, RNA, DNA and phosphorus than healthy tissue. These substances promote growth and establishment of bacterial pathogens. Histological observations revealed the presence of bacteria-like inclusions within giant cells in those plants which received nematodes three to four weeks before inoculation with bacteria. Histological observations of eggplant roots inoculated with *M. javanica* two or three weeks before *R. solanacearum* biotype 3 inoculation showed extensive cavities of broken cortical and endodermal cells (Sitaramaiah and Sinha, 1984a). The nematode-inoculated roots contained pronounced hyperplastic and hypertrophic regions characterized by 'giant cells' predisposed to bacterial invasion and colonization.

Modified host substrate is also claimed to favour bacterial pathogens. *Meloidogyne*-modified tissue acts as a more favourable substrate. De Moura *et al.* (1975) showed that on tomato cultivars, susceptible and resistant to *Corynebacterium michganense*, the presence of *M. incognita* increased the bacterial canker only when the nematode was inoculated prior to the bacterium. This suggested that the physiological changes in the host substrate caused by the nematode is important to develop synergistic relationships.

6.8.4. Nematode Breaks Resistance to Bacterial Pathogens

Failure of some bacterial disease-resistant crop cultivars in the presence of plant parasitic nematodes has drawn the keen attention of both plant breeders and nematlogists to breed crop cultivars resistant to nematodes and also to discover the mechanism of resistance. The increase in growth of disease-resistant crop cultivars in fumigated soils over those in non-fumigated soils can be great, indicating clearly the importance of breeding cultivars resistant to nematodes. Resistance breaking to bacterial pathogens implies that there is a possibility that plant parasitic nematodes could be able to supply something which is obviously lacking in the resistant cultivar in the absence of nematode invasion, or the nematode could alter cell permeability, making a particular cultivar susceptible to bacterial colonization.

It has been observed in several instances that a particular cultivar resistant to some bacterial pathogens becomes susceptible in the presence of *Meloidogyne* spp. *Meloidogyne* spp. as primary pathogens can bring about physiological changes favouring the bacterial pathogens to overcome the resistance. Parvatha Reddy *el al.* (1979) reported that when the eggplant cultivar Pusa Purple Cluster (highly resistant to *R. solanacearum*) was inoculated with a combination of the bacterium and *M. incognita*, a greater number of plants wilted when the organisms were inoculated simultaneously than when one was inoculated before the other. Field resistance of potato to *R. solanacearum* was broken down when the plants were infected with *M. incognita acrita* (Jatala and Martin, 1977a, b).

Napiere (1980) and Napiere and Quinio (1980) found that wilt disease development occurred earlier and with a higher mortality rate in both wilt-resistant and susceptible tomato cultivars grown in *R. solanacearum* and *M. incognita*-

infested soil. The susceptibility of young fresh plum trees to *Pseudomonas syringae* sub sp. *insidiosum* increased when they were infected with *Mesocriconemella xenoplax*.

To obtain maximum benefit from wilt-resistant cultivars it has been suggested to control the nematodes responsible for breaking the resistance to the bacterial pathogens. Resistance to vascular wilts caused by bacteria lies in the endodermis which contains large quantities of substances like phenols, naphthols and anthrols which in later stages give rise to benzoquinones, naphthoquinones and anthroquinones, some of which are highly antibacterial (Goodman *et al.*, 1967). Invasion of root-knot nematodes results in pericycle origin of syncytia and galls, and the endodermis is not properly differentiated in the affected areas. The absence of endodermis removes the barrier against invasion by vascular wilt bacteria.

6.8.5. Nematode Root Infection Increases Foliar Bacterial Diseases

Root-knot and lesion nematode infections have a systemic effect which could lead to an increase in susceptibility of other plant parts not actually infested by the nematode. Nematode-bacterium interactions have also been reported in the case of foliar bacterial diseases. Root infection with *M. javanica* greatly increased the severity of gladiolus scab caused by *Pseudomonas marginata* (El-Goorani *et al.*, 1974). Bacterial canker of tomato induced by *Clavibacter michiganense* was also increased when the roots were infected with *Meloidogyne* spp. (De Moura *et al.*, 1975). These studies show the systemic nature of physiological changes brought about by root-knot nematodes in favouring infection of foliar bacterial plant pathogens.

Chapter 7

NEMATODE - RHIZOBIA INTERACTIONS

The association of rhizobia with plant nematodes in the rhizosphere, and the beneficial effect of rhizobial symbiosis on plant nutrition and growth, led to investigations into the potential role of nematode parasitism on nodulation, and consequently on symbiotic nitrogen fixation.

One of the biological factors affecting nodule formation or dysfunction of existing nodules is the presence of nematodes in the rhizosphere. Several plant-parasitic nematodes with different modes of parasitism cause a reduction in nodulation on leguminous plants. Semi-endoparasites – *Rotylenchulus reniformis* on cowpea; and endoparasites – *M. incognita* on cowpea and Wando pea, *Meloidogyne* spp. on peas, *Heterodera cajani* on cowpea – all inhibit nodulation.

7.1. ROOT-KNOT NEMATODE - RHIZOBIA INTERACTIONS

Barker and Hussey (1976) used high inoculum level of *M. incognita* to suppress nodulation of Wando pea. In French bean, the root-knot nematode, *M. incognita* reduced the number of *Rhizobium* nodules in the root system (Singh and Reddy, 1981). Nodules containing root-knot nematodes deteriorate more rapidly than those free of nematodes. Thus the concurrent nematode infection reduces the total benefit from the nitrogen-fixing bacteria.

Sharma (1984) observed greater reduction in number of nodules when both *M. incognita* and *Rhizobium* were inoculated together or when the nematode had already established before the inoculation of *Rhizobium* as compared when *Rhizobium* had established before the inocuclation of nematodes. He suggested that the cause of reduced nodulation was nutritional interference by the nematode infestation and overall reduction of root system. In a few cases, developing females were found in the nodular tissue. The nematode infection interfered with the symbiotic nitrogen fixation and reduced nitrogen content of shoot and root. Combined inoculation of *M. incognita* and *H. cajani* adversely affected the root nodulation and nitrogen fixation in cowpea (Sharma and Sethi, 1976). *M. incognita* reduced nitrogen content to a greater extent than *H.* cajani. The root-knot nematode also reduced the leghaemoglobin content of the cowpea root nodules (Sharma and Sethi, 1985).

Ali *et al.* (1981) studied the antagonistic interaction between *M. incognita* and *Rhizobium leguminosarum* on cowpea. They observed that *M. incognita* reduced nodulation and inhibited nitrogen fixation (by about 63%) in the nodular tissue. Infected nodules contained different developmental stages of the nematode, but the nematodes did not alter the structure of nodules. However, infected nodules deteriorated earlier than non-infected ones. The nematode inoculation prior to rhizobia resulted in maximum reduction in nodules.

Root-knot (*Meloidogyne* spp.) nematodes are recognized as a major constraint to bean farming, causing up to 60% yield losses in heavily infested fields. Infection by root-knot nematodes is also known to suppress nodulation and hence nitrogen fixation in leguminous plants.

Numbers of nodules on bean plants infected with *M. incognita* were significantly ($P \leq 0.05$) lower than in control. Reduction in nodulation ranged from 35.4% in line

Table 7.1: Numbers of nodules and percentage of effective nodules in *Meloidogyne*- infected and non-infected bean plants

Bean genotype	***Nodule numbers in***		***Effective nodules (%) in***	
	Inoculated[1]	***Control***[2]	***Inoculated***	***Control***
GLP-24	90.0	166.8	70.0	98.0*
GLP-1004	20.0	58.6	14.0	50.0
NOB	111.2	172.2	72.0	90.0
KK8	70.2	132.2	36.0	78.0
KK14	88.6	161.8	38.0	88.0
KK15	77.4	157.6	40.0	80.0
KK22	38.0	124.6	32.0	62.0
E1	42.6	92.8	20.0	74.0
E3	19.4	121.4	8.0	40.0
E4	0	54.0	0	76.0
M14	1.4	30.2	0	68.0
M24	45.4	99.8	22.0	62.0
M26	41.6	92.4	10.0	58.0
M28	25.4	39.2*	20.0	70.0
M29	27.6	59.4	46.0	74.0
M30	9.2	28.0	8.0	38.0
L31	32.2	116.6	36.0	70.0
L32	32.4	107.8	34.0	66.0
L40	4.6	40.0	12.0	44.0
L45	19.0	46.2	18.0	50.0
LSD ($P \leq 0.05$)	**19.9**	**24.7**	**15.2**	**12.6**

NOB to 100% in line E4. No nodules were observed on the roots of bean line E4 inoculated with nematodes as compared to more than 111 nodules on roots of bean line NOB. *Meloidogyne* infection caused significant ($P \leq 0.05$) reduction in proportions of effective nodules in the bean genotypes tested (Kimenju *et al.*, 1999) (Table 7.1).

The number of nodules was significantly ($P \leq 0.05$) higher in bean cv. GLP-24 plants inoculated with *Rhizobia* alone than in plants inoculated with combinations of *Rhizobia* and *M. incognita* (Table 7.2). Plants inoculated with nematodes at emergence and with *Rhizobia* 10 days after emergence (DAE) had the lowest number of nodules on their roots. The percentage of effective nodules was significantly ($P \leq 0.05$) higher in bean plants inoculated with *Rhizobia* alone than those inoculated with *Rhizobia* and *M. incognita* (Table 7.2). Plants inoculated with nematodes at emergence had lower ($P \leq 0.05$) proportions of functional nodules than those inoculated with nematodes 10 DAE (Kimenju *et al.*, 1999).

Table 7.2: Effect of *Meloidogyne* infection on bean nodulation and functioning of nodules in bean cv. GLP-24 plants[1]

Treatments	*Nodule number/plant*[2]	*Effective nodules (%)*
Rhizobium alone at emergence	96.5[a]	83.0[a]
Nematode + Rhizobium (both at emergence)	74.4[b]	48.8[c]
Nematode (at emergence) + Rhizobium (10 DAE)	54.2[c]	47.4[c]
Rhizobium (at emergence) + Nematodes (10 DAE)	78.2[b]	66.0[b]

DAE = Days after emergence, [1]Data are means of 10 replications. [2]Means followed by the same letter along the columns are not significantly ($P \leq 0.05$) different by least significant difference test.

Stimulation of nodule formation on leguminous plants by plant-parasitic nematodes has also been observed. *M. incognita* on pea was observed to stimulate nodule formation (Verdejo *et al.*, 1988). However, nodules on plants with increased nodulation were smaller than nodules formed on plants free of nematode.

The cause for reduced nodulation has been postulated. According to Masefield (1958), the nematode galls on the roots may affect nodulation by causing nutrient deficiency in host plants and by occupying space on the root system, a suggestion which was supported later by Malek and Jenkins (1964). A competition phenomenon between nematode juveniles and root nodule bacteria was also postulated as a cause of reduction (Epps and Chambers, 1962).

Reduction in number and size of nodules and early degeneration of nodules because of nematode infection caused reduced nitrogen fixation in legumes (Taha and Raski, 1969).

7.2. CYST NEMATODE - RHIZOBIA INTERACTIONS

Combined inoculation of *M. incognita* and *H. cajani* adversely affected the root nodulation and nitrogen fixation in cowpea (Sharma and Sethi, 1976). *M. incognita* reduced nitrogen content to a greater extent than *H.* cajani (Sharma and Sethi, 1985).

Oostenbrink (1966) noticed that pea plants infected with *Heterodera goettingiana* possessed few nodules and exhibited poor growth.

7.3. RENIFORM NEMATODE - RHIZOBIA INTERACTIONS

Inoculation of *Rotylenchulus reniformis* reduced number and weight of *Rhizobium leguminosarum* nodules, dry shoot weight of pea plants. The nematode infection also interfered with the symbiotic nitrogen fixation and reduced the nitrogen content of shoot. Nematode infection and multiplication was reduced significantly in the presence of *Rhizobium* (Table 7.3) (Vats and Dalal, 1988).

Table 7.3: Effect of *Rotylenchulus reniformis* and *Rhizobium leguminosarum* on growth attributes of pea and nematode multiplication

Treatment	*No. of nodules/ plant*	*Dry shoot wt (g)*	*N content of shoot (mg/plant)*	*No. of females/root system*	*No. of nemas/100 g soil*
Uninoculated check	30	0.995	22.636	0 (1.00)	0 (1.00)
R. reniformis	17	0.705	11.068	173 (17.68)	664 (25.78)
R. leguminosarum	58	1.840	47.380	0 (1.00)	0 (1.00)
R. reniformis + *R. leguminosarum*	35	1.510	38.882	114 (10.71)	479 (21.90)
CD (P = 0.05)	**10.52**	**0.216**	–	**(0.85)**	**(0.35)**

Figures in parentheses are $(n+1)^{1/2}$ values

Rotylenchulus reniformis infection may deplete root hairs through which rhizobial infection could take place (Taha and Kassab, 1980). Also, suppression of lateral root formation by *R. reniformis* may cause reduction in the number of sites for nodule initiation, since lateral roots bear their own sites (Oteifa and Salem, 1972).

7.4. LESION NEMATODE - RHIZOBIA INTERACTIONS

Romaniko (1958) observed that *Pratylenchus penetrans* parasitized the nodules of peas and beans. *P. globulicola* also caused early destruction of nodules in pea. Green (1984) reported that *P. thornei* inactivated nodules so that fully formed nodules lacked leghaemoglobin on pea plants.

7.5. STUBBY ROOT NEMATODE - RHIZOBIA INTERACTIONS

Devitalization of root tips by *Trichodorus christiei* and the resulting lack of formation of root hairs were primarily responsible for reduced nodulation in hairy vetch (Malek and Jenkins, 1964).

7.6. STEM AND BULB NEMATODE - RHIZOBIA INTERACTIONS

Green (1984) reported that *Ditylenchus dipsaci* inactivated nodules so that fully formed nodules lacked leghaemoglobin on pea plants.

Table 7.4: Effect of nematodes on nodulation and nitrogen fixation in Rhizobium–legume interactions

Leguminous host	*Rhizobium species*	*Nematode species*	*Effect*		*Reference(s)*
			Nodulation	*N_2-fixation*	
Cowpea	*Rhizobium leguminosarum*	*M. incognita*	–	–	Ali *et al*., 1981: Sharma & Sethi, 1976
		M. javanica	–		Taha & Kassab, 1980
		R. reniformis	–		Taha & Kassab, 1980
		H. cajani	–	–	Sharma & Sethi, 1976
Broad bean	*Rhizobium* sp.	*M. incognita*	–	–	Yousif, 1979
Pea	*Rhizobium leguminosarum*	*M. incognita*	–	–	Barker & Hussey, 1979; Yousif, 1979
		P. penetrans	–		Barker & Hussey, 1979
		B. longicaudatus	–		Barker & Hussey, 1979
		Acrobeloides buetschlii	–	–	Wescott & Barker, 1976
French bean	*Rhizobium leguminosarum*	*M. incognita*	–	–	Singh & Reddy, 1981

7.7. MECHANISMS OF INTERACTION

In general, the benefits obtained by the plants through symbiotic association between plant roots and rhizobia are adversely affected by the involvement of the nematode as a pathogen in the system. Plant parasitic nematodes affect the system of biological nitrogen fixation by rhizobia at various stages from its establishment to efficient functioning. Survival of root nodule bacteria in the rhizosphere and colonization in the rhizoplane are influenced by root exudates.

Alterations in root exudates in nematode-infected plants may influence the establishment of rhizobia on legumes.

Most investigators have reported that plant nematodes irrespective of mode of parasitism cause reduced nodulation on leguminous plants. Nutrient depletion (Masefield, 1958), competition between nematode juveniles and root nodule bacteria (Epps and Chambers, 1962), devitalization of root tips as in the case of *Trichodorus christiei* (Malek and Jenkins, 1964) and suppression of lateral root formation as in *Rotylenchulus reniformis* infection (Oteifa and Salem, 1972) were suggested as possible causes of reduced nodulation.

Nematodes also damage root nodules by direct invasion. Species of *Meloidogyne, Heterodera, Pratylenchus, Rotylenchulus reniformis* and *Belonolaimus longicaudatus* invade root nodules on legumes. *Meloidogyne* spp. induce histological changes in nodular tissue and giant cells develop inside the nodules. Nodules also develop on root galls induced by the nematode. Cyst nematodes induce development of syncytia in nodules. Infected nodules show abnormal histology and degenerate earlier than healthy ones. The reduction in size and number of nodules and early degeneration of nodules on nematode infected leguminous plants are considered as two possible reasons for adverse effects on the nitrogen fixing capability of the roots.

Nodules from nematode-infected plants showed a higher Lb*c*/Lb*a* ratio, which indicated that nodule development was impaired. Significant reduction in overall Lb content, however, suggested that nodules were senescent. Reduction in nitrogen-fixing efficiency results either from reduced number and size of nodules, or their impaired development or early senescence or all act jointly, and the overall effect is deleterious for the crop.

Chapter 8

NEMATODES AS VECTORS OF BACTERIAL PLANT PATHOGENS

Plant parasitic nematodes assist plant pathogenic bacteria as carriers (Table 8.1). Some foliar nematodes act as specific vectors of bacterial plant pathogens and an obligate etiological relationship is established between the two, resulting in manifestation of a new disease syndrome in host plants (Taylor, 1990). The cauliflower disease of strawberry is such an example.

Table 8.1: Nematode vectors of plant pathogenic bacteria

Nematode	*Bacterial plant pathogen*	*Host*	*Disease*	*Reference(s)*
Aphelenchoides ritzemabosi or *A. fragariae*	*Rhodococcus fascians*	Strawberry	Cauliflower	Crosse & Pitcher, 1952; Pitcher & Crossc, 1958
Ditylenchus dipsaci	*Pseudomonas fluorescens*	Garlic	'Café au lait' bacteriosis	Caubel & Samson, 1984
Globodera pallida	*Pseudomonas solanacearum*	Potato	Wilt	Jensen, 1978

8.1. CAULIFLOWER DISEASE OF STRAWBERRY

The disease has been reported to occur in the UK and other European countries as well as in North America. Ritzema-bos (1891) proposed the term 'cauliflower' to describe malformation of the inflorescence and shortening and thickening of the aerial parts of strawberry plants. He considered an ectoparasitic nematode, *Aphelenchoides fragariae* as the causal agent of the disease since this nematode was consistently found in the diseased plants. Studies by Lacey (1936, 1942) provided evidence that a bacterium, *Rhodococcus fascians* was also associated with the disease. Research by Crosse and Pitcher (1952) and Pitcher and Crosse (1958) provided ample evidence of the obligatory association of the two pathogens for full cauliflower syndrome. According to Pitcher (1965) cauliflower disease is a complex requiring the nematode and certain strains of the bacterium and both organisms actively participate in the development of the full disease syndrome. Diseased plants are stunted with deformed stems, leaves and flowers which in advanced stages resemble tiny cauliflowers.

8.1.1. The Nematodes

Two species of *Aphelenchoides, A. ritzemabosi* and *A. fragariae*, act as vectors of *R. fascians* and play an obligate role in the cauliflower disease complex on strawberry. *A. ritzemabosi* is an obligate parasite and feeds endoparasitically on mesophyll cells and ectoparasitically on buds and growing points. Strawberry is an important host on which it is usually found with *A. fragariae* (Siddiqi, 1974). *A. fragariae* is an obligate parasite of above-ground parts of plants. The nematode feeds ectoparasitically or endoparasitically. It causes crimp or spring dwarf in strawberry and feeds ectoparasitically, living within folded crown and runner buds (Franklin, 1950). Twisting and puckering of leaves, discoloured areas with a rough surface, dwarfed leaves with crinkled margins, reddening of petioles, short internodes of runners, reduced flower tissues and death of crown buds are shown by the infected plants of strawberry.

8.1.2. The Bacterium

The cauliflower bacterium, *Rhodococcus fascians* (= *Corynebacterium fascians*) is widespread and remains in healthy strawberries as saprophyte, becoming pathogenic when the plants are invaded by the nematodes. Only a few strains are capable of inducing cauliflower symptoms, and they need to be carried to the meristem of a crown bud by the nematode for the development of cauliflower disease.

8.1.3. Symptoms

The disease manifests in the form of malformation of the inflorescence, shortening and thickening of the aerial parts of strawberry plants. *A. fragariae* and *A. ritzemabosi* act as vectors of *R. fascians* and play an obligatory role in the cauliflower disease complex on strawberry. It is believed that the bacterium attaches to the cuticle of the nematodes and is transported to the crown tissue of strawberry plants (Pionar and Hansen, 1986; Taylor, 1990). Strawberry plants infected with both *A. ritzemabosi* and certain strains of *R. fascians* resemble small cauliflowers. In extreme cases plants are stunted and flowers are deformed. The petals fail to develop or become small and greenish. The sepal becomes much enlarged. Stamens and receptacles are malformed so that fruits do not develop. Auxiliary buds are continually produced in the crowns. Therefore, both the organisms actively contribute to the cauliflower syndrome. The bacterium attaches to the cuticle of the nematodes and is transported to the crown tissue of strawberry plants.

8.1.4. Management Methods

(i) *Regulatory methods*: Selecting nematode-free propagation material is of paramount importance to reduce the impact of bud and leaf nematode. Whenever possible, certified stocks should be used.

(ii) *Physical methods*: Hot water treatment of strawberry runners at 50°C for 5 min or 48°C for 15 min or 46°C for 20 min eliminates the nematode infection.

(iii) *Cultural methods*: Uprooting and destroying infected plants, and propagation through healthy planting stock are recommended for the control of nematodes.

Regenerating plants from clean, dormant, excised axillary buds is effective for eliminating *A. fragariae* from strawberry.

(iv) *Chemical methods*: Foliar sprays with phenamiphos at 4-6 ml/3.8 litres of water gave good control of *A. fragariae.* Chemicals used for disinfestation of strawberry plants include parathion as a foliar spray and soil application of aldicarb in granular form.

(v) *Host resistance*: In Poland, the strawberry varieties “George Soltwedel”, “Regina” and “Talizman” are comparatively resistant as are “Saksonka” and “Festival'naya” in USSR.

Chapter 9

NEMATODE - NEMATODE INTERACTIONS

Plant parasitic nematodes occur in polyspecific communities (two or more genera, species or races) because of their persistence, polyphagous feeding habits, wide distribution and weak interspecific competition. Interaction of one population on another may enhance or inhibit nematode reproduction, which is ultimately the driving force in interspecific competition. Most investigations indicate that nematode-nematode interactions are often antagonistic to at least one of the species.

The presence of a particular species in a location is related to its method of dissemination, host suitability, host range, interaction with other organisms, cropping history and other factors. Some of the nematodes that occur in polyspecific communities may be relics from previous host crop species, but these species may also be competing for the same host. Competition is usually strongest between organisms that are most alike with respect to their physiological demands on the host. A single species may dominate if it is better adapted to the host or to the ecological conditions. Species usually coexist if they utilize different ecological niches or have dissimilar life cycles.

Species of plant parasitic nematodes interact with each other ecologically (as competitors) and etiologically (in their effect on plant growth and crop production). These two aspects of nematode-nematode interactions are interrelated since the amount of disease is often correlated with population density. Ecological interactions affect the reproduction capacity of the individual nematode populations. Etiological interactions alter development of the disease (Eisenback, 1985; Eisenback and Griffin, 1987). Most investigations indicate that nematode-nematode interactions are often antagonistic to at least one of the species.

9.1. ECOLOGICAL INTERACTIONS

Ecological interactions affect the reproduction capacity of the individual nematode populations. They may be beneficial or detrimental to one or all interacting nematodes or there may be no effect on either. Competition is the key for all the ecological interactions. The interacting nematodes compete for food and space and consequently the nature of competition is either physical or physiological or both. Host suitability, pathogenicity, mode of parasitism, nematode population density and time are important determinants of the competition.

9.1.1. Ectoparasites

Population density and environmental factors may alter the interaction between ectoparasites. *Paratylenchus neoamblycephalus* suppressed low numbers of *Criconemella xenoplax* at 20°C, and high numbers of *C. xenoplax* suppressed low numbers of *P. neoamblycephalus* in plum. High populations of *P. neoamblycephalus* were antagonistic to low populations of *C. xenoplax* at 26°C, and high populations of *C. xenoplax* greatly suppressed low populations of *P. neoamblycephalus*. Species were more competitive at higher densities even though the optimum temperature for *C. xenoplax* reproduction was 26°C and 20°C for *P. neoamblycephalus* (Braun *et al.*, 1975).

Table 9.1: Interactions between ectoparasites

Test host	*Nematode combination*	*Dominant species*	*General response*	*Reference(s)*
Plum	High nos. of *Paratylenchus neoamblycephalus* + low nos. of *Criconemella xenoplax*	*P. neoamblycephalus*	Antagonistic	Bruan *et al.*, 1975
Plum	Low nos. of *Paratylenchus neoamblycephalus* + high nos. of *Criconemella xenoplax*	*P. neoamblycephalus* at 26°C *C.xenoplax*at 20°C	Antagonistic Antagonistic	Bruan *et al.*, 1975

9.1.2. Ectoparasites and Migratory Endoparasites

Pinochet *et al.* (1976) reported that concomitant inoculation of *Pratylenchus vulnus* and *Xiphinema index* caused greater growth reduction of grape than caused by any one of them singly. Moreover, the population of *X. index* was reduced in the presence of *P. vulnus*.

Table 9.2: Interactions between ectoparasites and migratory endoparasites

Test host	*Nematode combination*	*Dominant species*	*General response*	*Reference(s)*
Grape	*Xiphinema index* + *Pratylenchus vulnus*	*P. vulnus*	Antagonistic	Pinochet *et al.*, 1976

9.1.3. Ecto- and Sedentary Endoparasites

Ectoparasites feeding on the preferred penetration sites of the infective stages of sedentary endoparasites may be a limiting factor in the success of the sedentary endoparasite. Ectoparasites may also damage the root system and thus indirectly reduce the number of feeding sites available for the sedentary endoparasites. *Meloidogyne incognita* was inhibited by *Tylenchorhynchus brassicae* on vegetables (Khan and Khan, 1986), by *T. brassicae* on cauliflower (Khan *et al.*, 1987) and by *Criconemella ornata, Hoplolaimus indicus* and *T. nudus* on brinjal (Misra and Das, 1979). Likewise, *Criconemella xenoplax* inhibited *M. hapla* on Concord grape (Santo and Bolander, 1977) and *Paratrichodorus minor* suppressed *M. javanica* on tomato (Van Gundy and Kirkpatrick, 1975). Cucumbers inoculated with *Aphelenchus avenae* and *M. incognita* had fewer galls than when inoculated with root-knot nematodes alone (Choi *et al.*, 1988; Ishibashi and Choi, 1991).

Table 9.3: Interactions between ectoparasites and sedentary endoparasites

Test *host*	***Nematode combination***	***Dominant species***	***General response***	***Reference(s)***
Grape	*Criconemella xenoplax + Meloidogyne hapla*	*C. xenoplax*	Antagonistic	Santo & Bolander, 1977
	M. hapla + C. xenoplax	*C. xenoplax*	Stimulatory	Santo & Bolander, 1977
Tomato	*Paratrichodorus minor + M. javanica*	*P. minor*	Antagonistic	Van Gundy & Kirkpatrick, 1975
	M. javanica + Hemicycliophora arenaria	None	Mutually antagonistic	Van Gundy & Kirkpatrick, 1975
	R. reniformis + T. brassicae	None	Mutually antagonistic	Khan & Khan, 1986
	M. incognita + Tylenchorhynchus brassicae	None	Mutually antagonistic	Khan & Haq, 1979
Brinjal	*T. brassicae + M. incognita*	None	Antagonistic	Khan *et al.*, 1986b
	T. nudus + M. incognita	*T. nudus*	Antagonistic	Misra & Das, 1979
	Criconemella ornata + M. incognita	*C. ornata*	Antagonistic	Misra & Das, 1979
	M. incognita + C. ornata	None	Mutually antagonistic	Misra & Das, 1979
	Hoplolaimus indicus + M. incognita	*H. indicus*	Antagonistic	Misra & Das,1979
Cucumber	*Aphelenchus avenae + M. incognita*	*A. avenae*	Antagonistic	Choi *et al.*, 1988; Ishibashi & Choi, 1991
Vegetables	*M. incognita + T. brassicae*	*T. brassicae*	Antagonistic	Khan & Khan, 1986

Sedentary endoparasites can suppress ectoparasites even though they are separated by plant tissue, probably by physiological mechanisms. *M. incognita* suppressed *T. brassicae* on tomato (Khan and Haq, 1979). *M. incognita* and

Criconemella ornata on brinjal (Misra and Das, 1979) and *M. javanica* and *Hemicycliophora arenaria* on tomato (Van Gundy & Kirkpatrick, 1975) were mutually antagonistic.

Interactions between ectoparasites and sedentary endoparasites may be beneficial to one or both species. *M. hapla* enhanced the reproduction of *C. xenoplax* on Concord grape (Santo and Bolander, 1977). The mechanisms involved are not known but may be physiological and/or physical.

9.1.4. Migratory Endoparasites

Migratory endoparasites generally utilize the same feeding sites and are probably very competitive with each other, although it is not uncommon to find concomitant populations. Field observations that *P. coffeae* suppressed *Scutellonema bradys* on guinea yam were supported in greenhouse tests (Acosta Ayala, 1976). *P. coffeae* and *R. similis* were mutually suppressive on citrus, but fine-textured soil favoured *P. coffeae* and coarse-textured soil favoured *R. similis* (O'Bannon *et al.*, 1976).

Table 9.4: Interactions between migratory endoparasites

Test host	*Nematode combination*	*Dominant species*	*General response*	*Reference(s)*
Citrus	*Radopholus similis* + *Pratylenchus coffeae*	None	Mutually antagonistic	O'Bannon *et al.*, 1976
Yam	*Scutellonema bradys* + *P. coffeae*	*P. coffeae*	Antagonistic	Acosta & Ayala, 1976

9.1.5. Migratory Endo- and Sedentary Endoparasites

Sedentary endoparasites may penetrate the host faster and inhibit the penetration by the migratory species. Prior inoculation by *M. incognita* inhibited penetration by *P. brachyurus* in tomato, a better host for *M. incognita* than *P. brachyurus* (Gay and Bird, 1973).

Differences in host tissue preference and nematode feeding mechanisms may limit interaction between migratory and sedentary endoparasites. *Ditylenchus dipsaci* and *M. hapla* were not affected by each other, probably because they occupied different sites in the host – tomato (Griffin, 1987).

Nematode interactions can be affected by the timing of inoculations. Reproduction of *P. vulnus* was greatly inhibited after 125 days when *M. incognita* was inoculated one month prior to *P. vulnus* in grapevine, but in simultaneous inoculations the inhibition was delayed until 250 days (Chitamber and Raski, 1984).

Migratory endoparasites are less advanced than sedentary endoparasites which establish a complex relationship with the host and alter plant physiology. This change in physiology often affects the suitability of the host for the migratory endoparasites. *M. incognita* induced a translocatable factor in split-root experiments that greatly inhibited the reproduction of *Pratylenchus penetrans* in tomato. *P. penetrans* also suppressed *M. incognita* but a translocatable factor was not involved. The inhibition may have been caused by disruption of feeding sites created by migratory endoparasites (Estores and Chen, 1970, 1972). On cowpea, *M. javanica* limited penetration by *P. sefaensis* (Egunjobi *et al.*, 1986).

Table 9.5: Interactions between migratory and sedentary endoparasites

Test host	*Nematode combination*	*Dominant species*	*General response*	*Reference(s)*
Pineapple	*Pratylenchus major* + *Meloidogyne* spp.	*P. major*	Antagonistic	Gueroount, 1968
Citrus	*P. coffeae* + *Tylenchulus semipenetrans*	None	Mutually antagonistic	Kaplan & Timmer, 1982
Grape	*M. incognita* + *P. vulnus*	*M. incognita*	Antagonistic	Chitamber and Raski, 1984
Tomato	*M. incognita* + *Pratylenchus brachyurus*	*M. incognita*	Antagonistic	Gay & Bird, 1973
	M. incognita + *P. penetrans*	None	Mutually antagonistic	Estores & Chen, 1970, 1972
		M. incognita	Antagonistic	
	Ditylenchus dipsaci + *M. hapla*	None	None	Griffin, 1987
Cowpea	*R. reniformis* + *P. sefaensis*	*R. reniformis*	Stimulatory	Egunjobi *et al.*, 1986
Bean	*Ditylenchus dipsaci* + *M. hapla*	None	None	Griffin, 1987
Black pepper	*R similis* + *M. incognita*	None	Mutually antagonistic	Sheela & Venkitesan, 1981

The species that predominates in antagonistic interactions between sedentary and migratory endoparasites is usually determined by climatic or edaphic factors. *P. coffeae* and *T. semipenetrans* on citrus (Kaplan and Timmer, 1982), *R. similis* and *M. incognita* on black pepper (Sheela and Venkitesan, 1981) were mutually suppressive.

9.1.6. Sedentary Endoparasites

Sedentary endoparasites are highly specialized parasites and have a long-lasting relationship with their host. Competition between species is generally mutually suppressive because they utilize the same sites for feeding. *Meloidogyne* species may inhibit, have no effect, or stimulate *Heterodera* species. *M. incognita* and *H. cajani* on cowpea did not interact (Sharma and Sethi, 1978).

Heterodera species may inhibit, or have no effect on *Meloidogyne* species. *M. hapla* was inhibited by *H. schachtii* on tomato (Griffin and Waite, 1982).

Table 9.6: Interactions between sedentary endoparasites

Test host	*Nematode combination*	*Dominant species*	*General response*	*Reference(s)*
Grape	*M. incognita* + *Rotylenchulus reniformis*	None	Mutually antagonistic	Rao & Seshadri, 1981
	T. semipenetrans + *R. reniformis*	None	Mutually an tagonistic	Taha & Sultan, 1977
Tomato	*M. hapla* + *Heterodera schachtii*	*H. schachtii*	Antagonistic	Griffin & Waite, 1982
	M. hapla + *Heterodera schachtii*	*M. hapla*	Antagonistic	Griffin, 1985
	Low nos. of *M. incognita* + *R. reniformis*	*R. reniformis*	Antagonistic	Kheir & Osman, 1977; Winoto & Lim, 1972
	Low nos. of *M. incognita* + *R. reniformis*	*R. reniformis*	Antagonistic	Kheir & Osman, 1977; Winoto & Lim, 1972
	High nos. of *M. incognita* + *R. reniformis*	*M. incognita*	Antagonistic	Kheir & Osman, 1977; Winoto & Lim, 1972
	M. incognita + *R. reniformis*	None	Antagonistic	Khan *et al*., 1985, 1986a
Brinjal	*M. incognita* + *R. reniformis*	None	Mutually antagonistic	Khan *et al*., 1986b
Cowpea	*M. incognita* + *Heterodera cajani*	None	Neutral	Sharma & Sethi, 1978
	M. incognita + *H. cajani*	*M. incognita*	Antagonistic/ Mutually antagonistic	Sharma & Sethi, 1978
	M. javanica + *R. reniformis*	None	Neutral	Rao & Prasad, 1971; Taha & Kassab, 1979; 1980
	M. incognita + *R. reniformis*	None	Stimulatory, cv. S-448 lost resistance to *R. reniformis* in presence of *M. incognita*	Khan & Husain, 1989

Table 9.6: (*Contd...*)

Table 9.6: (*Contd...*)

Test host	*Nematode combination*	*Dominant species*	*General response*	*Reference(s)*
	M. incognita + *R. reniformis*	None	Stimulatory, *M. incognita* increased the reproduction of *R. reniformis*	Khan & Husain, 1989
Pigeon pea	*M. incognita* + *R. reniformis*	*R. reniformis*	Antagonistic	Pathak *et al.*, 1985
Cauli-flower	*M. incognita* + *R. reniformis*	None	Mutually antagonistic	Khan *et al.*, 1987
Vegetables	*M. incognita* + *R. reniformis*	None	Antagonistic	Khan & Khan, 1986
Sweet potato	Low nos. of *R. reniformis* + *M. incognita*	*R. reniformis*	Antagonistic	Thomas & Clark, 1980, 1981, 1983a, 1983b
	High nos. of *R. reniformis* + *M. incognita*	*M. incognita*	Antagonistic	Thomas & Clark, 1980, 1981, 1983a, 1983b

The relationship between the root-knot and cyst nematodes may be density dependent. *M. incognita* was detrimental to *H. cajani* on cowpea when established first, but prior infection by *H. cajani* was detrimental to *M. incognita* (Sharma and Sethi, 1978).

Interactions between *Meloidogyne* and *Rotylenchulus* species can be mutually antagonistic in tomato (Khan *et al.*, 1985) and grape (Rao and Seshadri, 1981) or suppressive to just one species in pigeonpea (Pathak *et al.*, 1985). *R. reniformis* generally limited reproduction of *M. incognita* on tomato, except high populations of root-knot nematode inhibited reproduction of reniform nematode (Kheir and Osman, 1977; Winoto and Lim, 1972). *R. reniformis* initially inhibited *M. javanica* on cowpea, but the root-knot nematode was more competitive over time (Taha and Kassab, 1979, 1980). *M. incognita* was inhibited by low levels of *R. reniformis* on sweet potato, and low levels of *M. incognita* had no effect on the reniform nematode. High levels of *M. incognita* suppressed *R. reniformis*. Each species was capable of dominating, depending upon initial density (Thomas and Clark, 1980, 1981, 1983a, 1983b).

Interaction between *T. semipenetrans* and *R. reniformis* on grape were mutually antagonistic. Populations resulting from concomitant inoculations were significantly lower than populations from monospecific inoculations (Taha and Sultan, 1977).

9.1.7. Interactions between Species of the Same Genus

Reports about interactions between two species of the same genus are most common in *Meloidogyne, Heterodera* and *Globodera*. *M. javanica* and *M. hapla* were dominated by *M. incognita* at high temperatures, but *M. javanica* dominated *M. incognita* and *M. hapla* at low temperatures (Minz and Strich-Harari, 1959). At high temperatures, simultaneous inoculations of *M. incognita* and *M. hapla* resulted in a population of 90% *M. incognita* and 10% *M. hapla*. At low temperatures, the population was only 57% *M. incognita* (Chapman, 1965). Competition between species can be affected by factors other than temperature. *M. javanica* dominated *M. hapla* at 20°C, even though low temperatures generally favour *M. hapla* (Minz and Strich-Harari, 1959).

Table 9.7: Interactions between species of the same genus

Test host	***Nematode combination***	***Dominant species***	***General response***	***Reference(s)***
Tomato	*M. javanica* + *M. incognita*	*M. javanica* or *M. incognita*	Antagonistic	Minz & Strich-Harari, 1959
	M. javanica + *M. hapla*	*M. javanica*	Antagonistic	Chapman, 1965; Kinloch & Allen, 1972
	M. javanica + *M. incognita*	*M. incognita* race 2	Antagonistic	Khan & Haider, 1991
	M. incognita + *M. hapla*	*M. incognita*	Antagonistic	Chapman, 1965
	M. hapla (Diploid) + *M. hapla* (Polyploid)	*M. hapla* (Diploid)	Inhibitory	Triantaphyllou, 1991
Potato	*Globodera rostochiensis* + *G. pallida*	*G. rostochiensis*	Antagonistic	Marshall, 1989
	G. rostochiensis + *G. pallida*	*G. pallida*	Antagonistic	Parrot *et al.*, 1976
	G. rostochiensis + *G. pallida*	More competition between individuals of *G. pallida*	–	Seinhorst & Oostrom, 1989
	G. rostochiensis + *G. pallida*	Fertile hybrids produced	Hybridization	Miller, 1983

Globodera rostochiensis and *G. pallida* virtually occupy the same ecological niche and cannot coexist (Parrot *et al.,* 1976). They may occur together in varying proportions for different periods of time, but eventually one species dominates (Kort and Bekker, 1980). In New Zealand, *G. rostochiensis* was antagonistic to *G. pallida*, but at very low populations *G. pallida* was more competitive and was probably able to persist at very low densities (Marshall, 1989). However, in England, *G. pallida* was more competitive than *G. rostochiensis* (Parrot *et al.,* 1976). *G. rostochiensis* may hatch more freely and survive less efficiently at lower soil temperatures. Conversely, in The Netherlands, more competition occurred between individuals of *G. pallida* than with mixtures of *G. rostochiensis* (Seinhorst and Oostrom, 1989).

9.1.8. Interactions between Races of the Same Species

Cytological forms of a single species can interact with each other. The frequency of a tetraploid isolate of *M. hapla* was reduced from 50% to about 9% by competition from a diploid isolate after six generations on tomato (Triantophyllou, 1991).

9.2. ETIOLOGICAL INTERACTIONS

Etiological interactions between nematode species alter the course of the disease and consequently plant damage. Because nematodes rarely occur in monospecific communities, they may interact with each other to alter the course of the disease. If the amount of disease caused by both nematodes is less than the combined effect of each alone, the interaction is negative (antagonistic); if it is more, the interaction is positive (synergistic); and if it is same, there is no interaction (Burrows, 1987).

9.2.1. Negative Interactions

Less disease occurs when there is strong competition between nematode species. The less pathogenic species reduced the number of infection sites for the more pathogenic species. Negative interactions of nematode species usually involve very strong, pathogenic competitors. Either of the two species can predominate depending upon the inoculum level and other factors, but the two species usually

Table 9.8: Nematode-nematode interactions that affect disease expression

Nematode combination	*Test host*	*Plant growth response*	*Reference(s)*
Negative interactions			
M. incognita + *P. penetrans*	Tomato	*M. incognita* suppressed growth more by itself than when combined with *P. penetrans*	Estores and Chen, 1970
M. incognita + *R. reniformis*	Grape	Growth was less in combined inoculations than single inoculations	Rao & Seshadri, 1981
Positive interactions			
M. hapla + *H. schachtii*	Tomato	Combined inoculations suppressed root growth by 65, 64 and 61% below that of controls and single inoculations	Griffin & Waite, 1982

cannot coinhabit. *M. incognita* suppressed the growth of tomato more by itself than in combination with *P. penetrans* (Estores and Chen, 1970). Similar response occurred when *M. incognita* and *R. reniformis* occurred together on grape (Rao and Seshadri, 1981).

9.2.2. Positive Interactions

Nematodes may predispose a plant to attack by another organism. The mechanism involved is a change in host physiology that makes it easier to penetrate and establish a parasitic relationship. Interruption of defence mechanisms may also be involved. All of the positive interactions that have been reported for concomitant populations of nematodes involve a species of *Meloidogyne* and some other endoparasite.

H. schachtii and *M. hapla* suppressed growth of tomato by 65%, 64% and 61% below that of uninoculated controls and single inoculations of either *M. hapla* or *H. schachtii*, respectively (Griffin and Waite, 1982).

9.3. INTERACTIONS WITH OTHER PATHOGENS

Nematodes often interact with other organisms to cause plant disease. When two nematodes are present with another organism they may interact to make the disease even more severe. The mechanical damage and the physiological changes made by the nematodes may be involved in the mechanism of the interaction.

M. incognita, Belonolaimus longicaudatus and *Pythium aphanidermatum* caused more suppression of the growth of chrysanthemum than did any pathogen alone or any single nematode and fungus combination (Johnson and Littrell, 1970).

Table 9.9: Nematode-nematode interactions that affect disease expression with an additional pathogen

Test host	*Nematode combination*	*Additional pathogen*	*Plant growth response*	*Reference(s)*
Tomato	*P. penetrans* + *P. coffeae*	*Fusarium oxysporum*	All 3 cause greatest inhibition of growth & more severe wilt	Hirano & Kawamura, 1972
Cowpea	*M. incognita* + *R. reniformis*	*Rhizoctonia solani*	All 3 cause more disease than any organisms combined	Khan & Husain, 1989
Chrysanthemum	*M. incognita* + *B. longicaudatus*	*Pythium aphanidermatum*	All 3 cause more growth suppression than any organisms combined	Johnson & Littrell, 1970
Black pepper	*R. similis* + *M. incognita*	*Fusarium* sp.	All 3 cause severe growth suppression & yellowing than any organisms combined	Mustika, 1984

M. incognita and *R. reniformis* plus the fungus *Rhizoctonia solani* on cowpea caused more plant disease than either pathogen alone or any two organisms combined (Khan and Husain, 1989). *P. penetrans* and *P. coffeae* interacted with Fusarium to produce more wilt symptoms and suppression of plant growth than did either nematode alone or in combination with the fungus (Hirano and Kawamura, 1972). Reproduction of *P. coffeae* was stimulated by the presence of the fungus.

9.4. MECHANISMS OF INTERACTION

Competition is the key factor of all the ecological interactions between nematodes. The interacting nematodes compete mainly for food and space and consequently the nature of competition is either physical or physiological or both. Host suitability, pathogenicity, mode of parasitism, nematode population density and time are important determinants of the competition. The antagonistic interaction among surface feeders is more intense than between a surface feeder and sub-surface feeder, probably because their spatially separated feeding sites. Destruction of penetration sites of infective stages of sedentary endoparasites by feeding of ectoparasites lead to suppression of the endoparasites. Sedentary endoparasites suppress ectoparasites by physiological mechanisms (Norton, 1969). In general, competition between nematodes of similar feeding habits is more intense because of similar physical and physiological demands. Mechanical destruction of penetration sites and root tissue, physical occupation of feeding sites or altered host physiology are attributed mechanisms of ecological interactions between nematodes in capacitance.

Etiological interactions between nematode species alter the course of the disease and consequently plant damage (Duncan and Ferris, 1983). In relation to disease intensity this interaction may be negative (antagonistic) or positive (synergistic) or neutral (additive) (Burrows, 1987; Wallace, 1983). Synergistic interaction is rather rare. Intense competition between nematode species is generally antagonistic in relation to disease intensity. Physical damage to infection sites and physiological changes have been suggested as possible mechanisms of such interactions (Duncan and Ferris, 1983; Eisenback and Griffin, 1987). Changes in host physiology that facilitate penetration and establishment of the other nematode species is a possible mechanism of positive interactions (Griffin and Waite, 1982).

9.5. CONCLUSIONS

Concomitant populations of nematodes can interact with each other to affect their reproductive ability and they can alter the etiology of disease. The nature of the host-paraiste relationship, differences in pathogenicity, and environmental effects undoubtedly play an important role in nematode-nematode interactions. The following general conclusions are listed in order to summarize the complexity of interactions of nematodes in cohabitance.

- Nematode-nematode interactions can be studied from both ecological and etiological view points.
- Interactions can be stimulatory or inhibitory to one or both species.
- Competition between nematode species is generally weak.
- Competition between species can restrict the distribution of some species.
- Competition is more intense between species with similar feeding habits.
- Competitive advantage increases as the host-parasite relationship becomes more complex. Hence, endoparasites are more competitive than ectoparasites, and sedentary endoparasites are more competitive than migratory endoparasites.
- Competition between species may be density and time dependent.
- Environmental factors can modify the effects of competition.
- Host suitability is a key factor in interactions and is often responsible for one species dominating another.
- Mechanisms of competition may include mechanical destruction or physical occupation of feeding sites or induced physiological changes in the host's suitability or attractiveness.
- Generally, the amount of disease caused by two or more nematode species is additive or antagonistic, but sometimes it can be synergistic.
- Strong competitors cause less disease when in combination with another species than when alone.
- Most synergistic nematode-nematode interactions involve a sedentary endoparasite.
- Nematode-nematode interactions can become more complex when other organisms are also involved in the etiology of the disease.
- Interactions among nematodes are important in nature, and more precise experiments are needed if their effects on the variability of crop growth are to be quantified.

Chapter 10

NEMATODE – INSECT INTERACTIONS

There are less than ten reported parasitic and phoretic associations between phytoparasitic nematodes and insects (Poinar, 1975). Nematodes are aquatic metazoans with very limited power of dispersion. They are greatly benefitted by a synchronized association with an insect host for increased mobility and protection during travel to an insect's breeding or feeding sites.

One of the most important wilting diseases of trees in the world (red ring disease of coconut and other palms) is caused by insect-transmitted aphelenchoidid nematode. Nematodes in such interactions can be phoretically associated with the insect and parasitic on the plant or on both hosts. In cases where the nematode is parasitic on both hosts the interaction can be dicyclic; where the nematode alternately completes its life cycle in the insect then the plant host, or phytocyclic/ insect parasitic; where the nematode completes its life cycle on the plant host with one parasitic stage in the insect.

10.1. RED RING DISEASE OF COCONUT

The red ring disease was first reported from Trinidad by Hart in 1905. It is limited in distribution to the Neotropics, where it is one of the most important diseases of coconut and African oil palm (Griffith and Koshy, 1990). The losses in affected coconut and oil palm plantations are commonly in the range of 10-15%. The 'red ring' disease of coconut is caused by *Bursaphelenchus cocophilus* which is very destructive disease on some islands of Caribbean which threatens the existence of this crop. The extensive loss caused by this nematode is a major catastrophe to natives of these areas. The nematode is transmitted by *Rhynchophorus palmarum*. The nematodes are probably deposited during weevil oviposition into the leaf bases or wounds in a susceptible palm tree, where they invade, feed on parenchyma cells, and can cause a lethal wilt in as little as 2-4 months (Griffith, 1987). It has not been reported from India so far.

At present, most Latin American and Caribbean countries show losses ranging from less than 1% to more than 20% of 1 to 10 year old trees. Thirty five per cent mortality of young coconut palms has been reported in Trinidad and 80% loss in single plantation of Tobago (Esser, 1969).

10.1.1. Distribution

The red ring nematode is co-distributed with its insect vector, *Rhynchophorus palmarum*, in the southern Antilles, Mexico southward through Central America into South America, where it has been reported from Colombia, Surinam, Ecuador, Venezuela and Brazil.

10.1.2. Symptoms

Classical red ring symptoms in coconut often include premature nut fall (except for mature nuts), withering and necrosis of inflorescences, and yellowing, bronzing and death of progressively younger leaves. Yellowing of leaves usually starts at the tips of pinnae and moves inward to the rachis and then to the base of the petiole. The disease is present most commonly in young coconut trees (3-10 years old) with maximum incidence in trees of 4-7 years old. Coconut palms 3-10 years old usually die within several months of infection. Chlorosis first appears at the tips of the oldest leaves and spreads towards their bases. The brown lower leaves may break across the petiole or the lower part of the rachis or they may become partly dislodged at the base and hang down (Fig. 10.1). Premature shedding of the nuts may occur simultaneously with the development of the leaf symptoms or slightly before. As a result of the severe damage caused by the palm weevil internally, the crown often topples over about 4 to 6 weeks after the appearance of the first symptoms. The trunk then remains standing in the field for several months and finally it decays (Griffith and Koshy, 1989).

The most characteristic symptom of red ring disease is the internal lesions. In the beginning, about 3-5 cm beneath the stem surface, scattered reddish dots of about 1 mm diameter appear, which later coalesce to form an orange red ring of about 3 cm in width (Fig. 10.1). The ring extends the whole length of the stem and roots and in petioles it assumes a crescent-like shape. Large numbers of juveniles are seen in the centre of the discoloured areas and adults in the periphery. Dauer juveniles can be harvested from the discoloured tissue of the ring (up to 11,000 nematodes/g of tissue), from leaf petioles, or roots to confirm disease diagnosis in coconut palm (Blair, 1969). In longitudinal section, discolouration is usually continuous throughout the length of the stem, appearing as two bands which unite at the base and form discontinuous lesions near the crown. Severe damage to the crown of red ring diseased coconut palms is done by larval feeding of the weevil vector, often causing the palm crown to fall over under its own weight. Little leaf symptoms in coconut palm can be caused by the red ring nematode (Hoof and Seinhorst, 1962) and may be more common in older palms (Kraaijenga and Ouden, 1966).

Fig. 10.1: Red ring nematode (*Bursaphelenchus cocophilus*) on coconut. A-Infected plant. B-Cross section of trunk showing red ring symptoms (Courtesy: Sasser, 1971)

B. cocophilus causes little leaf disease of coconut and oil palm in Surinam and Guyana. A thermostable phytotoxin was produced due to breakdown of coconut tissue. Water uptake of infested coconut palms was less due to occlusion of xylem vessels. In coconut roots, the nematodes attack cortical tissues.

10.1.3. The Vector

The vector of the red ring nematode is the palm weevil (*Rhynchophorus palmarum*). The adult palm weevil acts as the carrier. The female weevil makes punctures in the softer tissue of the stem of the coconut palm with the help of its proboscis to lay its eggs. Normally the punctures are made in the soft axils to a depth of 5-8 cm and the eggs are placed deep in the puncture. The nematodes are injected into the tissues of the coconut palm when the insect deposits its eggs. Larvae of the palm weevil while feeding in diseased trees get infested with the nematodes and the adult beetles which develop from them in turn become the carriers.

10.1.4. Management Methods

(i) *Cultural methods*: Phytosanitation is currently the best method of red ring disease management. This strategy is directed at reducing the vector population as well as the number of sources for nematode inoculum. 'Cut and burn' the infested trees is one of the oldest and most effective practices used to prevent further spread of the disease.

(ii) *Chemical methods*: Trees should be sprayed with an insecticide (*e.g.* methomyl) and killed with 100-150 ml (48.3% *a.i.*) of the herbicide Monosodium Acid Methanearsonate (MSMA) which is injected or placed into the trunk (Griffith and Koshy, 1990). Once the tree is dry it should be cut and sectioned to make sure that weevils are not present, palms that are heavily infested with weevils should be cut and treated with insecticides such as methomyl, trichlorfon, monocrotophos, carbofuran, carbaryl or lindane or burned with kerosene.

More recent methods involve the destruction of palms using arboricides, herbicides or phytotoxic nematicides applied to holes made in tree-trunks. Hoyle (1968) recommended the use of cacodylic acid ('Silvisar') to poison trees. The chemical prevented weevil breeding, killed the diseased trees within two weeks, but had no nematicidal effect. Victoria *et al.* (1970) used sodium and potassium arsenate to kill both diseased trees and nematodes.

Adult weevils are attracted to anaerobic fermentation products such as ethyl alcohol, n-butyl alcohol and to certain volatile esters extracted from the diseased palm tissue (Griffith, 1987). This behaviour can be exploited to attract weevils to poisoned baits at 25/ha.

The trap or guard baskets protect plantations by attracting and killing the palm weevils which may enter the plantations from nearby diseased trees. The guard baskets are made of 2 cm mesh wire. They are cylindrical, 1 m high and 0.3 m in diameter. These baskets are filled with chunks of fresh tissue from diseased coconut trees to attract the beetle. The guard baskets are sprayed completely with about 4.5 litres of methomyl solution and distributed on the ground in the plantations at 2.5 baskets/ha of young coconut trees. This procedure is especially recommended in the dry season when the weevils are most active in the cool nights. Guard baskets remain for about two weeks, after which the tissue and insecticide in the basket should be burnt. Fresh tissue should be placed in the basket and treated as previously described.

If the vector population is effectively managed, the chances of nematode infestation will be minimized to a great extent. The palm weevil population was successfully managed when tree crowns were sprayed with 1-10% palmoral (Blair, 1969). A 0.5% emulsion of insecticides like fenitrothion, fenthion or carbaryl if sprayed at 2.3 litres/tree during May-June, can effectively reduce the vector population (Mamiya, 1984).

Injections of systemic nematicides, such as fenamiphos, oxamyl or carbofuran, into little leaf symptomatic palms cause apparent palm recovery in some of the trees tested. The recovery can take between 6-8 months because of the damage to the very young leaves in little leaf palms. Multiple applications of endrin (at 50 and 70-day intervals during the rainy and dry seasons, respectively) to the leaf axils of healthy coconut palms were effective at reducing red ring disease prevalence in Trinidad (Hagley, 1963).

(iii) *Biomanagement*

(a) *Antagonistic bacteria*: Griffith (1977) reported the pathogenicity of a bacterium, *Micrococcus agilis* to the palm weevil and suggested its potential use as a biocontrol agent.

(b) *Entomopathogenic nematodes*: The rhabditid nematodes of the genera *Steinernema* and *Heterorhabditis* can heavily parasitize the weevils by maintaining a high population of nematodes in the environment (Griffith and Koshy, 1989).

(c) *Parasitoids and pathogens*: A tachnid parasitoid, *Parabillaea rhyncophorae* from Bolivia (Candia and Simmonds, 1965), a cytoplasmic polyhedrosis virus (CPV) from *R. palmarum* from India (Gopinadhan *et al*., 1990) may hold some promise for the management of the vector.

10.2. INTERACTIONS BETWEEN *FUSARIUM* AND BEAN STEM MAGGOT

Studies on the interaction between *Fusarium* wilt and bean stem maggot at Mulungu showed that incidence and severity of *Fusarium* wilt were significantly reduced in plots where bean stem maggot was controlled (Table 10.1). This implies that the presence of bean stem maggot results in an increase in incidence and severity of *Fusarium* wilt. This means therefore that in the development of management practices to control *Fusarium* wilt, the possible influence of bean stem maggot has to be considered.

Table 10.1: Number of bean stem maggot pupae, severity of *Fusarium* wilt and yield of the climbing bean G 2333 grown on soils amended with farmyard manure and Tithonia

Treatment	*No. of pupae**	*Severity of Fusarium wilt*	*Yield (kg/ha)*
Control	2.8[a]	4.1[a]	204.0[b]
FYM	2.5[a]	3.3[b]	610.3[a]
Tithonia	2.4[a]	3.2[b]	475.1[ab]
C.V. (%)	**6.4**	**13.7**	**30.3**

* Values in the same columns followed by the same letter are not significantly different at P = 0.05

10.3. FIG WASPS AND *SCHISTONCHUS* SPP.

Schistonchus caprifici, the first described nematode member of the family Aphelenchoididae, was reported to be a phoretic associate of the fig wasp, *Blastophaga psenes* (Agaonidae) (Poinar, 1975). New evidence suggests that it is an obligate parasite of fig wasp and that it also parasitizes florets in the fig syconia of *Ficus caraica* (Vovlas *et al*., 1992).

Adult females of the *B. psenes* carry 200-400 juveniles and adults of the nematode *S. caprifici* in their hemocoels to fig syconia (caprifigs or edible figs). The nematodes are apparently deposited during oviposition by the fig wasp into pistillate florets, where they develop and reproduce, causing necrosis and cavities in the cortical parenchyma (Vovlas *et al*., 1992). Both male and female florets can be attacked and female florets in caprifigs are better for nematode reproduction than female florets of edible figs. Nematodes from a floret infect the fig wasp larva in its flower gall and apparently grow and reproduce within the insect as it goes through metamorphosis. Evidently, the wingless males are not used as nematode hosts (Vovlas *et al.*, 1992). The male fig wasps emerge first and mate with female wasps prior to emergence from their gall flowers. Female fig wasps emerge and gather pollen from mature staminate (male) florets and exit through an emergence hole cut by the male fig wasps. The female disperses to a new caprifig or an edible fig to repeat the cycle.

The fig wasps *Pegoscapus assuetes* and *P. jimenizi* (Agaonidae) from native *Ficus* spp. in southern Florida (*F. citrifolia* and *F. aurea*) are associated with new species of *Schistonchus*. The biology of the associations with these new species of *Schistonchus* are different from previously reported *Schistonchus* – *Ficus* – agaonid associations (Vovlas *et al.*, 1992), but the nematodes do appear to parasitize the fig florets during part of the interaction.

Chapter 11

MANAGEMENT OF DISEASE COMPLEXES

11.1. PHYSICAL METHODS

Soil solarization with 0.03 mm thick polyethylene transparent sheets in moist and well-cultivated soil continuously for 30-50 days has been found effective against soil-borne fungi and nematodes. *Ditylenchus dipsaci, Pratylenchus thornei, P. penetrans, Rotylenchus incultus* and *Parartrichodorus lobatus* among nematodes and *Fusarium* spp., *Verticillium dahliae* and *Phytophthora cinnamoni* among fungi were suppressed. In such a soil the growth of tomato and grapevine was significantly enhanced. Significant reduction in root galling on tomato by *Meloidogyne* spp. occurred in soil exposed to summer heat for about 9 weeks.

11.2. CULTURAL METHODS

(i) *Fruit crops*: In France, to save grapevines from attack by *Xiphinema index* and subsequently infection by grapevine fan leaf virus, a 7-year rotation with cereal and alfalfa crops has been recommended (Lamberti, 1981).

(ii) *Vegetable crops*: In fields infested with root-knot nematodes and *V. dahliae,* Scholte (1990) found that a five-year rotation of maize-sugar beet-barley-barley-potato had significantly higher potato yields than continuous potato.

The beneficial effects of neem cake were observed not only against *M. incognita* and the fungus *Macrophomina phaseolina* individually, but also when both the pathogens formed a disease complex on French bean.

(iii) *Plantation crops*: Phytosanitation is currently the best method of red ring disease management. This strategy is directed at reducing the vector population (*Rhynchophorus palmarum*) as well as the number of sources of nematode inoculum. As soon as coconut palms with red ring disease or *Bursaphelenchus cocophilus*-induced little leaf symptoms have been detected, they should be destroyed.

11.3. CHEMICAL METHODS

Dibromochloropropane (DBCP) in addition to nematicidal action possesses fungicidal properties also. Its twin potential in combination with fungicides like sodium azide is synergized.

(i) *Fruit crops*: Pentachloronitrobenzene (PCNB) has been found to effectively check the population increase of nematodes like *Longidorus elongatus* and *Xiphinema diversicaudatum*. This efficacy would eventually influence the spread of viruses transmitted by them (raspberry ring spot, strawberry latent ring spot, peach rosette mosaic) (Murant and Taylor, 1965; Taylor and Gordon, 1970).

Grapevine: Dichloropropane-Dichloropropene (DD) at 250 gallons/acre was a suitable recommendation for heavy soils against *Xiphinema index*, a vector of grape fan leaf virus, while only 200 gallons/acre was found to be effective in lighter soils (Rankin *et al.*, 1971).

Strawberry: Soil application of ethylene dibromide (EDB) exhibited dual efficacy against *M. hapla* – *V. dahliae* disease complex in strawberry (Meagher and Jenkins, 1970).

ii) *Vegetable crops*

Potato: Methylisothiocyanate has been successfully used for soil fumigation to reduce the populations of soil-borne fungus, *Verticillium dahliae* and alleviate the effects of disease complex involving wilt-inducing fungus (*V. dahliae*) and the lesion nematode (*Pratylenchus penetrans*) on potato (Rowe *et al.*, 1987).

Soil fumigation with methyl bromide resulted in yield increases of 68% for potato fields infested with *Globodera* spp. and *V. dahliae* (Hide and Corbett, 1973).

In *V. dahliae* – *Globodera rostochiensis* disease complex on potato, a mixture of methyl bromide and chloropicrin was found to have comparatively more fungicidal and nematicidal efficacy (Hide and Corbett, 1974).

Tomato: Application of metham sodium through drip irrigation was found to be effective against root-knot, *Pythium ultimum* and *Fusarium* sp. complex of tomato (Roberts *et al.*, 1988).

Jones and Overman (1976) noticed an improvement in yield of tomato affected by nematodes like *Belonolaimus longicaudatus, Trichodorus christiei, M. acrita* and wilt fungi such as *F. o.* f. sp. *lycopersici* race 2 and *Verticillium albo-atrum*, when carbofuran was applied at a higher soil pH (6.5 to 7.5).

Pentachloronitrobenzene (PCNB) has been found to effectively check the population increase of nematodes like *Longidorus elongatus* and *Xiphinema diversicaudatum*. This efficacy would eventually influence the spread of tomato blacking (Murant and Taylor, 1965; Taylor and Gordon, 1970).

Okra: Significant reduction in root-knot – *Fusarium* wilt disease complex of okra was obtained by soil fumigation with methyl bromide (Good, 1964).

Carrot: Application of metham sodium through drip irrigation was found to be effective against root-knot, *Pythium ultimum* and *Fusarium* sp. complex of carrot (Roberts *et al.*, 1988).

Cucumber: Benomyl application inhibited feeding of *X. americanum* on cucumber, which was reflected in a marked reduction of its transmission ability of tobacco ring spot virus (McGuire and Good, 1970).

11.4. BIOLOGICAL METHODS

(i) *Vegetable crops*: *Paecilomyces lilacinus* can be effectively used against *Rhizoctonia solani – M. incognita* complex of okra (Shahzad and Ghaffar, 1989).

11.5. INTEGRATED METHODS

(i) *Fruit crops*

Banana: Soil application of neem cake + *T. viride* + carbendazim was found effective in reducing the burrowing nematode (*R. similis*) and wilt (*F. o.* f sp. *cubense*) disease complex and in increasing the banana fruit yield (15,147 kg/ha). This treatment also gave minimum lesion index (1.1) and root-knot index (1.0) as compared to control (4.0) with cost: benefit ratio of 1:2.72 (Ravi *et al.*, 2001).

Grapevine: Treatment of grapevines with *P. fluorescens* at 100 g/vine + FYM at 20 kg/vine gave highest reduction in final soil nematode population of *M. incognita* (56.9%), least root gall index (1.8) and least wilt disease (*Fusarium moniliforme*) incidence (24.8%). The bunch weight of grapevine increased by 155.4% compared to untreated control (Senthil Kumar and Rajendran, 2004).

Papaya: Khan *et al.* (1997) have reported that *P. lilacinus* gave better over all protection of papaya plants against *M. incognita – F. solani* disease complex than *T. harzianum.* They reported that application of both the biocontrol agents further limited the damage caused by *M. incognita* or *F. solani* and gave a 35% increase in plant growth compared to individual applications.

(ii) *Vegetable crops*: *Pseudomonas aeruginosa* and *Paecilomyces lilacinus* when used together significantly reduced infection of *Meloidogyne javanica* and root infecting fungi (*Macrophomina phaseolina*, *Rhizoctonia solani*, *F. solani* and *F. oxysporum*) on pumpkin (*Cucurbita pepo*), guar (*Cyamopsis tetragonoloba*), chilli (*Capsicum annuum*) and watermelon (*Citrullus lanatus*). *P. aeruginosa* was more effective than *P. lilacinus* in reducing nematode infection. Combined use of *P. lilacinus* and *P. aeruginosa* was more effective in reducing nematode infection in guar; *M. phaseolina* and *F. oxysporum* on pumpkin; and *F. solani* on guar and watermelon, than either used alone (Perveen *et al.*, 1998).

Brinjal: Results related to root-knot and wilt disease complex management in eggplant under field conditions revealed that the combination formulation of *T.*

harzianum and *P. chlamydosporia* performed very well. The yield was 2.024 kg/ $3m^2$ followed by combination of *P. fluorescens* + *P. chlamydosporia* where the yield was 1.782 kg and *T. harzianum* + *P. fluorescens* (1.680 kg). There was a significant difference between these two treatments. Root gall index (RGI) and wilt disease incidence due to *Ralstonia solanacearum* (WDI) was minimum in combination treatment of *T. harzianum* + *P. chlamydosporia* followed by *P. fluorescens* + *P. chlamydosporia* and *T. harzianum* + *P. fluorescens* where the RGI was 1.8, 1.81 and 1.93, respectively and that of WDI was 22.18%, 25.13% and 27.28%, respectively (Naik, 2004).

Capsicum: In capsicum, based on the field performance of bio-agents, Naik (2004) found that combination of *T. harzianum* along with *P. fluorescens* increased the yield (3.320 kg/$3m^2$ plot) followed by combination of *T. harzianum* and *P. chlamydosporia* (2.960 kg/$3m^2$). The least gall index was present where the combination of *T. harzianum* and *P. fluorescens* was used (1.5) followed by *T. harzianum* + *P. chlamydosporia* (1.7) and *P. fluorescens* + *P. chlamydosporia* (1.7). But among combination treatment of bio-agents, all the treatments were on par (*P. fluorescens* + *T. harzianum*, *P. fluorescens* + *P. chlamydosporia*, *T. harzianum* + *P. chlamydosporia*). Results related to the percent bacterial wilt disease (*Ralstonia solanacearum)* incidence also showed similar trend in which *T. harzianum* + *P. fluorescens* performed very well and the incidence was 16.90% followed by *T. harzianum* + *P. chlamydosporia* (17.20%).

Field application of neem based formulation of fungal (*P. chlamydosporia*) and bacterial (*P. fluorescens*) bioagents resulted in significant reduction of the root-knot (*M. incognita*) and bacterial wilt (*Ralstonia solanacearum*) disease complex to the tune of 70% in capsicum and increased the crop yield by 37% (Rao *et al.*, 2002).

Cabbage and cauliflower: The bioformulation mixture of *P. fluorescens*, *T. viride* and chitin effectively reduced club root and root-knot disease complex in cabbage and cauliflower both under greenhouse and field conditions (Samiyappan, 2003).

(iii) *Ornamental crops*

Gladiolus: Gladiolus plants treated with *P. lilacinus* + *T. harzianum* + neem cake and *P. lilacinus* + *T. viride* + neem cake combinations not only controlled *M. incognita* infection but also *Fusarium* wilt (*F. oxysporum* f. sp. *gladioli*) till the harvest of flower spikes. The corms and cormels obtained from plants treated with these combinations were free from *Fusarium* infection. Fungal colonization of galled roots was maximum in *P. lilacinus* + *T. viride* combination (94%) followed by *P. lilacinus* + *T. harzianum* combination (69%). Similarly, the parasitization of eggs was maximum in *P. lilacinus* + *T. viride* combination (44%) followed by *P. lilacinus* + *T. harzianum* combination (38%), while egg mass parasitization was same in both the combinations (58%) (Nagesh *et al.*, 1999).

Carnation: The studies carried out to evaluate combination of bio-agents for the biological control of wilt (*F. o.* f. sp. *dianthi*) and root-knot nematode (*M. incognita*) disease complex in carnation revealed that *P. chlamydosporia* + *P. lilacinus* gave the best control of wilt and root knot disease complex (Shylaja, 2004).

Tuberose: In course of the experiments carried out to evaluate combination of bio-agents for the biological control of wilt (*F. o.* f. sp. *dianthi*) and root-knot nematode (*M. incognita*) disease complex in tuberose, the best result was found in plants treated with *P. chlamydosporia* + *T. harzianum* (Shylaja, 2004).

(iv) *Spice crops*

Black pepper: Integrated management of foot rot (*Phytophthora capsicii*) and nematodes (*Meloidogyne incognita* and *Radopholus similis*) on black pepper was achieved by i) mixing AMF and *Trichoderma harzianum* in solarized nursery mixture to raise healthy and robust seedlings, ii) application of *T. harzianum* and FYM in planting pit, iii) field application of neem cake at 1 kg/vine mixed with 50 g of *T. harzianum* during August.

Cardamom: *P. lilacinus* in combination with *Trichoderma* spp. suppressed *Meloidogyne* spp. and rhizome rot disease (*Rhizoctonia solani*) complex when incorporated in solarized cardamom nursery beds (Eapen and Venugopal, 1995).

Table 11.1: Integrated management of disease complexes in horticultural crops

Crop	*Disease complex*	*INM components*	*Reference(s)*
Banana	*R. similis* + *F.o.* f. sp. *cubense*	*T. viride* + Neem cake + Carbendazim	Ravi *et al.*, 2001
Papaya	*M. incognita* + *F. solani*	*P. lilacinus* + *T. harzianum*	Khan *et al.*, 1997
Grapevine	*M. incognita* + *F. moniliforme*	*P. fluorescens* + FYM	Senthil Kumar & Rajendran, 2004
Tomato	*M. incognita* + *F.o.* f. sp. *lycopersici*	*T. harzianum* + Vydate/ Nemacur	Stephen *et al.*, 1996
	M. incognita + *Pythium aphanidermatum*	*T. harzianum* + *T. viride* + Press mud/ Carbofuran	Sundaram & Tangaraju, 2001
Brinjal	*M. incognita* + *R. solanacearum*	*T. harzianum* + *P. chlamydosporia*	Naik, 2004
Potato	*Meloidogyne* spp. + *V. dahliae*	5 yr. rotation of maize – sugar beet – barley – barley – potato	Scholte, 1990
Capsicum	*M. incognita* + *R. solanacearum*	*P. fluorescens* + *P. chlamydosporia*/ *T. harzianum*	Rao *et al.*, 2002

Table 11.1: (*Contd...*)

Table 11.1: (*Contd...*)

Crop	*Disease complex*	*INM components*	*Reference(s)*
		T. harzianum + *P. fluorescens*	Naik, 2004
Cabbage & Cauliflower	*M. incognita* + *Plasmodiophora brassicae* (Club root)	*T. viride* + *P. fluorescens* + Chitin	Samiyappan, 2003
Pumpkin, Cluster bean, Chilli, Water melon	*M. javanica* + *M. phaseoli*/*R. solani*/ *F. solani*/ *F. oxysporum*	*P. aeruginosa* + *P. lilacinus*	Perveen *et al.*, 1998
Pigeonpea	*M. incognita* + *F. udum*	Seed treat. with *T. harzianum*/*T. viride* at 4 g/ kg seed + carbofuran/captan	Gowder & Kulkarni, 1999
	H. cajani + *F. udum*	*T. harzianum* + *P. chlamydosporia*/ *G. mosseae*	Siddiqui & Mahmood, 1996
Tuberose	*M. incognita* + *F.o.* f. sp. *dianthi*	*P. chlamydosporia* + *P. lilacinus*, *P. chlamydosporia* + *T. harzianum*	Shylaja, 2004
Gladiolus	*M. incognita* + *F.o.* f. sp. *gladioli*	*P. lilacinus* + *T. harzianum*/*T. viride* + Neem cake	Nagesh *et al.*, 1999
Carnation	*M. incognita* + *F.o.* f. sp. *dianthi*	*P. chlamydosporia* + *P. lilacinus*, *T. harzianum* + *P. lilacinus*	Shylaja, 2004
Black pepper	*M. incognita* + *R. similis* + *P. capsici*	*T. harzianum* + Endomycorrhizae	Anandraj *et al.*, 2001
Cardamom	*M. incognita* + *R. solani*	*P. lilacinus* + *Trichoderma* spp.	Eapen & Venugopal, 1995

Section II

Disease Complexes in Horticultural Crops and their Management

Chapter 12

HORTICULTURAL CROPS

12.1. IMPORTANCE OF HORTICULTURAL CROPS

Agriculture is the mainstay of the Indian economy, and horticulture is a crucial component thereof. Horticulture development has been accorded high priority in the XIth Five Year Plan. The impact of enhanced investment in horticulture has been highly encouraging in terms of vastly improved quality production and export potential-an increase from 96.1 million MT in 1990-91 to 275.4 million MT during 2007-08.

India is the second largest producer of fruits next only to China contributing 12.0% of the total world production. Total production of fruits during 2007-08 has been estimated at 63.503 million MT from 5.775 million ha (Bijay Kumar, 2009) (Table 12.1). India is the largest producer of mango, banana, and papaya. India produces 41.0% of the world mangoes, 29.0% of the bananas and 30.0% of papayas. In grapes, India has recorded the highest productivity per unit area in the world (26 MT/ha). The per capita consumption of fruits has increased from 40 to 85g/day within the last one decade.

Table 12.1: Area, production and productivity of horticultural crops in India (2007-08)

Horticultural crops	*Area ('000 ha)*	*Production ('000 MT)*	*Productivity (MT/ha)*	*% Share of hortil. crops*
Vegetables	7803	125887	16.1	60.84
Fruits	5775	63503	11.0	30.69
Plantation crops	3226	12045	3.7	5.82
Spices	2603	4103	1.6	1.98
Flowers	161	870	5.4	0.42
Aromatic & Medicinal plants	386	325	0.8	0.16
Nuts	132	177	1.3	0.09
Mushrooms	–	37	–	–
Honey	–	65	–	–
Total	20087	207012	10.3	

India is the second largest producer of vegetables next only to China contributing 13.0% of the total world production. Vegetables occupy an area of 7.803 million ha with a production of 125.887 million MT (Bijay Kumar, 2009). India produces 18.0% of the world onions. In vegetables, India occupies the first position in the production of peas and okra, second in onion, brinjal, cabbage, cauliflower and third in potato and tomato. The per capita consumption of vegetables has increased from 96 to 175 g/day within the last one decade.

Floriculture is estimated to cover an area of 161,000 ha with production of 870,000 MT of loose flowers. The area under cut flowers has increased significantly in recent years and the production has reached 4341 million numbers.

1ndia has emerged as the largest producer of spices, tea, cashew and areca nut. Coconuts sustain an area of 1.940 million ha with a production of 108.94 million MT. Areca nut is cultivated in 0.386 million ha with a production of 0.476 million MT. Cashew nuts are grown in an area of 0.868 million ha with a production of 0.665 million MT. The annual production of spices is 4.103 million MT from 2.603 million ha.

Horticulture has emerged as an important sector for diversification of agriculture. Presently, horticulture has established its credibility in improving income through increased productivity, generating employment and in enhancing exports besides providing protective food, household nutritional security, precious raw material for various processing and ancillary industries and finally in enhancing the aesthetic aspects of life. The focused attention on investment in horticulture during the last decade has been rewarding in terms of increased production and productivity of horticultural crops with manifold export potential. Horticulture has played a significant role in women empowerment, providing employment opportunities through mushroom cultivation, floriculture, processing, nursery raising, vegetable seed production etc.

12.2. EXPORT OF HORTICULTURAL PRODUCE

Horticulture in India is fast emerging as a major commercial venture, because of higher remuneration per unit area and the realization that consumption of fruits and vegetables is essential for health and nutrition. In the last one decade, export potential of horticultural crops has significantly increased attracting even multinationals into floriculture, processing and value added products.

The present annual growth rate of horticulture is more than 6.5% which contributes 24.5% from the mere 8% of area to the GDP and 54.55% to the export earnings in the agriculture sector (Tables 12.2 and 12.3).

The major destinations for export of different horticultural products and potential international markets for fresh and processed products are presented in Table 12.4.

Table 12.2: Export of horticultural produce from India (2007-08) (Bijay Kumar, 2009)

Horticultural product	***Export quantity (MT)***	***Value (Rs. Crores)***	***% Share of value***
Floriculture & seeds			
Floriculture	42016.60	64983.50	13
Fruit & vegetable seeds	8031.09	12146.68	2
Total	**1654282.66**	**159332.63**	
Fruits & vegetables			
Fresh onions	1378373.17	116330.57	25
Other fresh vegetables	275909.49	43002.06	9
Walnuts	5062.06	11803.06	2
Fresh mangoes	79060.88	14193.95	3
Fresh grapes	85897.79	30192.45	6
Other fresh fruits	159569.71	25643.64	5
Total	**549949.22**	**173163.18**	
Processed fruits & vegetables			
Dried & preserved vegetables	118787.46	42406.20	9
Mango pulp	156835.52	50582.79	10
Pickle & Chutneys	145216.00	29359.48	6
Other processed fruits and vegetables	129110.24	50814.71	10
Grand Total	**2754181.00**	**505659.00**	

Table 12.3: Export of horticultural produce from India (2007-08) (Bijay Kumar, 2009)

Horticultural produce	***Quantity in MT***	***Value in lakh (Rs.)***
Apple	32665	3330
Banana	16662	2607
Oranges	29261	2726
Grapes	96723	31706
Gauva	2495	417
Litchi	161	63
Mango	54350	12741
Papaya	10879	1123
Pineapple	4194	339
Sapota	2150	503
Brinjal	338	191
Cauliflower	70	2
Onion	100860648	103577
Peas	814	270
Tomato	134845	15290
Potato	78450	4142
Sweet potato	346	35

Table 12.4: Major destinations for export of different horticultural products(Bijay Kumar, 2009)

Horticultural crops	***Major destinations for export***
Fruits	
Fresh grapes	UK, UAE, The Netherlands, Bangla Desh, Germany, Sri Lanka, Norway, Belgium, Nepal, Oman, Russian Republic
Pomegranate	UAE, Saudi Arabia, Sri Lanka, Kuwait
Fresh guava	Saudi Arabia, Netherlands, UAE, Kuwait, Yaman Arab Republic, The Netherlands, Nepal, Sudan, Canada, Oman, Japan
Fresh gooseberries	UAE, Qatar
Fresh mango	UAE, UK, Bangladesh, Nepal, Saudi Arabia, Kuwait, USA, Bahrain, Singapore, Germany
Fresh apple	Bangla Desh, Nepal, China, USA, Saudi Arabia, Germany, UAE
Fresh banana	UAE, Saudi Arabia, Bahrain, Kuwait, Qatar, Nepal, Oman, Maldives, UK, Bangladesh
Fresh orange	Bangladesh, Nepal, USA, Canada, The Netherlands, Oman, Spain, China, UAE, Malaysia
Fresh litchi	Nepal, UAE, UK, Bahrain, Bahamas, Canada, Oman
Fresh papaya	UAE, Saudi Arabia, Qatar, Bahrain, France, Nepal, Kuwait, Bahamas, Germany, The Netherlands
Fresh pineapple	Nepal, UAE, Maldives, Saudi Arabia, Spain, Pakistan, Kuwait, USA, Bangladesh, Armenia
Fresh sapota	UAE, UK, Bahrain, Saudi Arabia, Qatar, Canada, Bahamas, USA, Oman, Kuwait
Mango pulp	UK, Canada, Bangladesh, Bahrain, Netherlands, Japan, Saudi Arabia
Mango slices	UK, Saudi Arabia, Japan, USA
Dried grapes	UK, Malaysia, Netherlands, UAE
Mango pickles	Saudi Arabia, UAE, UK, Canada
Mango chutney	Oman, USA
Guava jam & jelly	UK, Germany, Hong Kong, USA, Netherlands, Saudi Arabia
Dried tamarind	Egypt, UAE, Pakistan, Saudi Arabia, UK
Dried figs	Nepal, Bangladesh
Vegetables	
Fresh brinjal	UK, France Saudi Arabia, Germany, Hong Kong, Portugal, Sweden, Canada, Belgium, UAE
Fresh cauliflower	Nepal, UAE, Maldives, Canada
Fresh onion	UAE, Singapore, Sri Lanka, Bangla Desh, Malaysia, Nepal, Bahrain, Saudi Arabia, Mauritius, Oman
Fresh peas	Saudi Arabia, UAE, Kuwait, USA, Qatar, Nepal, Comoros, Oman, Singapore, Liberia
Fresh tomato	Pakistan, UAE, Bangladesh, Nepal, Maldives, Taiwan, China, Oman, Kuwait, Tanzania Republic

Table 12.4: (*Contd...*)

Table 12.4: *(Contd...)*

Horticultural crops	***Major destinations for export***
Fresh potato	Nepal, Mauritius, Sri Lanka, Maldives, Seychelles, UAE, Iran, Bangladesh, Pakistan, Malaysia
Fresh sweet potato	UAE, Maldives, Israel, France, Kuwait, Oman, Bahrain, Canada, Qatar
Gherkin	France, USA, Spain, Belgium
Dried vegetables	UAE, Sri Lanka, USA, UK, Singapore
Dehydrated onion	Germany, Romania, Sri Lanka, USA, Malaysia, Netherlands, UK
Vegetable seeds	USA, The Netherlands, France, Italy, Japan, Pakistan
Flowers	
Flowers	USA, Japan, The Netherlands, UK, Italy

Horticulture today is not merely a means of diversification but forms an integral part of food and nutritional security, as also an essential ingredient of economic security. Adoption of horticulture both by small and marginal farmers brought prosperity in many regions of the country of which Maharashtra, Karnataka and Andhra Pradesh are the prime examples. A Golden Revolution through significant production increases in horticultural crops is already on the horizon.

Since India is now a signatory member of WTO, it will require its produce and products to be competitive both in the domestic and export markets. This would call for the use of improved horticultural practices such as use of HYV/F_1 hybrids, recommended agro-techniques and plant protection measures, hi-tech intervention and scientific post-harvest management.

After the success of 'Green Revolution' in agriculture, India is now moving towards the **'Golden Revolution'** comprising fruits, vegetables and flowers. Commercialization of horticulture along with post harvest management, processing and marketing assume great significance and add value not only in the context of farmers' income, but also its contribution to the agriculture GDP. Commercialization of horticulture is largely dependent upon use of appropriate technology and market oriented production.

Chapter 13

FRUIT CROPS

Fruits are amongst the most cherished bounties of nature. They are the major sources of Vitamins 'A' and 'C' and minerals like calcium and iron. Fresh fruits have low calorific values and are low in fats. A large variety of fruits are grown in India. Of these, banana, citrus, pineapple, papaya, sapota, jack fruit and avocado (tropical fruits); mango, grape, guava, litchi and loquat (subtropical fruits); apple, pear, peach, plum, apricot, almond and walnut (temperate fruits) and amla, ber, pomegranate, annona, fig and phalsa (arid zone fruits) are important. The leading fruit growing states are Maharashtra, Andhra Pradesh, Tamil Nadu, Kerala, Gujarat, Karnataka, Bihar, Uttar Pradesh, West Bengal and Orissa.

India is the second largest producer of fruits next only to China contributing 12.0% of the total world production. The present area under fruit crops is estimated to be about 5.775 million hectares with an estimated production of 63.503 million MT (Tables 13.1 to13.3). Fruit crops and nuts contribute 31.81% share to the total production of horticultural crops. The nut crops are grown in 0.1209 million hectares with an annual production of 0.1335 million MT (Table 13.4). India is the largest

Table 13.1: Major fruit producing countries of the world (2007-08) (Bijay Kumar, 2009)

Country	*Area (in '000 ha)*	*Production (in '000 MT)*	*Productivity (in MT/ha)*	*% Share of world production*
China	9587	94418	10	19
India	5775	63503	11	12
Brazil	1777	36818	21	7
USA	1168	24962	21	5
Italy	1246	17891	14	3
Spain	1835	15293	8	3
Indonesia	846	11615	14	2
Mexico	1100	15041	14	3
Iran	1256	12102	10	2
Turkey	1049	12390	12	2
Others	22841	208036	9	42
Total (World)	48481	512070	13.09	

producer of mango, banana and papaya. India produces 41.0% of the world mangoes, 29.0% of the bananas and 30.0% of world papaya. In grapes, India has recorded the highest productivity per unit area (26.0 MT/ha) in the world. The per capita consumption of fruits has increased from 40 to 85 g/day within the last one decade.

Table 13.2: Crop-wise area and production of fruit crops in India (2007-08)

Fruit Crops	*Area ('000 ha)*	*Production ('000 MT)*	*Productivity (MT/ha)*	*% Share of India production*
Apple	264	2002	7.6	3.2
Banana	647	23205	35.9	36.5
Citrus	843	7574	9.0	11.9
Grapes	64	1677	26.0	2.6
Guava	178	1975	11.1	3.1
Litchi	69	418	6.0	0.7
Mango	2205	13792	6.3	21.7
Papaya	80	2686	33.4	4.2
Pineapple	80	1216	15.1	1.9
Pomegranate	122	858	7.0	1.4
Sapota	150	1238	8.2	1.9
Others	1071	6862	6.4	10.8
Total	5775	63503	11.0	

Table 13.3: State-wise area, production and productivity of fruit crops (2007-08) (Bijay Kumar, 2009)

State	*Area ('000 ha)*	*Production ('000 MT)*	*Productivity (MT/ha)*	*% Share of India production*
Maharashtra	1432.3	11047.6	7.7	18
Andhra Pradesh	853.0	10772.3	12.6	17
Tamil Nadu	292.3	7530.0	25.8	12
Gujarat	306.9	5849.7	19.1	9
Karnataka	279.0	4588.0	16.4	7
Uttar Pradesh	316.0	3757.5	11.9	6
Bihar	286.2	3252.4	11.4	5
West Bengal	194.3	2766.7	14.2	4
Kerala	323.3	2579.8	8.0	4
Jammu & Kashmir	194.9	1435.9	7.4	2
Assam	122.7	1410.8	11.5	2
Orissa	265.3	1275.1	4.8	2
Madhya Pradesh	46.7	1237.0	26.5	2
Punjab	61.6	1055.4	17.1	2
Others	800.5	3924.8	4.7	8
Total	5775.0	62503.0	11.0	

India has been exporting fruits and fruit products since several decades. However, in the recent past export promotional activities received greater attention. Mango and grapes are the main items in fresh fruit export and some other fruits like pomegranate, sapota, banana, citrus, litchi, apple and strawberry also entered fresh fruit export market in limited quantities.

Table 13.4: Crop-wise area, production and productivity of nut crops in India (2007- 08) (Bijay Kumar, 2009)

Nut crop	*Area ('000 ha)*	*Production ('000 MT)*	*Productivity (MT /ha)*
Almond & walnut	132	177	1.3

13.1. BANANA, *MUSA* SPP.

Globally, banana (*Musa* spp.) is one of the maximum distributed fruit crop grown in more than 130 countries in 4.435 million hectares with an annual production of 82.702 million tonnes (Table 13.5). It is the fourth most important food crop after rice, wheat and maize. It is a dessert fruit for millions and is used in different regions of the world as staple food. As a diet, banana is filling, easy to digest, nearly fat free, rich source of carbohydrate with calorific value of 67 calories/100 g and is free from sodium, making it salt-free diet and also has zero cholesterol. Hence, regular use of it is also considered to be good for heart. Banana also contains several minerals such as calcium, magnesium, potassium and phosphorus. Banana fruit is particularly rich in vitamin-C and contains significant amount of other vitamins such as vitamin-A.

Table 13.5: Major banana producing countries of the world (Bijay Kumar, 2009)

Country	*Area (in '000 ha)*	*Production (in '000 MT)*	*Productivity (in MT/ha)*	*% Share of world production*
India	647	23205	36	29
China	306	7325	24	9
Philippines	430	7000	16	8
Brazil	509	6972	14	8
Ecuador	210	6130	29	7
Indonesia	310	5000	16	6
Coasta Rica	43	2240	52	3
Mexico	75	2200	29	3
Thailand	153	2000	13	2
Colombia	65	1800	28	2
Others	1688	18830	11	23
Total	4435	82702	24.36	

Banana is an important fruit crop having great socio-economic significance in India, as it supports livelihood of millions of people. India is the largest producer of banana in the world with an annual production of 23.205 million tonnes from an area of 0.647 million ha (Table 13.6). Although, banana occupies 10.3 per cent of the area under fruits it contributes 31.8 per cent of the total fruit production in the country. The major banana producing states include Tamil Nadu, Maharashtra, Gujarat, Andhra Pradesh, Karnataka, Bihar, West Bengal, Madhya Pradesh, Assam and Kerala.

Table 13.6: Area, production and productivity of leading banana growing states of India (2007-08) (Bijay Kumar, 2009)

Country	***Area (in '000 ha)***	***Production (in '000 MT)***	***Productivity (in MT/ha)***	***% Share of India production***
Tamil Nadu	114.1	6116.5	53.6	26.4
Maharashtra	80.0	4962.9	62.0	21.4
Gujarat	57.7	3157.7	54.8	13.6
Andhra Pradesh	75.1	2254.3	30.0	9.7
Karnataka	59.9	1513.3	25.3	6.5
Bihar	30.5	1329.4	43.6	5.7
West Bengal	37.4	892.2	23.9	3.8
Madhya Pradesh	15.2	788.2	51.9	3.4
Assam	44.1	610.9	13.8	2.6
Kerala	61.5	493.9	8.0	2.1
Others	71.4	1085.4	15.2	4.7
Total	646.9	23204.8	35.9	

The national productivity of banana is very impressive (36 tonnes per ha). Its productivity in commercial zones especially in Maharashtra, Gujarat, Tamil Nadu, Karnataka, Andhra Pradesh and Bihar is highly impressive and comparable to the productivity in other countries. Although there is a regional preference for cultivars, "Dwarf Cavendish" and "Robusta" (AAA) are the major cultivars of commerce having substantial area. Other important banana cultivars grown in the country are Poovan, Rasthali, Virupaksha, Nendran (AAB), Chakrakeli (AAA) and Monthan (ABB).

Research and development efforts made in the last two decades have resulted in commendable increase in productivity (144 per cent in the last decade) and overall production of banana in the country. The improved production technologies include higher plant densities (4500 plants per ha), improved crop geometries coupled with adoption of drip irrigation, more judicious fertilization management through balanced use of organic and inorganic fertilizers, appropriately timed split applications of inorganics, use of proper sized rhizomes and well timed plantings, commercial production of *in vitro* plants and development of effective integrated pest management strategies. By adoption of these improved technologies, productivity of several individual farmers in different parts of the country has resulted in substantial increase (60 to 65 tonnes per ha).

13.1.1. Burrowing Nematode, *Radopholus similis* and Panama Wilt, *Fusarium oxysporum* f. sp. *cubense* Disease Complex

Incidence and losses due to Panama wilt caused by *F. oxysporum* f. sp. *cubense* is enhanced in association with the burrowing nematode, *R. similis* under high nematode population. This clearly indicates the existence of synergistic interaction between the burrowing nematode and Fusarium wilt pathogen in banana (Fig. 13.1).

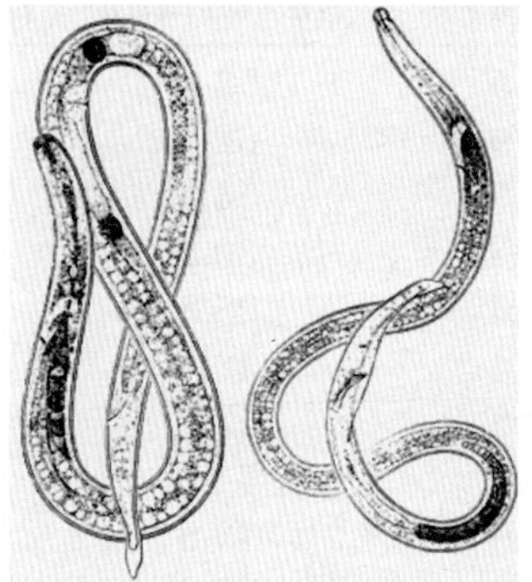

Fig. 13.1: Burrowing nematode and Panama wilt disease complex in banana

Newhall (1958) showed that the incidence of Panama wilt in banana caused by *F. oxysporum* f. sp. *cubense* was doubled in the presence of *R. similis* during the experimental period of three months. When Gros Michel bananas infected with *R. similis* were inoculated with *F. oxysporum* f. sp. *cubense*, the period between inoculation and the onset of wilt was also considerably shortened (Loos, 1959).

Rishbeth (1960) suggested that nematodes breakdown resistance to Panama wilt in Lacatan bananas.

Lesions formed after inoculation with both *R. similis* and *F. oxysporum* f. sp. *cubense* were more extensively necrotic and increased in size more rapidly than when *R. similis* alone was used (Blake, 1966).

Stover (1966) suggested that *F. solani* may contribute to the extension of lesions on banana roots caused by *R. similis* and even to the death of roots.

F. oxysporum f. sp. *cubense* readily establishes itself in the feeder roots of banana when they are invaded by the nematode *R. similis*, but the fungus has seldom been recovered from nematode-free roots (Blake, 1966).

Management methods

(i) *Integrated management*

a) *Bioagents and botanicals*: Soil application of neem cake + *T. viride* + carbendazim was found effective in reducing the burrowing nematode (*R. similis*) and wilt (*F. o.* f. sp. *cubense*) disease complex and in increasing the banana fruit yield by 53.55% (15.147 MT/ha as compared to 9.887 MT/ha in control). This treatment also gave minimum lesion index (1.1) and root-knot index (1.0) as compared to control (4.0) with cost: benefit ratio of 1:2.72 (Ravi *et al.*, 2001).

13.1.2. Nematodes (*Meloidogyne incognita, Pratylenchus coffeae*) and Panama Wilt, *Fusarium oxysporum* f. sp. *cubense* Disease Complex

The root-knot and lesion nematodes along with Fusarium wilt are known to interact and cause disease complex in banana. The plants receiving inoculum of either *M. incognita* or *P. coffeae* or both inoculated before inoculation of *F. oxysporum* f. sp. *cubense* resulted in the early onset and increased severity of wilt disease complex and maximum reduction in plant growth parameters (Table 13.7) (Sundararaju and Thangavelu, 2009).

Table 13.7: Effect of *M. incognita, P. coffeae* and *F. oxysporum* f. sp. *cubense* on plant growth and disease complex in banana

Treatment	*Plant height (cm)*	*Root-lesion index (1-5 scale)*	*Root-knot index (1-5 scale)*	*Wilt index (1-5 scale)*
P. coffeae (Pc)	110	4	1	1
M. incognita (Mi)	115	1	5	1
F. oxysporum f. sp. *cubense (Foc)*	118	1	1	4
Pc + Foc	105	4	1	3
Mi + Foc	110	1	4	5
Pc + Mi + Foc	90	5	3	5
Control	125	1	1	1

Management methods

Botanicals

Application of compost and neem cake was most effective in reducing the disease intensity by *Meloidogyne* sp. and *F. oxysporum* f. sp. *cubense* on banana (Roy *et al.*, 1999).

13.1.3. Root-knot Nematode, *Meloidogyne javanica* and Root Rot, *Macrophomina phaseolina, Rhizoctonia solani, Fusarium* spp. Disease Complex

Association of *M. javanica* and root infecting fungi such as *M. phaseolina, R. solani* and *Fusarium* spp. on tomato was reported by Siddiqui *et al.* (2000).

13.1.4. Root-knot Nematode, *Meloidogyne incognita* and Wilt, *Fusarium oxysporum* f. sp. *cubense, Ralstonia solanacearum* Disease Complex

Association of *M. incognita, F. oxysporum* f. sp. *cubense* and *Ralstonia solanacearum* has been reported causing severe damage on banana crop (Fig. 13.2).

Fig. 13.2: Root-knot nematode and bacterial wilt disease complex

Management

Dipping of banana suckers in 0.2% bavistin solution for 45 min (Roy *et al*., 1998a); flooding and application of inorganic fertilizers at N-225, P-75, K-60 kg/ha (Roy *et al*., 1998a); and soil application of bavistin at 0.2% (Roy *et al*., 1998b) were effective in increasing plant growth, reducing nematode population, and in increasing fruit yield (Table 13.8).

Table 13.8: Management of *M. incognita, F. oxysporum* f. sp. *cubense* and *R. solanacearum* disease complex

Treatment	***Dose***	***Results***	***Reference(s)***
Bavistin	Dipping of suckers for 45 min	Increased plant growth, reduced nematode population, increased fruit yield	Roy *et al*., 1998a
Flooding, appln. of inorganic fertilizers	N-225, P-75, K-60 kg/ha	Increased plant growth, reduced nematode population, increased fruit yield	Roy *et al*., 1998a
Bavistin	Soil appln. of 0.2% 5, 7 & 9 months	Highest fruit yield of 17.4 kg/plant	Roy *et al*., 1998b

13.1.5. Burrowing, Spiral and Root-knot Nematode Complex

Radopholus similis, Helicotylenchus multicinctus and *Meloidogyne incognita* cause nematode complex in banana.

Management methods

Integrated methods

Integration of paring and hot water treatment (at 55°C for 20 min) of banana suckers along with application of carbofuran 3G (at 16.6 g/plant) and neem cake (at 1 kg/plant) at the time of planting reduced the soil and root population of nematodes besides increasing growth, development and yield of banana (Table 13.9) (Shreenivasa *et al.*, 2005).

Table 13.9: Management of nematode complex in banana

*Treatment**	*Fruit yield/ plant (kg)*	*Cost: benefit ratio*	*Nema popn/200 ml soil***			*Nema popn/ 5 g roots*		
			Rs	*Hm*	*Mi*	*Rs*	*Hm*	*Mi*
Untreated check	4.2	1: 1.49	213	150	133	62	51	3
Paring + hot water treat. + carbofuran	5.8	1: 2.01	97	78	40	30	19	21
Paring + hot water treat. + carbofuran + neem cake	7.6	1: 2.12	74	53	48	20	11	10
CD (P = 0.05)	**0.6**	–	**10.6**	**10.6**	**13.9**	**3.0**	**3.5**	**3.3**

* Hot water treatment at 55°C for 20 min, Carbofuran 3G at 16.6 g/plant, Neem cake at 1 kg/plant.

** *Rs – Radopholus similis, Hm – Helicotylenchus multicinctus, Mi – Meloidogyne incognita.*

13.1.6. Burrowing Nematode, *Radopholus similis* and Root Rot, *Cylindrocladium* sp. Disease Complex

Cylindrocladium sp. was reported for the first time from Martinique on banana root systems which were highly damaged. *Cylindrocladium* sp. was proved to be highly pathogenic even when it was inoculated without nematodes. When it was inoculated concurrently with *Radopholus similis* the necrosis of roots was highly accelerated.

Cylindrocladium sp. has been reported on banana roots in Martinique, Costa Rica, Guadeloupe, Ivory Coast and Cameroon. It is likely to be present in all main banana production areas of the world where it could be one of the most damaging agents in association with nematodes (Sarah, 1994).

13.2. CITRUS, *CITRUS* SPP.

Citrus fruits (*Citrus* spp.) rank third in area and production after banana and mango with an estimated production of 7574 million tonnes from an area of 0.843

million hectares (Table 13.10). There are many commercially grown *Citrus* spp. such as mandarins (*Citrus reticulata*), sweet orange (*C. sinensis*), acid lime (*C. aurantifolia*), lemon (*C. limon*), grape fruit (*C. paradisi*), pumello (*C. grandis*), etc. Although one or the other species of the citrus fruit is grown in almost all the states of India, but the major citrus growing states are Andhra Pradesh, Maharashtra, Punjab, Gujarat, Madhya Pradesh, Rajasthan, Karnataka, Orissa, Assam and Uttarakhand. The productivity of citrus in India is low with national average of 9 tonnes per hectare, as compared to 25-30 tonnes per hectare in other citrus growing countries.

Table 13.10: Area, production and productivity of leading citrus growing states of India (2007-08) (Bijay Kumar, 2009)

States	*Area (in'000 ha)*	*Production (in '000 MT)*	*Productivity (in MT/ha)*	*% Share of India production*
Andhra Pradesh	236.5	2997.9	12.7	39.6
Maharashtra	261.3	1627.7	6.2	21.5
Punjab	35.4	618.2	17.4	8.2
Gujarat	34.6	360.2	10.4	4.8
Madhya Pradesh	17.9	286.4	16.0	3.8
Rajasthan	14.4	253.4	17.6	3.3
Karnataka	11.7	250.3	21.4	3.3
Orissa	26.8	211.7	7.9	2.8
Assam	19.0	163.4	8.6	2.2
Uttarakhand	26.8	127.4	4.8	1.7
Others	158.8	677.7	4.3	8.9
Total	843.2	7574.4	9.0	

13.2.1. Burrowing Nematode, *R. citrophilus* and Wilt *Fusarium* spp. Disease Complex

Combined infection of grapefruit seedlings by *R. citrophilus* and *Fusarium* spp. resulted in more severe damage than from the nematode or fungus alone (Feder and Feldmesser, 1961).

13.2.2. Citrus Nematode, *Tylenchulus semipenetrans* and Root Rot Fungi Disease Complex

The role of root-rot fungi in the disease syndrome caused by *T. semipenetrans* is significant. Misra and Edward (1977) reported severe root damage to citrus when the nematode was accompanied by secondary invaders like *Fusarium oxysporum* and *Macrophomina phaseolina*. The nematode helps weakly pathogenic fungus like *Fusarium* to cause secondary infection of citrus roots (Chhabra, 1972). In the presence of *T. semipenetrans*, significantly more citrus plants become infected by *Fusarium oxysporum* and *F. solani* (O'Bannon *et al.*, 1967) and the combined activity of the nematode and *F. solani* is thought to be a factor in the debilitation of citrus. *Fusarium* spp. seems to cause more damage in combination with the citrus nematode in cool, wet soils and with *Phytophthora* spp. in warm, wet soils.

Under greenhouse conditions, the citrus nematode apparently interacts with *Fusarium solani* to reduce citrus seedling growth, and the effect of the pathogen combination is noticeably greater than that of either pathogen alone. This may be important in subsequent root rot of citrus (Van Gundy and Tsao, 1963).

13.2.3. Burrowing Nematode, *Radopholus citrophilus* and Root Rot Fungi Disease Complex

Spreading decline in citrus tree is due to the combined activity of the burrowing nematode, *R. citrophilus* and other root rot organisms that are always found in the parasitized roots. Although it is recognized that the nematode is the primary cause of disease, pathogenic fungal species such as *Fusarium, Sclerotium* and *Thielaviopsis* are commonly found in lesions soon after nematode invasion. Subsequently, bacteria and cellulose destroying fungi such as *Penicillium* and *Aspergillus* invade the root, causing additional destruction. As the nature of the root tissue changes, the nematodes leave their habitat and migrate to new healthy roots (DuCharme, 1968).

13.3. PAPAYA, *CARICA PAPAYA*

Papaya (*Carica papaya*) is one of the most important fruit crops valued for its rich source of vitamins and nutrient content. It is native to tropical America and its place of origin is said to be in Southern Mexico and Costa Rica. It is believed to have been introduced into India during the 16th century. It is grown both in tropical and sub-tropical countries of the world like Brazil, Australia, USA, Taiwan, Puerto Rico, Peru, Pakistan, Bangladesh, Malaysia, India and various parts of Central and South Africa. The ripe papaya fruits are used for table purpose, raw fruits are cooked and used as vegetable. Immature fruits are used for the extraction of proteolytic enzyme papain, which is mainly used as a drug in digestive disorders and in the treatment of ulcers and diphtheria. It has also varied industrial uses *viz.*, in tanning industry, degumming of silk, etc.

India is the largest producer of papaya in the world. It is cultivated on a commercial scale in Andhra Pradesh, Gujarat, West Bengal, Karnataka, Chhattisgarh, Assam, Kerala, Tamil Nadu and Madhya Pradesh. Papaya is grown in 81,000 ha with annual production of 2.686 million tonnes (Table 13.11). Productivity of papaya is the highest (33.4 MT/ha) (Table 13.12) among the fruit crops, which has attracted the growers for its commercial cultivation resulting in the spectacular increase in area and production in last few decades. In the last decade, phenomenal increase in area and production is recorded from Southern states owing to demand of fruits for industrial use.

Table 13.11: Major papaya producing countries of the world (2007-08) (Bijay Kumar, 2009)

Country	*Area (in '000 ha)*	*Production (in '000 MT)*	*Productivity (in MT/ha)*	*% Share of world production*
India	81	2686	33	30
Brazil	37	1898	52	21
Mexico	20	800	40	9
Nigeria	93	765	8	9
Indonesia	8	645	81	7
Ethiopia	11	230	21	3
Congo	13	215	17	2
Peru	14	175	13	1
Philippines	10	164	17	2
Venezuela	10	151	16	2
Others	77	1194	15	13
Total	372	8923	28.45	

Table 13.12: Area, production and productivity of leading papaya growing states (2007-08) (Bijay Kumar, 2009)

States	*Area (in '000 ha)*	*Production (in '000 MT)*	*Productivity (in MT/ha)*	*% Share of India production*
Andhra Pradesh	14.4	1123.4	78.0	41.8
Gujarat	11.2	489.0	43.7	18.2
West Bengal	10.7	308.6	28.9	11.5
Karnataka	3.7	248.9	67.4	9.3
Chhattisgarh	7.4	136.1	18.4	5.1
Assam	7.0	107.1	15.4	4.0
Kerala	18.5	80.6	4.4	3.0
Tamil Nadu	0.4	66.4	181.8	2.5
Madhya Pradesh	0.8	28.3	33.9	1.1
Others	5.0	67.3	13.1	2.5
Total	80.3	2685.9	33.4	

On an average, 1000 g of ripe papaya fruit contains 2500 I.U. of Vitamin A, 8 I.U. of Vitamin B, 33 Sherman units of Vitamin B_2 and 70 mg of Vitamin C. Besides, it contains papain, a proteolytic enzyme which acts as pepsin and is useful in digestive disorders.

13.3.1. Root-knot Nematode, *Meloidogne incognita* and Wilt, *Fusarium solani* Disease Complex

Increasing inocula of *M. incognita*, whether present alone or together with *F. solani*, increasingly decreased seedling emergence of papaya. The combination of highest inocula of both pathogens (3000 nematodes and 3g culture of fungus) caused maximum inhibition of seedling emergence and also increased post-emergence damping-off of papaya seedlings (Fig. 13.3) (Khan and Hussain, 1990).

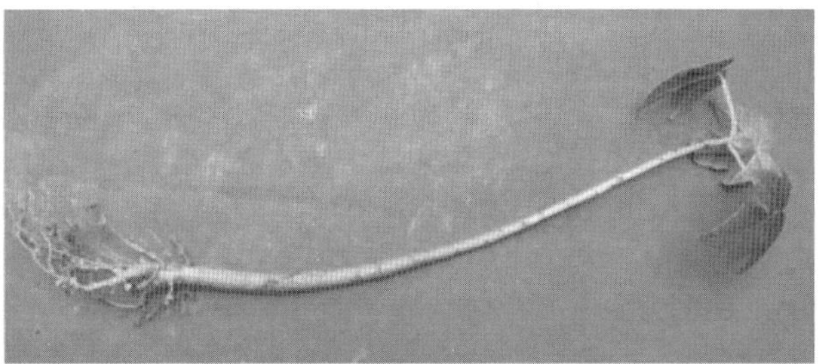

Fig. 13.3: Papaya plant infected with the root-knot nematode

Significant reduction was observed in plant height, root length, shoot and root weight in plants inoculated with nematode and fungus simultaneously, and prior inoculation of nematode followed by fungus 12 days later. Root galling was highest in case of nematode alone followed by nematode inoculation 12 days prior to fungus, and simultaneous inoculation of nematode and fungus (Table 13.13) (Kishore *et al.*, 2005).

Table 13.13: Effect of *Meloidogne incognita* and *Fusarium solani* on disease complex in papaya

Treatment	*Damping-off (%)*	*Plant height (cm)*	*Shoot weight (g)*	*Root weight (g)*	*Galls/g roots*
M. incognita	12	98.7	58.2	10.8	74
F. solani	48	106.3	63.3	13.3	0
M. incognita + *F. solani* (simultaneously)	64	65.3	53.6	9.2	43
M. incognita (prior) + *F. solani* (later)	04	73.2	60.3	9.7	65
F. solani (prior) + *M. incognita* (later)	52	83.7	76.6	15.6	35
Uninoculated Control	0	125.7	85.8	70.6	0
CD (P = 0.05)	**7.15**	**12.66**	**5.12**	**2.37**	-

Management methods

i) ***Integrated management***

(a) *Two bioagents*: Khan *et al.* (1997) have reported that *P. lilacinus* gave better over all protection of papaya plants against *M. incognita* – *F. solani* disease complex than *T. harzianum.* They reported that application of both the biocontrol agents further limited the damage caused by *M. incognita* or *F. solani* and gave a 35% increase in plant growth compared to individual applications.

13.4. STRAWBERRY, *FRAGARIA VESCA*

Strawberry is an attractive, luscious, tasty and nutritious fruit with a distinct and pleasant aroma and delicate flavour. It is rich in vitamin C and iron and is mainly consumed as fresh. Jam and syrup are also prepared from strawberry. It is cultivated in tropical and sub-tropical areas round the year. It is commercially cultivated in Himachal Pradesh, Uttar Pradesh, Maharashtra, West Bengal, Nilgiri Hills, Delhi, Haryana, Punjab and Rajasthan.

13.4.1. Cauliflower Disease, *Aphelenchoides fragariae* and/or *A. ritzemabosi, Rhodococcus fascians*

The disease has been reported to occur in the UK and other European countries as well as in North America. Ritzema-bos (1891) proposed the term 'cauliflower' to describe malformation of the inflorescence and shortening and thickening of the aerial parts of strawberry plants. He considered an ectoparasitic nematode, *Aphelenchoides fragariae* as the causal agent of the disease since this nematode was consistently found in the diseased plants. Studies by Lacey (1936, 1942) provided evidence that a bacterium, *Rhodococcus fascians* was also associated with the disease. Research by Crosse and Pitcher (1952) and Pitcher and Crosse (1958) provided ample evidence of the obligatory association of the two pathogens for full cauliflower syndrome. According to Pitcher (1965) cauliflower disease is a complex requiring the nematode and certain strains of the bacterium and both organisms actively participate in the development of the full disease syndrome. Diseased plants are stunted with deformed stems, leaves and flowers which in advanced stages resemble tiny cauliflowers.

(i) *The nematodes*: Two species of *Aphelenchoides, A. ritzemabosi* and *A. fragariae*, act as vectors of *R. fascians* and play an obligate role in the cauliflower disease complex on strawberry. *A. ritzemabosi* is an obligate parasite and feeds endoparasitically on mesophyll cells and ectoparasitically on buds and growing points. Strawberry is an important host on which it is usually found with *A. fragariae* (Siddiqi, 1974). *A. fragariae* is an obligate parasite of above-ground parts of plants. The nematode feeds ectoparasitically or endoparasitically. It causes crimp or spring dwarf in strawberry and feeds ectoparasitically, living within folded crown and runner buds (Franklin, 1950). Twisting and puckering of leaves, discoloured areas with a rough surface, dwarfed leaves with crinkled margins, reddening of petioles, short internodes of runners, reduced flower tissues and death of crown buds are shown by the infected plants of strawberry (Fig. 13.4).

Fig. 13.4: Left – Strawberry healthy plant, Right – Infected with *A. fragariae*

(ii) *The bacterium*: The cauliflower bacterium, *Rhodococcus fascians* is widespread and remains in healthy strawberries as saprophyte, becoming pathogenic when the plants are invaded by the nematodes. Only a few strains are capable of inducing cauliflower symptoms, and they need to be carried to the meristem of a crown bud by the nematode for the development of cauliflower disease.

(iii) *Symptoms*: The disease manifests in the form of malformation of the inflorescence, shortening and thickening of the aerial parts of strawberry plants. *A. fragariae* and *A. ritzemabosi* act as vectors of *R. fascians* and play an obligatory role in the cauliflower disease complex on strawberry. It is believed that the bacterium attaches to the cuticle of the nematodes and is transported to the crown tissue of strawberry plants (Pionar and Hansen, 1986; Taylor, 1990). Strawberry plants infected with both *A. ritzemabosi* and certain strains of *R. fascians* resemble small cauliflowers (Fig. 13.5). In extreme cases plants are stunted and flowers are deformed. The petals fail to develop or become small and greenish. The sepal becomes much enlarged. Stamens and receptacles are malformed so that fruits do not develop. Auxiliary buds are continually produced in the crowns. Therefore, both the organisms actively contribute to the cauliflower syndrome.

Fig. 13.5: Cauliflower disease affected strawberry plant

(iv) *Management methods*

(a) *Regulatory methods*: Selecting nematode-free propagation material is of paramount importance to reduce the impact of bud and leaf nematode. Whenever possible, certified stocks should be used.

(b) *Physical methods*: Hot water treatment of strawberry runners at 50°C for 5 min or 48°C for 15 min or 46°C for 20 min eliminates the nematode infection.

(c) *Cultural methods*: Uprooting and destroying infected plants, and propagation through healthy planting stock are recommended for the control of nematodes.

Regenerating plants from clean, dormant, excised axillary buds is effective for eliminating *A. fragariae* from strawberry.

(d) *Chemical methods*: Foliar sprays with phenamiphos at 4-6 ml/3.8 litres of water gave good control of *A. fragariae.* Chemicals used for disinfestation of strawberry plants include parathion as a foliar spray and soil application of aldicarb in granular form.

(e) *Host resistance*: In Poland, the strawberry varieties “George Soltwedel”, “Regina” and “Talizman” are comparatively resistant as are “Saksonka” and “Festival’naya” in USSR.

13.4.2. Black Root Rot, *Pratylenchus penetrans, Rhizoctonia fragariae*

Black root rot is a complex disease of perennial strawberry caused by *Rhizoctonia fragariae* and the lesion nematode *Pratylenchus penetrans* (Fig. 13.6). Black root rot is caused by a complex interaction of fungi, nematodes and environmental factors. Several fungi are implicated in the disease including *Rhizoctonia* spp., *Pythium* spp., and *Fusarium* spp. When root lesion nematodes are present in the planting material, the disease is often more severe. This indicates that root feeding by the nematodes may predispose the roots to infection by the fungi. The disease most commonly occurs in low areas, on compacted soils or where organic matter is low.

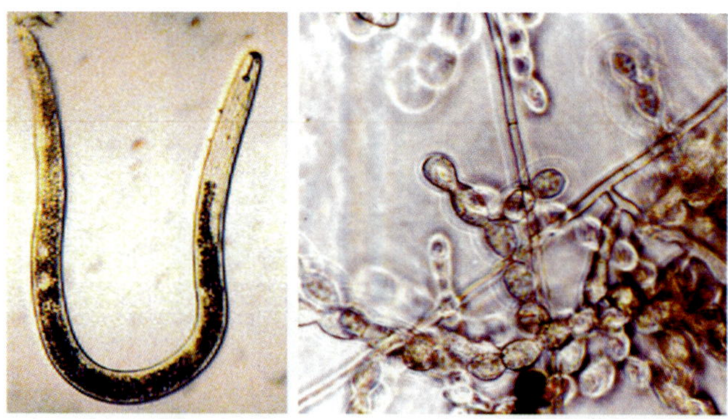

Fig. 13.6: *Pratylenchus penetrans* and *Rhizoctonia fragariae* responsible for black root rot disease complex in strawberry

One or more of the fungi listed above is usually present in the field at the time of planting. Thus, black root rot is not usually introduced into a new planting through nursery stock or contaminated equipment. The exception to this would be recontamination of fungi after a field is sterilized by chemical fumigation. If an environmental factor or other factor (*e.g.*, nematode feeding) is present, the disease complex may be triggered. Factors that favour black root rot include poorly drained soils, winter injury, nutrient imbalances and herbicide damage.

Strawberry plants exhibited black root-rot symptoms when grown in soil infected with *Rhizoctonia fragariae*. *Pratylenchus penetrans* added to soil at 170, 90 and 17 nematodes/cm of soil increased severity of root rot (LaMondia and Martin, 1989).

An economic analysis of lesion nematode populations in *R. fragariae* - infested field soils was conducted based on the regression of yield data with *P. penetrans* populations in small plots at the Connecticut Agricultural Experiment Station Valley Laboratory in Windsor, Connecticut. A "Strawberry Profit Spreadsheet Template" model developed by DeMarree and Riekenberg (1998) was used to analyze the effects of nematodes on profitability. Based on four years of projected fruiting from a planting, strawberry profit expressed as a percentage of gross sales was predicted to be 33%, 30%, 18%, or 0% (operation at a cumulative loss) over four harvest years at initial densities of 0, 12, 50, or 125 *P. penetrans* per g root, respectively. Half of the samples from surveyed growers' fields in Connecticut had populations in excess of 125 nematodes per g root.

Symptoms

Above ground: Symptoms of black root rot usually begin in the first fruiting year. The injury will be most noticeable in low or soil compacted areas of a field where drainage is poor. Strawberry plants with black root rot show a general lack of vigor with poor runner growth and small berries (Fig. 13.7). Plants may collapse when water demand is high such as during spring growth, during or after fruiting, or during drought stress. Aboveground symptoms may resemble symptoms caused by other root disorders; therefore, roots also need to be examined.

Fig. 13.7: Variable plant growth in a black root rot-affected strawberry field

Below ground: To diagnose black root rot, strawberry plants should be sampled in April or May. When taking a plant sample, dig, rather than pull the plant from the ground. Examine the white roots of plants that are showing signs of decline. In healthy plants, the interior of woody roots is yellowish-white. Also, there are abundant fleshy white roots and lateral roots with white cores. If black root rot is present, roots may be rotted at their tips or may appear mottled with black lesions along the white roots. The root core in early stages of black root rot is

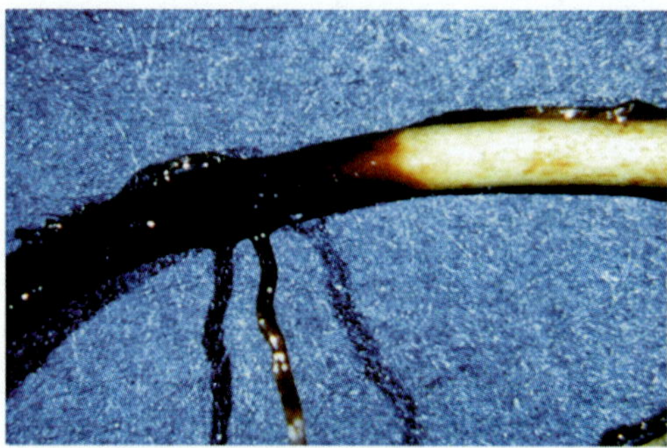

Fig. 13.8: Cortical root rot symptoms on a strawberry structural root

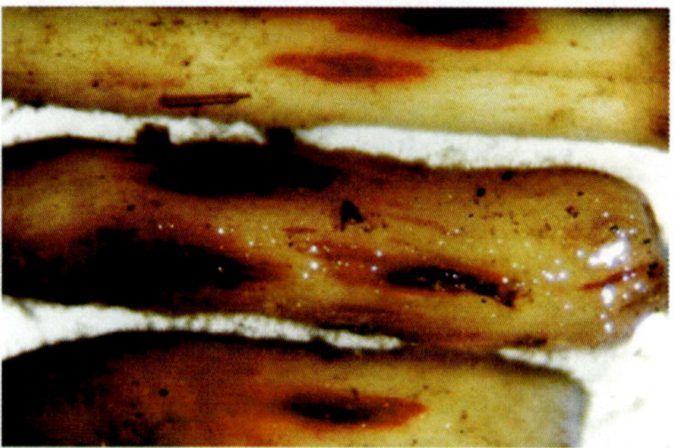

Fig. 13.9: Elliptical cortical lesions resulting from *Pratylenchus penetrans* infection

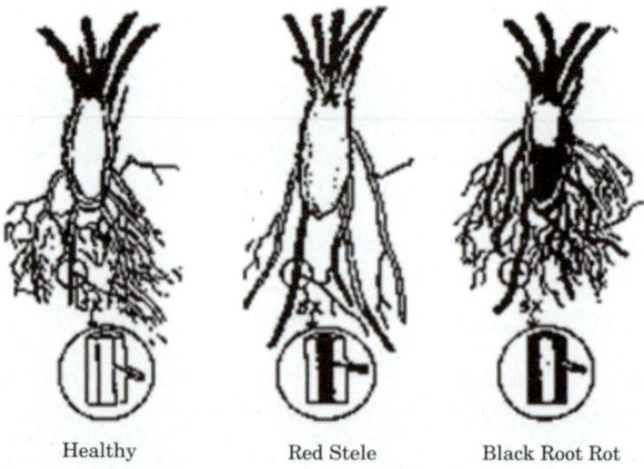

Fig. 13.10: Comparison of root structures of healthy plants and those infected with red stele or black root rot

white, rather than red (which indicates red stele disease) (Figs. 13.8 to13.10). If plants are severely affected, both the core and outer tissue will be black.

Management

Physical methods

Soil solarization: In the Pacific Northwest, soil solarization for two months significantly reduced strawberry root necrosis and root infection by a number of fungi, including *R. fragariae* (Pinkerton *et al.,* 2002). However, solarization did not eliminate the pathogens from the soil.

Cultural methods

Black root rot is favoured by wet soils and soils low in organic matter. As a result, proper site selection and preparation are both important management tools for this disease complex.

Site selection: The future planting site should be selected at least one year before planting. Soil drainage should be good. Avoid low-lying areas that have a tendency to be poorly drained.

Site preparation: If the site had been planted in strawberries previously, practice crop rotation by planting the area into cover crops or annual cash crops for at least two years, but preferably three to five years. Do not rotate with crucifers or legumes because these are alternate hosts for *Rhizoctonia* spp. Sweet corn or pumpkins might be a good choice.

Cover crops should be planted for at least one growing season, even if strawberries were not grown previously, to build up organic matter in the soil. Do not use a legume if black root rot is of concern. Cover crops include annual ryegrass, Sudan grass and sorghum/Sudan (Sudex).

If the site selected does not have good soil drainage, the strawberry planting should be established on raised beds. The raised beds will allow excess soil water to drain from the strawberry root system, creating an environment less favourable to the disease causing fungi. In addition, less soil compaction will occur near the root system.

Soil tests for fertility and pH should be conducted during site preparation. Fertilizer and/or lime should be added before or at planting based on the results.

Preplant soil fumigation is often recommended for control of soil fungi and nematodes. However, there have been mixed results after fumigation with regard to black root rot. In some instances, black root rot has been reported to build up more quickly in fumigated soils than in nonfumigated soils. Researchers suspect that the reason for this may be that when the soil is sterilized by fumigation, beneficial or competitive soil microorganisms are eliminated. In their absence,

Rhizoctonia and other fungi involved in the black root rot complex may increase more rapidly.

Planting: Use only healthy, white rooted plants which have been purchased from a reputable nursery. There are no strawberry varieties, at this time, that are resistant to black root rot. However, some varieties appear to be more tolerant than others. In varietal trials conducted on infested soils in New York, Raritan, Lester and Sparkle showed more tolerance than Blomidon and Honeoye.

After planting: Always use cultural practices which favour good plant growth and development. A soil and/or tissue analysis should be performed each year to determine optimum fertilizer applications. Try to minimize soil compaction created by equipment and cultural practices. Avoid over or under irrigation of the strawberry field. Strawberry plants should be mulched in late fall to prevent winter injury, which predisposes plants to black root rot. In particular, compacted, wet soils tend to heave more during the winter if mulching is insufficient.

Crop rotation: Rotation was one of the first management tactics suggested for black root rot (Zeller, 1932), and current recommendations continue to suggest rotation into small grains for two years (Pritts and Wilcox, 1990; Schroeder, 1988). Small grains have been suggested to suppress *Rhizoctonia solani* (Zeller, 1932), but some *R. solani* AG are pathogenic to small grains (Weller *et al.,* 1986) and the effects on *R. fragariae* are unknown. Rotation away from strawberry to unspecified crops reduced *R. fragariae* isolation from plants to about one third of that seen from continuous strawberry production (Martin, 1988). A dense planting of small grains may reduce broadleaf weeds, but the lesion nematode has a wide host range, including most small grains (Mai *et al.,* 1977), and rotation with grains has been associated with increased lesion nematode damage to potato (Florini and Loria, 1990). Growers in Connecticut have rotated to small grains and still observed poor strawberry growth and black root rot symptoms in replanted fields.

Nematode antagonistic rotation or cover crops, such as 'Saia' oat, sorghosudangrass, *Rudbeckia hirta,* pearl millet '101' and 'Polynema' marigold, have been reported to suppress lesion nematode populations (LaMondia, 1999b; LaMondia and Halbrendt, 2003), but many of these crops have serious drawbacks that limit their utilization, such as seed availability, difficulty in establishment, or cost. Additional plant species need to be evaluated for efficacy against *P. penetrans* and the black root rot complex, seed availability, low cost, and ease of establishment.

Mineral nutrition: Black root rot also may be reduced through the management of mineral nutrition. Ammonium forms of nitrogen nutrition resulted in less black root rot than when plants were fertilized with nitrate nitrogen. The application of ammonium rather than nitrate nitrogen forms to field plots (Elmer and LaMondia,

1995) and microplots (Elmer and LaMondia, 1999) reduced the severity of root rot disease by 10% to 20% and yields were increased by about 15%. Nitrogen form may influence plant mineral composition, the pH of the rhizosphere and the microbial ecology of the rhizosphere (Elmer and LaMondia, 1999), any of which may affect the development of disease.

Biological methods

Biological approaches to disease management also have been attempted. Entomopathogenic nematodes have been implicated in the reduction of root-knot nematode diseases (Bird and Bird, 1986), and *Trichoderma harzianum* (RootShield) has been an effective biological control of a number of fungi, including *Rhizoctonia* spp. (Yuen *et al.,* 1994). In small plots and field microplots, inundative application of entomopathogenic *Steinernema carpocapsae* or *S. feltiae* did not impact *P. penetrans* populations and *T. harzianum* did not affect the development of strawberry black root rot by *R. fragariae*.

Chemical methods

Chemical management of a disease complex caused by soil-borne fungal and nematode pathogens may be achieved by fumigation with a broad-spectrum biocide such as methyl bromide or chloropicrin (Wolfe *et al.,* 1990; Yuen *et al.,* 1991). In fact, virtually all of the strawberry acreage in California is produced on soil fumigated with one of these two compounds (Martin and Bull, 2002). But fumigation is not always successful for a variety of physical and biological reasons, and the application of metham-sodium did not increase root growth, health, or fruit yield (Yuen *et al.,* 1991). The only management tactic associated with nearly complete control of lesion nematodes in Connecticut was fumigation with methyl bromide.

The overall result of soil fumigation may be greatly reduced if the black root rot pathogens are reintroduced on the crowns transplanted into the field. *Rhizoctonia fragariae* consistently has been isolated from healthy nursery-produced planting stock (Ribeiro and Black, 1971) and lesion nematodes also may be extracted at low frequencies. Therefore, the use of healthy planting stock is of particular importance after fumigation or rotation, and the use of fumigation and other control tactics is especially important in nursery production.

In the absence of soil fumigation, alternative control tactics may be targeted at each of the pathogens in the complex. Application of nonfumigant nematicide, such as fenamiphos (Nemacur), may result in temporary nematode control and can increase strawberry vigour and yield. However, nematode population densities still may increase later in the season and may eventually be higher in nematicide-treated plants than in severely diseased plants, which lack sufficient roots to support nematode populations (LaMondia, 1999a). Repeated applications of nematicides or alternate methods for suppressing nematode populations may be required to maintain plant health and productivity over time.

The use of fungicides to suppress *R. fragariae* has been attempted and is currently under investigation, but no success has yet been reported. Fungicidal control of *R. fragariae* may best be targeted at eliminating non-symptomatic infections from planting stock.

13.4.3. Root-knot Nematode, *Meloidogyne hapla* and Wilt, *Verticillium dahliae* Disease Complex

(i) *Management methods*

(a) *Chemical methods*: Soil application of ethylene dibromide (EDB) exhibited dual efficacy against *M. hapla-V. dahliae* disease complex in strawberry (Meagher and Jenkins, 1970).

13.5. RASPBERRY

13.5.1. Tomato Ringspot Virus (TmRSV) in Raspberry

In British Columbia, the virus TmRSV (Fig. 13.11) and vector *Xiphinema americanum* occur only in clay loam soils in an area encompassing about 125 square kilometers and cause serious problem. Raspberries plantings so infected begin to decline about three years after planting. While there is decrease in the amount of vegetative growth each year the most serious problem to the grower is the effect on fruit quality. The virus causes a crumbly condition in the fruit which results in poor quality and difficulty in harvest. Replanting into such a field without adequate nematode control results in a reoccurrence of the problem and a necessity to replant within another 3-4 years.

Fig. 13.11: Raspberry ring spot virus

Oxamyl at 1.0 and 2.0 lb/100 gal applied as foliar spray beginning at full leaf in May 4 times to 4-yr-old raspberry planting reduced the vector population significantly for 2 years. The above treatments also significantly increased yields.

Oxamyl at 2.0 lb/100 gal resulted in an increase of 4246 lb/acre over non-treated plants and a gross return of $ 1132/acre.

13.6. CHERRY

13.6.1. Cherry Rasp Leaf Virus (CRLV)

The cherry rasp leaf virus disease occurs in the western United States and in British Columbia, Canada (Bodine *et al.*, 1951). It has been shown that the disease is caused by the sap-transmissible CRLV (Mansen *et al.*, 1974) and therefore differs from the English rasp leaf virus disease which is induced by the synergistic action of two viruses, prune dwarf virus and raspberry ringspot virus or arabis mosaic virus (Cropley, 1964). Enation symptoms caused by CRLV are very severe and typically restricted to the lower part of the tree. *Xiphinema americanum* transmits the virus. *X. americanum* was recovered from 43 of 64 samples collected and CLRV from 12 of those containing this species. Viruliferous nematodes were present from the soil surface layers to a depth of at least 60 cm. several common orchard weeds such as dandelion (*Taraxacum officinale*), plantain (*Plantago major*), and balsamroot (*Balsamorhizer sagittata*) are virus hosts and may serve as reservoirs from which nematodes can transmit the virus to cherry (Mansen *et al.*, 1974).

13.7. MULBERRY, *MORUS ALBA*

13.7.1. Root-knot Nematode, *Meloidogyne incognita* and Root Rot, *Macrophamina phaseolina* Disease Complex

Inoculation of *M. phaseolina* prior to *M. incognita* significantly reduced the nematode population. Inoculation of *M. phaseolina* simultaneously with *M. incognita* and 15 days after nematode inoculation caused disease incidence and significant reduction in plant growth compared to the inoculation of fungus or nematode alone and fungus inoculation prior to nematode (Muthulakshmi *et al.*, 2010).

13.8. GRAPEVINE, *VITIS VINIFERA*

Grapevine (*Vitis vinifera*) is one of the important commercial crops grown in different climatic zones of the world for table delicacies and wine preparation. It is fairly good source of minerals like calcium, phosphorus and iron and vitamins like B_1 and B_2. In India, it is grown in tropical states like Maharashtra, Karnataka, Tamil Nadu, Andhra Pradesh, Punjab, Mizoram, Madhya Pradesh, Haryana, Jammu and Kashmir and Himachal Pradesh. Grapevine is grown in 64,000 ha with annual production of 1.677 million tonnes (Tables 13.14 and 13.15). In grapes, India has recorded the highest productivity (26 MT/ha) in the world.

Currently, about 85,898 tonnes of fresh grapes are exported, accounting for 19.53% of the total production. Significant quantity of grapes is exported to the Middle East markets and the rest to U.K, Europe and few South-east Asian markets.

Table. 13.14: Major grape producing countries of the world (2007-08), (Bijay Kumar, 2009)

Country	*Area (in '000 ha)*	*Production (in '000 MT)*	*Productivity (in MT/ha)*	*% Share of world production*
Italy	770	8519	11	13
France	830	6500	8	10
China	504	6250	12	9
USA	380	6105	16	9
Spain	1200	6013	5	9
Turkey	540	3923	7	6
Iran	315	3000	10	5
Argentina	220	2900	13	4
Chile	182	2350	13	4
India	64	1677	26	3
Others	2498	19043	8	28
Total	**7503**	**66281**	**11.72**	

Table. 13.15: Area, production and productivity of leading grape growing states (2007-08) (Bijay Kumar, 2009)

States	*Area (in '000 ha)*	*Production (in '000 MT)*	*Productivity (in MT/ha)*	*% Share of India production*
Maharashtra	45.6	1290.0	28.3	76.9
Karnataka	11.1	209.0	18.9	12.5
Tamil Nadu	2.8	83.5	29.8	5.0
Andhra Pradesh	2.5	52.4	21.0	3.1
Punjab	1.0	26.7	26.7	1.6
Mizoram	0.7	8.3	12.4	0.5
Madhya Pradesh	0.1	3.5	35.0	0.2
Haryana	0.1	2.9	29.0	0.2
Jammu & Kashmir	0.2	0.5	2.5	0.0
Himachal Pradesh	0.1	0.1	1.0	0.0
Others	0.2	0.2	1.0	0.0
Total	**64.4**	**1677.1**	**26.0**	

Among these markets, U.K. is the most paying besides being reliable. Middle East markets are not dependable and are also less paying.

13.8.1. Root-knot, *Meloidogyne incognita* and Wilt, *Fusarium moniliformae* Disease Complex

Management methods

(i) *Integrated management*

(a) *Bioagents and botanicals*: Soil application of *Pseudomonas fluorescens* at 100 g and FYM at 20 kg/vine gave effective management of disease complex and improved the plant stand by reducing the final soil nematode population (56.9%), root gall index (1.8) and disease incidence (15.67%). This treatment also increased the number and weight of fruit bunches (17.83 and 155.40%, respectively) and fruit quality (more TSS – 13.53 Brix, TSS: acid ratio – 14.87, lower acidity – 0.91%). The bunch weight of grapevine increased by 155.4% compared to untreated control (Senthil Kumar and Rajendran, 2004).

13.8.2. Root-knot, *Meloidogyne incognita* and Root Rot, *Rhizoctonia solani* Disease Complex

A disease complex involving *M. incognita* and *R. solani* was associated with stunting of grapevines in a field nursery. Nematode reproduction was occurring on Colombard (*Vitis vinifera)* and Ramsey (*V. champinii*) cultivars of grapevine. Root rotting was higher when grapevines were inoculated with both *M. incognita* and *R. solani* and was highest when nematode inoculation preceded the fungus. Shoot weights were lower when vines were inoculated with the nematode 13 days before the fungus compared with inoculation with both the nematode and the fungus on the same day. It was concluded that both the *M. incognita* population and some *R. solani* isolates were virulent against both Colombard and Ramsey cvs. of grapes (Walker, 1997).

13.8.3. Reniform Nematode, *Rotylenchulus reniformis* and Crown Gall, *Agrobacterium tumefaciens* Disease Complex

Lele *et al*. (1978) described an interaction between *R. reniformis* and *A. tumefaciens* in grapevine (Fig. 13.12). Initial damage by the reniform nematode facilitated entry, establishment and disease development.

Fig. 13.12: Crown gall on grapevine

13.8.4. Grapevine Fanleaf Virus

Grapevine fanleaf virus degeneration is the oldest known disease of *Vitis vinifera*. Records in Europe date back 200 years and specimens in herbaria displaying symptoms predate the introduction of American rootstocks (Martelli and Sovino, 1988). The observation that *Xiphinema index* transmits grapevine fanleaf virus (GFLV) was the first record of virus transmission by a nematode (Hewitt *et al.*, 1958). Grapevine fanleaf virus (GFLV), which is transmitted by the ectoparasitic nematode *X. index*, is causing $1 billion losses to the grapevine industry in France (Fig. 13.13).

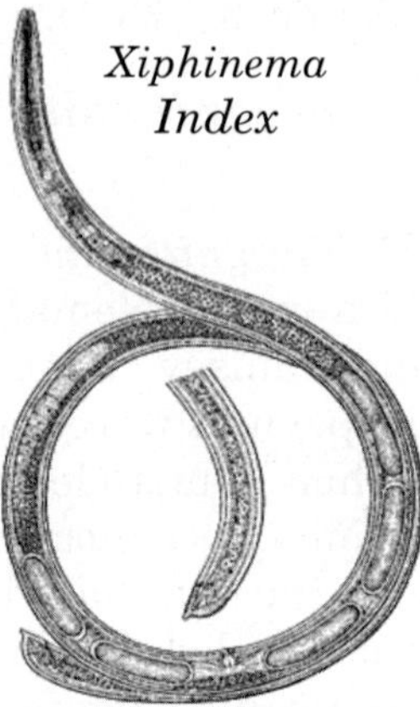

Fig. 13.13: *Xiphinema index*, vector of grape fanleaf virus

GFLV is naturally transmitted by a single nematode species, *X. index* (Andret-Link *et al.*, 2004). *X. index* transmits the fan leaf yellow mosaic vein banding virus (GVFV-GVYMV) disease of grape (Figs. 13.14 to 13.16). Leftover roots of dead and old plants in vineyards act as source of viral inoculum of dagger nematode year after year. *X. americanum* has been found transmitting tobacco mosaic virus, grape yellow vine virus, peach rosette mosaic virus and tomato ring spot virus.

Fig. 13.14: Grapevine plant and leaf infected with fanleaf virus disease

The virus is lost during the molt of the cuticle between life stages. The virus is not passed through the egg stage. Consequently, the virus is re-acquired by feeding of each vermiform life stage of the nematode. GFLV causes reduced vigour, lack of fruit set, and reduced yield of grapevines. Of great importance in the design of control programmes for the *X. index*/GFLV problem is that portions of the grape

Fig. 13.15: Fanleaf infected grapevine leaves. Left – Healthy, Right – Infected

Fig. 13.16: Fanleaf infected grapevine fruit bunches. Left – Healthy, Right – Infected

root system can remain alive and serve as a reservoir for virus and nematode for at least 5 years after vine trunks have been removed (Raski *et al.*, 1965).

The grapevine fanleaf virus is a Nepovirus with isometric particles of 30nm diameter. It is transmitted by *X. index* and *X. italiae*, with *X. index* the more efficient vector (Argelis, 1987; Martelli and Savino, 1988).

The virus can be acquired in less than 15 min of feeding on an infected plant and is retained up to eight months in the absence of feeding (Taylor and Robertson, 1975).

The retention site for GFLV in *X. index* is the cuticular lining of the esophagus, including the area surrounding the odontophore (Taylor and Robertson, 1975). As with other viruses vectored by longidorid and trichodorid nematodes, it is selectively and specifically adsorbed at the retention site, indicating a specific association between protein coat of virus and cuticular surface. Dissociation with cuticle probably occurs as glandular secretions pass forward into the plant cell (Taylor and Robertson, 1975).

The virus does not pass through the egg stage. It is not retained through a molt when the esophageal lining and odontostyle are also shed. In *Xiphinema*, although the lining of the esophagus is not shed at a molt, it undergoes structural changes and virus particles may pass into the intestine (Taylor and Robertson, 1975).

The virus does not replicate in the nematode vector.

Detection

From a management standpoint, it is important to determine whether *X. index* is present in a site intended for a vineyard, and whether the nematodes are viruliferous.

Virus particles can be detected in extracts from single nematodes by immunosorbent electron microscopy (Roberts and Brown, 1980). However, as indicated by Esmenjaud *et al.* (1993) the procedure is complex and not readily adapted to routine assays.

Enzyme-linked immunosorbent assay (ELISA) methods are routinely used to determine GVFL in plant tissue (Walker *et al.*, 1994). When applied to nematodes, the techniques appear sufficiently sensitive to detect the virus if more than ten viruliferous individuals are used, but are unreliable for single individuals (Esmenjaud *et al.*, 1993).

Symptoms

Martelli and Savino (1988) describe three distinct syndromes or symptoms of grapevine fanleaf degeneration:

Malformations: Leaves are variously distorted, asymmetrical, with wide petiole sinuses and abnormal vein arrangement that results in the appearance of an open fan. Leaf symptoms may also include chlorotic mottling. Foliar symptoms are most evident in the spring and through the vegetative growth period. Shoots show abnormal branching, with double nodes and short internodes, while forming angles at internodes. Bunches are small with poor fruit set and shot berries. They ripen irregularly.

Yellow mosaic: Bright yellow discoloration is observed on leaves, shoots, tendrils and inflorescences in the spring. Foliage and clusters are not malformed, but clusters are small.

Veinbanding: Yellow flecking along main veins and associated interveinal areas of mature leaves appear in mid to late summer. Leaves are not malformed, but fruit set is poor and yield is low.

Several closely related nepoviruses found in grapevines are apparently related to GFLV but are serologically different. They are sap-transmissible, and where vectors have been definitively established they are species of *Xiphinema* and *Longidorus*. They include arabis mosaic virus, raspberry ringspot virus, tomato blackring virus, grapevine chrome mosaic virus, Strawberry latent ringspot virus, artichoke Italian latent virus, and grapevine Bulgarian latent virus (Argelis, 1987; Stellmach and Goheen, 1988).

One or more of ten species of *Xiphinema* are present in all major grape-growing regions of the world. They are *X. algeriense*, *X. americanum*, *X. brevicolle*, *X. diversicaudatum*, *X. index*, *X. italiae*, *X. mediterraneum*, *X. pachtaicum*, *X. turcicum*, and *X. vuittenezi* (Raski, 1988). With *X. index* a notable exception, most species of *Xiphinema* have a wide host range and are adapted to a wide range of soil textures.

X. americanum is actually considered a group of some 39 species, which include *X. brevicolle* and *X. pachtaicum*. Molecular techniques (Vrain, 1993), biological evaluations (Halbrendt and Brown, 1993) and virus transmission studies (Brown *et al.*, 1993) are continuing to underscore the variability and similarities within the *X. americanum* group. These studies are adding to the body of knowledge that will be required for an eventual taxonomic revision. The group also includes *X. californicum* Lamberti and Bleve-Zacheo, which is probably associated with grape in California (Lamberti and Ciancio, 1993).

(iv) Management methods

Cultural methods

In France, to save grapevines from attack by *X. index* and subsequently infection by grapevine fan leaf virus, a 7-year rotation with cereal and alfalfa crop has been recommended (Lamberti, 1981).

Chemical methods

D-D at 250 gallons/acre was a suitable recommendation for heavy soil against *X. index*, a vector of grapevine fan leaf virus, while only 200 gallons/acre was found to be sufficient in lighter soil (Rankin *et al.*, 1971).

Management Guidelines for GVFLV - Pre-plant

Avoidance

The wide host ranges of ectoparasitic nematodes make pest avoidance difficult to achieve. One exception is *X. index* and GFLV. The use of virus-free and nematode-free rootings provide the best method of controlling the complex and certification programmes have reduced its dissemination through vegetative propagation (Esmenjaud *et al.*, 1993). The wide host range and current abundance of *X. americanum* in U.S. soils is an indication that quarantine of the nematode is not possible, but the certified virus-free nursery stock program continues to have merit. On-farm nurseries should be maintained nematode-free and located away from old vineyard sites.

Cultural controls

By their nature, ectoparasitic nematodes can survive in soil or persist on weed hosts for long periods of time. Fallowing will only reduce populations, not remove them. The persistence of old grape roots in replant sites is a problem. New grape roots tend to follow the channels in the soil left by old roots. GFLV can be moved from these reservoirs to healthy planting stock through exploratory feeding probes by *X. index.* It may be difficult to develop 'immune' rootstocks on which the nematode cannot reproduce and never even attempts to feed. Replanting of a virus-infected vineyard with a rootstock that is resistant to *X. index* within 6-7 years after removing a virus infected vineyard will not be successful in the long term unless the nematode does not attempt to feed. The site will probably not be safe until 1-2 years after the last old root is dead. Such time intervals may not be economically feasible. Resistance to GFLV will be an important solution to the problem (Esmenjaud, 1986).

Old roots are most easily killed before the old vine is removed by using systemic herbicides. Once the vine is removed the only reliable method of killing roots in the top 2 m of soil is by soil fumigation. Soil ripping to that depth on 30 or 60 cm centers may increase the speed of root death, but this tactic has not been field tested. Heat is a useful root killing agent but we are unable to deliver the required heat using current methods. Soil flooding for months will not kill old grape roots or nematodes. Selected rotation crops will reduce population levels of certain nematode species over time. This will have little value for endoparasites but should be tested with ectoparasites.

Certain *Sorghum bicolor* x *Sorghum sudanense* hybrids grown in summer have negative effects on ectoparasites but the number of years required to eliminate a nematode problem is unknown. Addition of manure, composts and ammonia does not kill old roots, but there are various reports of the benefits of ammonia as a nematode control agent (Mojtahedi and Lownsbery, 1976).

Chemical Control

Soil fumigation with methyl bromide or 1,3-Dichloropropene is effective for killing old roots 1.5 and 2 m deep in soil. Such treatments can also give 99.9% reduction of all nematode species in the top 1.5 to 2 m of soil when properly applied. Two to six years after such treatments, the nematodes do return unless resistant rootstocks are replanted. The use of high rates of 1,3-dichloropropene was halted in the state of California in spring 1990. Methyl bromide use is to be phased out by the year 2010, and adequate replacement fumigants must be environmentally safe. Environmental problems with nematicides have also occurred elsewhere (Rupp, 1990).

The future of all general biocides as a fumigation replacement is unclear. An approach we have taken is the integration of "softer" root killing agents or tools coupled with "softer" soil treatments which reduce soil-dwelling population levels. The practice of removing a vineyard and replanting within one year is not a current option in vineyard management where nematode problems occur. There will be much discovery in this area in the decades ahead. One treatment strategy worthy of study involves use of root killing and soil cleansing treatments followed by one or more years of non-host crops.

Management Guidelines for GVFLV - Established Vineyards

Avoidance

Movement of contaminated grape harvesting equipment and tractors from a site infected with *X. index* should be avoided.

Biological control

The addition of biological agents to soil has been inconsistent and usually ineffective. It may be possible, however, with the advent of drip irrigation systems, to apply microbe-produced toxins directly to the soil in irrigation water. Zoosporic fungal parasites have been isolated from *X. rivesi* and *X. americanum*. Such fungi require free water for their movement through soil. Their occurrence suggests research to investigate water management protocols that will enhance fungal parasitism of dagger nematodes.

Cultural control

The diseases may be prevented by planting vineyards with virus disease-free grapevines in clean soil, and by replanting diseased vineyards only after a long fallow period of more than ten years.

By reducing vine stress through more frequent irrigation the damage caused by nematodes can be reduced.

The use of grassy cover crops in vineyards infested with *X. index* should be studied. However, legume cover crops in vineyards should be monitored to assess population levels of *Mesocriconemella xenoplax,* which may increase. Populations of *X. americanum* will also increase on most cover crop selections. The difficulty in choice of cover crops is that the host range of most ectoparasites is quite broad.

Nematode species with long bodies tend to be in shallow and non-disturbed sites of a vineyard, so placement of any treatment is important. Tillage and soil disturbance can reduce population levels for short periods of time, but root surface area is also reduced.

Drip-irrigation-applied fertilizers that release ammonia may reduce population levels of ectoparasitic nematodes when applied repeatedly. Fifteen kg/ha of nitrogen in urea salt when re-applied three to five times at 30 to 45-day intervals can reduce population levels of most ectoparasites by half. More field testing of these strategies is necessary. Since grapevines do not have a high nitrogen requirement and some vineyards already have excess nitrogen, this technique should be tested and implemented with caution.

Chemical control

High-rate, deep-placement applications of the soil fumigant, 1,3-D, have been successful over a 3-year test period in controlling both the nematode vector, *Xiphinema index*, and the fanleaf-yellow mosaic virus disease of grapevines. A 250-gal-per-acre application rate appears to be necessary, especially on heavy soils. Recent commercial applications of methyl bromide under continuous polyethylene sheeting indicate a good potential for control of fanleaf virus-dagger nematode disease.

Organophosphate and carbamate nematicides are lethal to ectoparasitic nematodes when used as single treatments at high rates via drip irrigation. Treatments with currently-available commercial nematicides only reduce populations about 50% for 6 to 8 months after treatment. Multiple treatments with low rates of phenamiphos (1 kg/ha) are ineffective against ectoparasites. When multiple treatments are used against endoparasites for several years, population levels of ectoparasites such as *X. americanum* can be observed to increase above the non-treated population levels.

In the case of ectoparasites, the value of systemic nematicides will be minimal unless the toxicant is available during feeding or leaks out into the rhizosphere (Edwards, 1991).

Grapevine Rootstocks Resistant to GFLV

Grapevine rootstocks 039-16 (*V. vinifera* x *V. rotundifolia*) and 043-43 have been patented by the University of California and are available through grape rootstock nurseries for use in GFLV-infested vineyards (Table 13.16). Their long-term use is still under study, but preliminary data suggest that they be recommended for use and distribution as rootstocks providing field resistance to infectious degeneration. Resistance to infectious degeneration is complex and could act at several levels. It could result from the root's ability to prevent nematode feeding, or root cells' ability to limit or stop GFLV spread. An ideal rootstock would be resistant not only to the feeding of dagger nematode but also to GFLV, providing two levels of resistance. Then, even a chance probing and transmission of GFLV by the nematode would not result in disease because the rootstock would be able

Table 13.16: Performance of Cabernet Sauvignon scion on selected rootstocks growing on GFLV-infested site in NAPA Valley

Grapevine rootstocks	*Fruit yield/plant (lbs)*	*Resistance to GFLV*
039-16	26.4	Resistant
043-43	24.95	Resistant
AxR#1	16.0	Susceptible
Harmony	9.1	Susceptible
St. George	6.3	Susceptible

to prevent or limit spread of the virus. Researchers at University of California, Davis are continuing to look for resistance to GFLV with the hope of crossing GFLV-resistant plants with known sources of dagger nematode resistance to produce the next generation of rootstocks to combat infectious degeneration.

Future Research

Except for *X. index,* ectoparasitic nematodes have not received the same research attention as endoparasites. Long-term experiments on the damage potential and management of ectoparasitic nematodes in vineyards are needed.

The apparent increase in incidence of *X. index* associated with reduced use of conventional soil fumigants is important and should be monitored. Other problems with ectoparasitic nematodes may similarly emerge. New pre-plant control strategies will be needed in the near future. Resistance and tolerance of rootstocks will continue to be important areas of study.

Tomato Ringspot Virus in Grapevines

Tomato ringspot virus degeneration is also called grape yellow vein disease, tomato ringspot disease, and little berry disease. It is endemic in the northeastern U.S. and Canada, and occurs less frequently in California (Gonsalves, 1988). The nepovirus is transmitted by *X. americanum*, *X. californicum*, and *X. rivesi*. Particles are carried in the esophagus lumen adhering to the cuticular lining. Since there is considerable variability in the species complex that is loosely-termed *X. americanum*, Lamberti and Bleve-Zacheo (1979) contend that reports of *X. americanum* as a vector of TRSV represent a misidentification, or predate taxonomic revisions. They consider that *X. californicum* is the vector of TRSV in California. However, Griesbach and Maggenti (1989) found a wide range of vectoring capabilities of TRSV strains with California populations of *X. americanum sensu lato*.

Symptoms: Tomato ringspot virus degeneration differs geographically, with the disease more severe in colder regions. Leaves are mottled, with oak leaf patterns, new growth is weak and buds are susceptible to cold damage. Internodes are shortened, leaf area is reduced, and clusters are small. The grape yellow vein symptom, common in infections with this virus in California, consists of yellow flecking along the veins and associated lamina. There is reduced fruit set and slow decline of the vine (Gonsalves, 1988).

Peach Rosette Mosaic Virus in Grapevines

Peach rosette mosaic virus decline is a disease of *V. labrusca* grapevines that occurs only in Michigan in the U.S. The virus also causes a disease in peach, as indicated by the name (Ramsdell, 1988). This nepovirus is a strain of TRSV transmitted by *X. americanum* and *Longidorus diadecturus* with the characteristics of other nepoviruses transmitted by these genera. The virus may be endemic in vineyard weeds, which complicates exclusionary measures. Symptoms of Peach Rosette Mosaic Virus Decline include puckered leaves with flattened basal sinuses, sparse clusters, and crooked shoots. The vines are susceptible to winter injury (Ramsdell, 1988).

13.9. GUAVA, *PSIDIUM GUAJAVA*

Guava rank fifth in area and production with an estimated production of 1.975 million tonnes from an area of 0.178 million hectares (Table 13.17). Guava occupies 3.3 per cent of the area under fruits it contributes 3.1 per cent of the total fruit production in the country. The major guava producing states include Maharashtra, Bihar, Uttar Pradesh, Karnataka, West Bengal, Punjab, Andhra Pradesh, Gujarat, Tamil Nadu and Orissa.

Table. 13.17: Area, production and productivity of leading guava growing states (2007-08) (Bijay Kumar, 2009)

States	*Area (in '000 ha)*	*Production (in '000 MT)*	*Productivity (in MT/ha)*	*% Share of India production*
Maharashtra	32.0	250.5	7.8	12.7
Bihar	28.7	255.7	8.9	12.9
Uttar Pradesh	15.1	181.6	12.0	9.2
Karnataka	7.7	166.7	21.6	8.4
West Bengal	11.9	162.2	13.7	8.2
Punjab	8.2	155.5	19.0	7.9
Andhra Pradesh	9.7	136.8	14.1	6.9
Gujarat	8.6	131.1	15.2	6.6
Tamil Nadu	8.5	93.3	11.0	4.7
Orissa	14.0	93.4	6.7	4.7
Others	33.8	348.2	10.3	17.6
Total	178.2	1975.0	11.1	

13.9.1. Guava Wilt, *Fusarium oxysporum* f. sp. *psidii, Helicotylenchus dihystera*

It is a very destructive disease which is widely distributed in almost all guava growing regions of India. It is caused by the fungus, *Fusarium* along with spiral nematode in Chotanagpur region of Jharkhand. Guava wilt is prevalent in Uttar Pradesh, Delhi, West Bengal, Bihar, Madhya Pradesh, Andhra Pradesh, Haryana, Punjab and Rajasthan. The disease complex recorded 5-10% loss annually in 12 districts of Uttar Pradesh which was estimated around Rs. 0.8 to 1.0 million. In West Bengal, guava wilt reduced the yields by 80%, from 11.3 MT/ha in healthy plantations to about 1.82-2.27 MT/ha in affected plantations.

(i) *Symptoms*: Symptoms appear during rainy season. The leaves turn to brown colour, dry and drop. The infected leaves show deficiency of nitrogen and zinc.

The infected branches and twigs show typical die-back symptom. As the disease advances, the entire plant dies in three to four weeks. The cross section of root shows typical blackening of xylem vessels.

Management methods

(i) *Biomanagement*

(a) *Antagonistic bacteria*: *Bacillus subtilis* was found relatively effective by controlling 50% of the anthracnose disease on fruits at post-harvest stage.

(b) *Antagonistic fungi*: Biological control with *Aspergillus niger* AN 17 (Pusa Mrida), *Penicillium citrinum*, *Trichoderma harzianum, T. virens* and intercropping with marigold and turmeric gave good control of wilt. *A. niger* was found most effective which can be mass multiplied on FYM and applied every year as it controls wilt and also provide nutrition to guava plants (Table 13.18).

Table 13.18: Evaluation of bioagents against guava wilt

Bio-agent	*Average wilting (%)*	*% Control*
Penicillium citrinum	47.33	52.67
Aspergillus niger AN 17 (Pusa Mrida)	21.66	78.34
Trichoderma harzianum	24.66	75.34
Control	100.00	0.00

Trichoderma harzianum, which can be easily multiplied on rotten amla fruits and other cheaper substrates locally available with the orchardists, is a potential biocontrol agent against guava wilt.

Out of 5 AMF inoculants, *Glomus aggregatum* was found to show highest potential in colonizing the roots of guava and improve its performance. It also suppressed the population of *Fusarium oxysporum* f. sp. *psidii* in roots.

13.9.2. Root-knot Nematode, *Meloidogyne mayaguensis* and Root Rot, *Fusarium solani* Disease Complex

In Brazil, *M. mayaguensis* has become a menace for guava production. Nationwide, a fourth of the cultivated area is infested, for which the prospect is the decimation of the orchards. Because infected plants have their roots entirely rotten as the disease progresses, a study was set to investigate whether a soil borne pathogen could be involved. In the field, guava seedlings were inoculated with *M. mayaguensis* only or a combination of the nematode with one of 11 *Fusarium* spp. isolates. One hundred-fifty days later, half of the plants inoculated with *M. mayaguensis* and *Fusarium* sp. had started to decline. Variables related to the plants´ shoot and root growth were negatively affected by the fungus, especially when root rooting was intense. The six most damaging isolates were identified as *F. solani*. The extent of the root rotting and damage caused by *F. solani* when associated with *M. mayaguensis* indicates that a complex disease involving these organisms may be responsible for the widespread decimation of orchards throughout Brazil.

13.10. PEACH, *PRUNUS PERSICA*

Peach (*Prunus persica*) is a temperate fruit rich in proteins, sugar, minerals and vitamins. In India, peach is grown in Himachal Pradesh, Punjab, Haryana, Delhi, west Uttar Pradesh, Jammu and Kashmir and Assam.

13.10.1. Peach Tree Short Life

Peach Tree Short Life is a disease complex that is causally related to pathogenic nematodes (ring nematode, *Mesocriconemella xenoplax*; root-knot nematode, *Meloidogyne* spp.; and possibly others), other pathogens and various improper cultural techniques, such as pruning at the wrong time (Nyczepir and Lewis, 1984). Typically, symptoms are a sudden wilting of the tree shortly after bloom. It may involve one or two branches, or the whole tree may collapse during spring time (typically 3 to 7 years of age). When trees or limbs are dying, discolored tissues appear just beneath the bark. When the bark is cut away, a characteristic "sour sap odor" is present. Cold injury and *Pseudomonas syringae* (bacterial canker or blast) are usually considered the final killing factors, but predisposition by nematodes, other pathogens, and incorrect cultural techniques also contribute to death. Plant pathologists and nematologists are working together to search for microbial biocontrol agents (*e.g.*, fluorescent pseudomonads) against *M. xenoplax*.

In the southeastern United States, peach tree short life (PTSL) refers to the sudden spring collapse and death of young peach trees. Generally, trees 3 to 7 years old are affected. PTSL is not caused by a single specific factor, but rather by a complex of cold damage and bacterial canker, caused by *Pseudomonas syringae* pv. *syringae,* which act together in some years and independently in other years. In any case, the final result is the same (*i.e.*, tree death). Many other factors contribute to the PTSL complex, including time of pruning, rootstocks, orchard floor management, fertilization practices, and rapid fluctuation in late winter/early spring temperatures. The primary biotic factor responsible for predisposing peach trees to bacterial canker or cold injury or both is the ring nematode, *Mesocriconemella xenoplax* (Fig. 13.17). Cytospora canker often moves into and finally kills trees badly damaged by cold and/or bacterial canker. The role of cytospora canker in PTSL is, however, considered secondary.

Other problems such as *Armillaria* root rot, stem pitting, and borers have at times been thought to be associated with PTSL, but these problems have distinct symptoms and control measures.

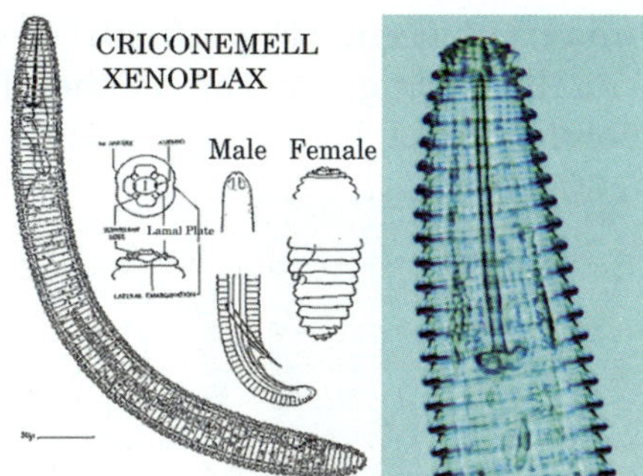

Fig. 13.17: The nematode, *Mesocriconemella xenoplax*

(i) *Symptoms*

The classic symptom of PTSL is the sudden spring collapse of peach trees that were apparently healthy the previous fall. Symptoms are similar to those of any plant deprived of an adequate root system. Cutting into the bark of these trees reveals internal browning extending to, but not below, the soil line (Fig. 13.18). A distinct sour sap odor is often present. Water soaking of bark, or leakage, sometimes occurs in association with this internal browning (Fig. 13.19). Individual

Fig. 13.18: Initially, internal browning due to peach tree short life extends only to soil line (knife blade)

Fig. 13.19: Water-soaked bark and trunk leakage often accompany internal damage to peach tree short life

bacterial cankers may or may not occur (Fig. 13.20). It is usually necessary to examine trees very early in the spring (*i.e.*, at or shortly before bud break) to determine the specific involvement of bacterial canker.

Fig. 13.20: Bacterial canker damage associated with peach tree short life

Bacterial canker of peach is generally associated with delay of bloom or leaf out of individual limbs, with such limbs usually dying by the end of summer. If the infection is severe, entire trees may collapse, with branches exhibiting alternate zones of darkened and healthy tissue having a sour sap odor. Dead trunk tissue usually does not extend beneath the soil line, thus leaving the primary root system alive. As a result, root suckers are usually produced at the base of the tree during summer (Fig. 13.21) when cold injury is present, trees will begin to leaf out until additional water is required by the tree. At that time, leaves collapse (Fig. 13.22) and the bark may crack and separate from the scaffold limbs and trunk (Fig. 13.23). Trees killed or severely damaged by PTSL are frequently invaded by *Cytospora* species. After trees have been dead for several weeks, it may not be possible to specifically separate bacterial canker from cold damage. If trees have been dead one month or more, it may not be possible to diagnose any specific cause.

Fig. 13.21: Peach tree short life damage is typically confined to the above-ground portion of the tree. The surviving root system will often sucker profusely

Fig. 13.22: Sudden collapse of developing leaves and flowers is typical of peach tree short life

Fig. 13.23: Typical peach tree short life cold injury with cracking and separation of bark from trunk and scaffold limbs

(ii) *Disease development*

Bacterial canker can cause sudden spring collapse of peach and other stone fruit trees if the trees have been weakened by factors such as ring nematode or cold damage. Even in the absence of cold damage, trees grown on Nemaguard rootstock are much more susceptible to bacterial canker than trees grown on Lovell, which in turn are more susceptible than trees grown on Guardian (BY520-9) rootstock. In the Southeast, the specific importance of bacterial canker as a factor independent of cold damage is difficult to determine. Cold damage is known to predispose trees to death from bacterial canker. Cold damage alone can also kill trees with similar symptoms. How many trees would have died without invasion of *P. syringae,* or how many would have been killed sooner or later by *P. syringae* without cold damage, cannot be determined and is of little practical importance. The important point is that PTSL is a disease complex, and the control program to be discussed later is aimed at this complex.

Observations indicate that the sudden spring-wilt syndrome of PTSL is related to vascular cambial death. In healthy trees the older wood of the xylem, the water conducting tissue, becomes nonfunctional through time. This nonfunctionality is marked by build-up of gum in the vessel elements. Peach trees are dependent upon the yearly production of new water-carrying vessel elements. New xylem tissue is the tree's main route of water movement. New water-conducting elements are generated annually each spring from a layer of cells called the vascular cambium. The vascular cambium lies just outside the xylem, immediately beneath the spongy bark or phloem. If production of new water-carrying vessels is prevented by cold injury to the vascular cambium, the sudden increase in water consumption due to bloom and/or leaf emergence may place the tree in an overwhelming water deficit that results in collapse and death.

Some injured trees in a PTSL site do not collapse immediately but are weakened and decline slowly during the summer. In these trees, injury to the vascular cambium results in loss of vessel element production insufficient to cause death, but sufficient to impair the tree's water-carrying capacity. This condition results in a weakened tree that undergoes a summer-long decline. Cytospora canker often kills PTSL-weakened trees and is usually very common 1 or 2 years after an injurious winter.

The following different theories of disease development have evolved over the years.

Hormone theory

Death of the vascular cambium and the resulting failure of that tissue to produce water-carrying vessel elements in the spring appears to be due largely to cold injury. Normally, during winter the vascular cambium is dormant and highly resistant to cold damage. After the chilling requirement of the tree has been satisfied and dormancy is broken, growth and production of new vessels begin with the return of warm weather. The growing, active cambium becomes very sensitive to cold damage and can be injured by temperatures that would not have affected it earlier.

The fact that tree death from cold injury is not uniform in PTSL sites suggests that certain trees are protected from cold injury and/or bacterial canker. In many plants, the best adaptive protection mechanism from cold injury is the dormant condition. It is widely accepted that dormancy is controlled by naturally occurring growth hormones, including auxins.

The auxin theory of cambial growth states that renewed cambial activity in the spring is initiated by auxin produced in the expanding portion of the crown (tips of peach branches, for example) and transported to the stem. A loss in cold hardiness accompanies the termination of dormancy. If the cambium is stimulated to become prematurely active during the time of year when cold temperatures are present, cold injury is likely. The question then becomes, what factor or factors influence dormancy? Nematode feeding and fall pruning are factors known to be injurious to peach trees. Both nematode feeding and pruning are wounds that stress the tree. Classically, a wound stress response is seen as an increase in cambial auxin level. Fall pruning stimulates auxin levels more than winter pruning. Indeed, winter pruning does not affect auxin levels at all. Trees growing in fumigated soil showed an immediate auxin increase after fall pruning but not after winter pruning. In nonfumigated soil, auxin levels were high initially, and subsequent early pruning did not further stimulate auxin levels in these trees. The shift toward higher auxin levels may have resulted in later entry of the vascular cambium into dormancy, earlier release from dormancy, or both. Either situation would predispose trees to cold injury.

Nematode injury alone can cause elevated auxin levels. It should be recognized that each tree and tree site varies in terms of vigour, nematode population, soil type, drainage, severity of pruning, and other factors. Such differences can affect hormone levels and release from dormancy. If the nondormant trees are then subjected to the fluctuating cold-warm-cold temperature patterns common to the Southeast in late winter, cold injury to the vascular cambium may result. Many of the most severe occurrences of PTSL take place when the completion of the chilling requirement and release from dormancy are followed by a period of warm weather, which is subsequently followed by a bout of subfreezing temperatures. Further evidence of hormonal disruption within peach trees was shown when levels of cytokinins, another group of plant hormones, were elevated in trees grown in ring nematode-infested soil as compared with noninfested soil. The trees in the ring nematode soil eventually exhibited characteristic PTSL symptoms and died.

Carbohydrate-partitioning theory

It has been demonstrated that Nemaguard peach rootstock (susceptible to ring nematode and PTSL) is more sensitive to ring nematode feeding than Guardian (tolerant to ring nematode and PTSL) peach rootstock. Recent work indicates that Nemaguard permits more carbohydrate reserves to be transmitted from shoot to root in response to ring nematode parasitism than does Guardian. This, in turn, depletes carbohydrate levels in the above-ground portion of trees on Nemaguard rootstocks. High carbohydrate reserves have been shown to be positively correlated with superior cold hardiness and stress resistance. Above-ground depletion of nutrients observed in trees on Nemaguard rootstock subjected to ring nematode feeding is believed to result in tree injury or death at levels of environmental and biological stresses that are innocuous to healthy trees.

(iii) *Economic importance*

Tree loss due to PTSL varies widely from year to year in the Southeast. Losses, sometimes catastrophic, may be localized (Fig. 13.24) or widespread; for example, nearly 200,000 peach trees were reported to have died in Georgia in the spring of 1962. It has been estimated that peach growers in South Carolina alone experience over $6 million loss per year as a result of this disease.

Fig. 13.24: A mixed orchard on Nemaguard (green leaf) and Nemared (red leaf) seedling rootstocks that has been severely afflicted by peach tree short life

(iv) *Management methods*

Control of PTSL in the Southeast and bacterial canker in California has been the subject of considerable research. Several practices have been shown to lessen the occurrence of this disease, and a program for control of PTSL tree losses has been developed. The various aspects of this program can be grouped into three phases of orchard management as follows: (A) site preparation, (B) planting stock selection, and (C) sound cultural practices.

Site preparation

PTSL control should begin during site preparation. Losses due to PTSL are commonly more severe on old peach land than on land newly planted to peaches. The factors that predispose virgin orchard sites to PTSL are not clear. However, root zone nutrient depletion, low pH, hardpan development, and ring nematode buildup all contribute to the occurrence of PTSL, and one or more of these factors occur in any orchard during its life. Ring nematode buildup seems to be an extremely important factor. The time to assess the need for corrective measures is before planting. Soil sampling will determine the need for nutrients and lime. Enough lime should be added to raise the soil pH to 6.5 at a depth of 16 to 18 inches. Nutrients can be incorporated during liming. Subsoiling to break hardpans is a second practice that improves tree survival. Sandy soils and/or dry soils will respond best to subsoiling. Wet clay soils will respond poorly, if at all.

The ring nematode, *M. xenoplax*, greatly increases severity of PTSL. Any land scheduled for planting to peaches should be tested for presence of ring nematode, and, if necessary, pre-plant fumigation should be carried out. Pre-plant crop rotation with wheat has been shown to be comparable to methyl bromide fumigation in suppressing the ring nematode population prior to establishing a peach orchard. Any site that cannot be sampled and prepared properly should not be planted to peaches.

Planting Stock Selection

Good trees alone cannot assure a good orchard, but bad trees almost always assure a bad orchard. Buy trees only from nurseries with a good reputation for producing vigorous, true-to-type, disease-free trees grown in fumigated soil. The major consideration in planting stock will be rootstock selection. Trees grown on Nemaguard rootstock are much more susceptible to PTSL than trees grown on Guardian or Lovell rootstock. Nemaguard rootstock should never be used on any site where ring nematode, *M. xenoplax*, is present. Sites with a history of PTSL or old orchard sites where nematode samples were not taken should not be planted with trees on Nemaguard rootstock. Guardian rootstock has significantly better resistance to PTSL than Lovell (Fig. 13.25) and especially Nemaguard. Its use is recommended on sites prone to PTSL. However, Guardian is not immune to PTSL and will only perform satisfactorily in conjunction with the best management practices for suppressing the disease. Guardian has root-knot nematode resistance comparable to that of Nemaguard and can be considered as an alternative to Nemaguard in those areas of the coastal plain where this nematode is a problem. In addition, it should provide superior resistance to PTSL, which appears to be a growing problem in this area as peach land is replanted.

Fig. 13.25: On severe peach tree short life sites, Guardian (center) provides superior performance compared with Lovell (flanking left - dead), Boone Co. (flanking right - dead), and other tested commercial rootstocks in background

Sound Cultural Practices

Time of pruning is a major cultural practice affecting the incidence of PTSL. Pruning directly affects the susceptibility of trees to cold damage. Trees pruned during October, November, December, and January are more susceptible to PTSL than trees pruned during any other time of the year. Pruning during these months should be avoided. If any pruning must be done during this period, select blocks least likely to have problems with PTSL. Choose blocks over 7 years old, blocks known to be free of *M. xenoplax*, and blocks on sites with no PTSL history. Do not prune trees on Nemaguard rootstock from October through January.

Another practice that affects the occurrence of PTSL is orchard floor management. Harrowing contributes to PTSL by damaging root systems. Use of herbicides in the tree row and sod between the rows reduces PTSL. A sound nutrition program based on annual soil and tissue sampling will help maintain a vigorous orchard and help reduce losses from PTSL. Weak, dying, or dead trees should be removed from the orchard. Several insects and diseases that may become a general nuisance can build up on dead or dying trees.

Obviously, PTSL has no single specific cause and control does not involve a single direct control practice. Rather, the whole package of sound, proven effective cultural practices, the details of which should make up any good orchard management program, must be considered.

Crop Rotation

A non-chemical approach, but one which requires several years of advance planning, is to use rotational crops. Preliminary research in the southeastern U.S. indicates that the crops grown prior to peaches can influence the number of root-knot and ring nematodes. Soybean has been shown to be a good host on which ring nematode can increase. Thus peaches should not be planted in a field immediately following soybean. In contrast, coastal Bermuda grass and small grains, especially wheat, are poor to non-hosts for ring nematode. Several consecutive years of growing coastal Bermuda grass or wheat on an anticipated peach orchard site may possibly reduce the necessity for using a preplant fumigant.

Chemical Methods

Preplant Nematode Control

Where root-knot and ring nematodes are present and PTSL has occurred, preplant soil fumigation is an important component for the establishment of a productive orchard. The soil should be fumigated in October to mid-November prior to planting trees the coming winter or early spring. Telone II (1,3-dichloropropene) at the broadcast rate of 30 gallons per acre is suggested. If nematode counts are low and the site has not been planted in peaches for at least 5 years, fumigation of a strip 6-8 feet wide centered where the tree row will be located may provide adequate nematode control. Because several months elapse before planting the trees, the fumigated areas should be marked. This can be done by planting the row middles with a cover crop such as rye or grass intended to be used for sod, then fumigate between these strips. This practice also limits soil erosion and helps prevent the movement of nematode-contaminated soil into the fumigated areas.

Telone II should be applied 10-12 inches deep using chisels spaced approximately 12 inches apart. The soil should be prepared thoroughly with the surface loose and level. All grass and other plant debris should have been worked into the soil and be well decayed prior to fumigating. The soil temperature should be 50-80°F at a 6-inch depth. Soil moisture should be optimal for planting a crop. A drag or similar type of equipment should be used behind the applicator to seal the soil surface.

Most rates and application schedules of fenamiphos improved peach tree longevity compared with no post-plant treatment. In two experiments, several of the fenamiphos treatments resulted in 100% productive trees by ninth year, while only one of 28 nontreated trees was productive. In a third experiment, the percent of productive trees depended upon the pre-plant treatment and the length of the interval before post-plant fenamiphos treatments were initiated. Discontinuation of post-plant fenamiphos treatments after two or three consecutive years caused *M. xenoplax* numbers to increase significantly and generally resulted in fewer productive trees four years later. The 6.42 kg/ha/application treatment applied as two fall plus one spring application provided the longest beneficial effect resulting in 88-100% productive trees by the beginning of ninth year (Ritchie, 1989).

Integrated Management

- Before planting, apply lime to adjust soil pH to 6.0-6.5 in the top 20 cm.
- Subsoiling during site preparation to break up the hardpan, thereby improving water infiltration, drainage, root growth, nutrient uptake, and diffusion of nematicides.
- In sandy soils where peach trees have been grown previously and in other soils where ring and root-knot nematodes are a problem, fumigate the soil before planting trees.
- Plant trees that have been grown in fumigated soil or in soil free of parasitic nematodes and other diseases.
- Plant trees propagated on Lovell or Halford rootstocks (both are very susceptible to root-knot nematodes; thus, preplant fumigation often is essential).
- Apply nutrients and lime as needed based on soil tests, foliar analysis, and local recommendations.
- Prune as late as possible, never before 1 January and preferably after 1 February. If earlier pruning is unavoidable, prune older trees first. Early pruning is especially hazardous for trees grown on locations where peaches were previously grown. Discontinue summer pruning (including topping and hedging) by 15 September.
- Use recommended herbicides for weed management. Mechanical cultivation, if used, should be shallow to avoid root injury.
- In sites where preplant fumigation was necessary, use a postplant nematicide if ring nematode populations increase. Assay soil for nematodes annually.
- Promptly remove from the orchard and destroy all dead and dying trees.

13.10.2. Root-knot Nematode, *Meloidogyne javanica* and Crown Gall, *Agrobacterium tumefaciens* Disease Complex

Crown gall of peach, caused by *Agrobacterium tumefaciens*, is increased by high populations of *M. javanica* (Nigh, 1966).

In an interaction study, Lovell seedlings were more susceptible to crown gall causing organism, *A. tumefaciens* in the presence of *M. javanica*. Not only there was significantly higher incidence of bacterial infection but also a larger number (27.5% increase) of nematode galls when the two organisms occurred together.

13.10.3. Lesion Nematode, *Pratylenchus* spp. and Root Rot, *Pythium* spp. Disease Complex

Nematodes of the genus *Pratylenchus* along with *Pythium* fungus has been associated with a large percentage of peach trees suffering from peach decline. *Criconemoides* spp. and *Tylenchorhynchus* spp. have also been associated with this condition (Hendrix *et al.*, 1965).

13.11. PLUM, *PRUNUS DOMESTICA*

13.11.1. Ring Nematode, *Mesocriconemella xenoplax* and Wilt, *Pseudomonas syringae* Disease Complex

The susceptibility of young fresh plum trees to *Pseudomonas syringae* increased when they were infected with *Mesocriconemella xenoplax*.

13.12. LITCHI, *LITCHI CHINENSIS*

Table 13.19: Area, production and productivity of leading litchi growing states (2007-08) (Bijay Kumar, 2009)

States	*Area (in '000 ha)*	*Production (in '000 MT)*	*Productivity (in MT/ha)*	*% Share of India production*
Bihar	29.8	223.2	7.5	53.3
West Bengal	8.1	77.8	9.6	18.6
Assam	4.8	34.0	7.0	8.1
Jharkhand	3.3	16.5	5.1	3.9
Tripura	2.6	14.4	5.4	3.4
Punjab	1.5	19.3	12.8	4.6
Uttarakhand	8.8	15.1	1.7	3.6
Orissa	4.0	12.8	3.2	3.1
Himachal Pradesh	3.9	2.5	0.6	0.6
Uttar Pradesh	0.3	0.9	3.4	0.2
Others	2.1	1.8	0.9	0.4
Total	69.2	418.4	6.0	

13.12.1. Litchi Decline

There is currently 70,000 ha of litchi in Vietnam. Litchi decline became a major disease problem in 1997, when 300 ha of litchi from the Bac giang and Hai duong provinces died and a further 1,000 ha were affected by the disease.

The fungi associated with this disease complex include *Fusarium oxysporum, Fusarium solani, Rhizoctonia* sp. and *Pythium* sp. It is likely that nematodes also have a role in this disease complex.

Symptoms: The first symptom may appear as dieback of the new shoots on the tree. The leaves initially become dull in appearance and eventually turn brown on one part of the canopy or the entire tree, depending on the degree of root infection. If only one part of the root system is necrotic, then only that side of the foliage will become brown (Fig. 13.26). The tree will die with the leaves still attached. The brown part of the canopy will not have any new growth or fruit. Infected roots become black and then necrotic. There may be pink discolouration of the collar region of the trunk. Adventitious roots develop from the base of the tree along the soil surface, which can aid in the trees survival for a short time. If the infection remains localized in one part of the canopy, the tree will survive under optimal conditions.

Fig. 26: Decline affected litchi plant

Geographical distribution: This disease has been recorded in all the litchi growing regions of northern Vietnam. It is more serious in Bac giang, Hai duong and Quang ninh provinces in Vietnam.

Host range: The 'vai thieu' cultivar of litchi is the most susceptible. The 'vai lai' and the 'vai chua' cultivars are more tolerant to the disease.

Epidemiology: The warm and wet conditions during the typhoon season promotes disease development. When there is inadequate soil drainage, the disease is more prevalent, especially in those trees growing in low lying areas.

Management

- Set up a drainage system prior to planting the litchi to ensure that there is no standing water around the base of the trees.
- Prune the tree to reduce evaporation and encourage root growth to help the tree to recover quickly.
- Add organic manure to improve the biological resilience of the soil.
- Do not propagate trees from cuttings in soil from diseased areas.

Chapter 14

VEGETABLE CROPS

China is the largest producer of vegetables contributing 47.0% of the total world production (Table 14.1). Vegetables occupy an area of 23.936 million ha in China with a production of 448.983 million MT. Productivity of vegetable crops is highest (33.0 MT/ha) in Spain.

India is the second largest producer of vegetables next only to China contributing 13.0% of the total world production. Vegetables occupy an area of 7.803 million ha with a production of 125.887 million MT (Bijay Kumar, 2009) (Tables 14.2 and 14.3). Vegetable crops contribute 58.93% share to the total production of horticultural crops. India produces 18.0% of the world onions. In vegetables, India occupies the first position in peas; second in cabbage, cauliflower, onion, brinjal; third position in potato and tomato. The per capita availability of vegetables has increased from 76 g in 1951 to 256 g in 2001 as against 280 g/day recommended by ICMR for a balanced diet. Increase in production is mainly due to increase in

Table 14.1: Major vegetable producing countries of the world (2007-08) (Bijay Kumar, 2009)

Country	*Area (in '000 ha)*	*Production (in '000 MT)*	*Productivity (in MT/ha)*	*% Share of world production*
China	23936	448983	19	47
India	7803	125887	16	13
USA	1333	38075	29	4
Turkey	996	24454	25	3
Russia	970	16516	17	2
Egypt	598	16041	27	2
Iran	641	15993	25	2
Italy	528	13587	26	1
Spain	379	12676	33	1
Japan	433	11938	28	1
Others	16957	222625	13	24
Total	**54573**	**946774**		

Source: FAO (except Indian data) (Indian data source: Indian Horticulture Database, 2008)

productivity. Productivity is still lower in India than the world average productivity. There is scope for increasing the productivity of vegetables in India further. The vegetable production target for 2020 has been fixed at 150 million MT.

Table 14.2: Crop-wise area, production and productivity of vegetable crops in India (2007-08) (Bijay Kumar, 2009)

Vegetable crops	*Area ('000 ha)*	*Production ('000 MT)*	*Productivity (MT/ha)*	*% Share of India production*
Potato	1786	34463	19.3	27.4
Onion	805	12157	15.1	9.7
Tomato	572	10261	17.9	8.2
Brinjal	566	9596	17.0	7.6
Tapioca	270	9054	33.5	7.2
Cabbage	265	5888	22.2	4.7
Cauliflower	321	5797	18.1	4.6
Okra	409	4193	10.3	3.3
Peas	314	2560	8.2	2.0
Sweet Potato	126	1146	9.1	0.9
Others	2370	30774	13.0	24.4
Total	**7803**	**125887**		

(*Source*: Indian Horticulture Database, 2008)

Table 14.3: State-wise area, production and productivity of vegetable crops in India (2007-08) (Bijay Kumar, 2009)

State	*Area ('000 ha)*	*Production ('000 MT)*	*Productivity (MT/ha)*	*% Share of India production*
West Bengal	1313.2	22456.8	17.1	17
Uttar Pradesh	960.9	19790.3	20.6	15
Bihar	823.7	14067.7	17.1	11
Orissa	660.8	1214.6	12.4	7
Tamil Nadu	262.7	7975.7	30.4	6
Gujarat	411.7	7402.9	18.0	6
Maharashtra	455.3	6454.8	14.2	5
Karnataka	406.0	5030.9	12.4	4
Andhra Pradesh	284.1	4769.6	16.8	4
Assam	328.9	4474.2	13.6	4
Jharkhand	238.9	3639.7	15.2	3
Kerala	166.9	3479.7	20.8	3
Haryana	274.7	3277.1	11.9	3
Others	915.2	2185.3	2.3	12
Total	**7803**	**125887**	**16.1**	

Potato, tomato, brinjal, chilli, cabbage, cauliflower, onion, peas, gourds, okra and melons are now grown on commercial scale with increasing demand, better output, change in production system and development of suitable cvs. These crops have high potential for foreign exchange earning. Today the emphasis on vegetable breeding is on value addition, disease resistance and better quality.

The important vegetables exported include onions-7.1%, cabbage-4%, cauliflower-4%, okra-3%, peas-3%. Traditional vegetables like onion, potato, okra, bottle gourd and chilli are exported to Malaysia, Singapore, Gulf, Sri Lanka, Bangladesh, Pakistan and Nepal. Non-traditional vegetables like asparagus, cabbage, capsicum, sweet corn, baby corn, green peas, French bean, cucumber, gherkin, cherry tomato, Brussels sprout, broccoli, processed and frozen French bean and lima bean are exported to European countries, Australia, Gulf and S.E. Asian countries. French bean, lima bean, sweet corn and capsicum are exported to Gulf. Some of the new products introduced in export market are tomato paste in bulk aseptic packs, processed button mushrooms, freeze dried and instant frozen vegetables. Dehydrated onion and tomato products have great potential as export commodities.

14.1. POTATO, *SOLANUM TUBEROSUM*

Potato (*Solanum tuberosum*) is one of the important vegetable crops grown in India. It is being cultivated in 1.786 million hectares producing 34.463 million tonnes of potatoes with an average yield of 19 tonnes per hectare (Bijay Kumar, 2009) (Tables 14.4 and 14.5). It thrives best in cool climate. Therefore it is a summer crop in the hills and a winter crop in the plains. The major potato growing

Table 14.4: Major potato producing countries of the world (2007-08)

Country	*Area (in '000 ha)*	*Production (in '000 MT)*	*Productivity (in MT/ha)*	*% Share of world production*
China	5002	72040	14	22
Russia	2863	36784	13	11
India	1786	34463	19	10
Ukraine	1452	19102	13	6
USA	456	17654	39	5
Germany	276	11605	42	4
Poland	570	11221	20	3
Belarus	413	8744	21	3
Netherlands	161	7200	45	2
France	145	6271	43	2
Others	6390	104835	16	32
Total	**19513**	**329919**	**25.90**	

Source: FAO (except Indian data) (Indian data source: Indian Horticulture Database, 2007-08)

Table 14.5: Area, production and productivity of leading potato growing states in India (2007-08)

State	*Area (in'000 ha)*	*Production (in '000 MT)*	*Productivity (MT/ha)*	*% Share of India production*
Uttar Pradesh	504.9	11094.9	22.0	32.2
West Bengal	400.8	9900.8	24.7	28.7
Bihar	315.5	6019.7	19.1	17.5
Gujarat	65.2	1493.9	22.9	4.3
Punjab	79.0	1477.3	18.7	4.3
Madhya Pradesh	50.0	650.4	13.0	1.9
Assam	79.3	514.6	6.5	1.5
Uttarakhand	23.7	483.6	20.4	1.4
Karnataka	61.0	398.4	6.5	1.2
Jharkhand	40.0	377.1	9.4	1.1
Others		2051.8	12.3	6.0
Total	**1786.1**	**34462.5**	**19.3**	

(*Source*: Indian Horticulture Database, 2007-08)

states include Uttar Pradesh, West Bengal, Bihar, Gujarat, Punjab, Madhya Pradesh, Assam, Uttarakhand, Karnataka and Jarkhand.

14.1.1. Potato Early Dying Complex, *Pratylenchus penetrans*, *Verticillium dahliae*

The lesion nematodes often interact with *V. dahliae* in development of potato early dying disease complex in potato. In the arid and semiarid regions of the world, potato early dying (also known as early dying complex or *Verticillium* wilt) is a common limiting factor in potato production (Rowe *et al.*, 1987). It is endemic in many potato production areas of North America.

Potato early dying is arguably the most economically damaging disease complex of potato in the USA when considering direct losses of yield and quality and the cost of control (Rowe and Powelson, 2002). In the Midwestern USA, the disease complex is most important in the French fry processing sector, which relies on long season potato cultivars such as Russet Burbank. *Verticillium dahliae* is generally regarded as the primary pathogen involved in this disease complex (Rowe and Powelson, 2002). This fungus survives as microsclerotia in the soil for extended periods of time (Davis and Huisman, 2001). *V. dahliae* is a variable fungus, separated into groups based on vegetative compatibility (Rowe and Powelson, 2002). Vegetative compatibility group (VCG) 4 is the most aggressive on potato, with subgroup 4A the most important (Joaquim and Rowe, 1991; Strausbaugh, 1993). Soil inoculum levels are known to influence disease severity (Ben-Yephet and Szmulewich, 1985). Previous research in the Midwestern USA has demonstrated that an economic threshold of 8 *V. dahliae* propagules/g (vdpg) of soil triggers the use of the soil fumigant metham sodium (Nicot and Rouse, 1987), although more recent work has cast doubt on the relationship of *V. dahliae* soil populations and yield of potato (Davis *et al.*, 2001). This more recent work suggests edaphic factors, such as organic matter and sodium concentration in the soil, influence disease severity, potato yield and perhaps inoculum efficacy.

Although *V. dahliae* is the most important pathogen in this complex, the root lesion nematode *Pratylenchus penetrans* and the black dot pathogen *Colletotrichum coccodes* are also involved (Otazu *et al.*, 1978; Davis and Huisman, 2001). Coinfection of *V. dahliae* and *C. coccodes* in the same plant hastens plant maturity and reduces plant development more so than either pathogen alone [Otazu *et al.*, 1978; Tsror (Lahkim) *et al.*, 1999]. Most potato cultivars are very susceptible to *V. dahliae* and *C. coccodes*. However, two cultivars, Alturas and Ranger Russet (Novy *et al.*, 2003), developed within the last decade, have resistance or tolerance to *Verticillium* wilt. They now account for nearly 10% of potato production for French fry processing. Unfortunately, the rest of this market sector relies on the use of metham sodium as a soil fumigant for the management of potato early dying. There have been numerous examples of control failure with metham sodium that required investigation, especially since this agrochemical is expensive and can potentially have significant negative impacts on the environment.

(i) *Economic importance*

When disease develops throughout a field midway through the growing season and then becomes severe during the period of maximum tuber bulking, a significant reduction in tuber size and total marketable yield can result (Rowe and Powelson 2002). In North America, yield reduction in moderately diseased fields can easily be 10 to 15 % and in severely diseased fields it can be as high as 30-50% (Powelson and Rowe 1993). The economic impact of potato early dying across the potato industry is significant because of the direct losses resulting from low yields (Rowe *et al.*, 1987). Early Dying complex of potato is a major issue in Manitoba.

On potatoes, the *Globodera–Verticillium dahliae* and *Pratylenchus–Verticillium dahliae* disease complexes have become particularly notorious. Early senescence or 'early dying' caused by *V. dahliae* and *V. albo-atrum* is accentuated by populations of *Pratylenchus* spp. (Martin *et al.*, 1981; Wheeler *et al.*, 1992; Bowers *et al.*, 1996; Hafez *et al.*, 1999), *G. rostochiensis* (Evans, 1987), and *G. pallida* (Hide *et al.*, 1984; Storey and Evans, 1987). In terms of yield, Martin *et al.* (1982) calculated that 15, 50 and 150 *P. penetrans* per 100 cm^3 soil in combination with *V. dahliae* would result in 36, 60 and 75% reductions in potato tuber weight, respectively. However, tuber weights were unaffected by the presence of the individual pathogens, except where nematode populations were high (150 *P. penetrans* per 100 cm^3), when a 12% reduction was found. Yield reduction from the *Pratylenchus–V. dahliae* complex has also been reported elsewhere (Botseas and Rowe, 1994), as have other damaging effects such as the disruption of photosynthesis, stomatal conduction and transpiration (Saeed *et al.*, 1997a; Saeed *et al.*, 1997b). However, the fundamental importance of *P. penetrans* in potato early dying is its ability to activate low populations of *V. dahliae* that would otherwise be inconsequential in disease development (Bowers *et al.*, 1996).

The early dying complex in the USA is exacerbated by the presence of *C. coccodes* – the black dot pathogen. *C. coccodes* can cause reductions in yield and quality by itself [Johnson, 1994; Tsror (Lahkim) *et al.*, 1999], but appears to be particularly important with coinfections of *V. dahliae* [Otazu *et al.*, 1978; Tsror (Lahkim) and Hazanovsky, 2001]. A great deal of attention has been given to tuber infections of *C. coccodes*, which causes a tuber blemish and much of what we know about the disease black dot is from this very important phase of the disease (Read and Hide 1995; Lees and Hilton 2003). However, foliar infections of *C. coccodes* are known to occur and to be an important component of the disease (Johnson and Miliczky, 1993; Johnson, 1994). It also appears that the fungus preferentially grows towards the roots and stolons rather than towards the apex (Nitzan *et al.*, 2006*b*). Whether or not this is the mechanism by which the fungus reaches the vascular tissue is not known. The importance of seed-borne (Johnson *et al.*, 1997) (Data taken from Taylor *et al.*, 2005) compared with soil-borne inoculum (Cullen *et al.*, 2002) is also not known, but there is information suggesting the importance of either source may be cultivar specific (Nitzan *et al.*, 2005). Unfortunately, metham sodium is ineffective in the management of black dot, presumably because the microsclerotia of *C. coccodes* are significantly larger than those of *V. dahliae* making them resistant to this biocide.

The concentrated efforts are underway in understanding the infection frequency of various plant parts relative to yield (Gudmestad *et al.*, 2005). Previous research in Europe has suggested that belowground plant parts, stems, stolons and roots are the most rapidly colonized by *C. coccodes* (Read and Hide, 1995; Andrivon *et al.*, 1998). It is demonstrated that aboveground stems are more rapidly colonized by this fungus and that yield loss is associated with the infection frequency of this plant part (Gudmestad *et al.*, 2005). Data obtained to date also suggest that soil-borne inoculum of 70 *C. coccodes* propagules/g of soil is sufficient to negatively affect tuber yield. The process of developing a duplex real-time PCR technique to quantify soil populations of *V. dahliae* and *C. coccodes* is underway. The *C. coccodes* primers used in the duplex are based on previously published studies (Cullen *et al.*, 2002) and the DNA sequences used to develop the *V. dahliae* were derived from studies characterizing this pathogen (Dobinson *et al.*, 2000). It has been able to correlate signal strength with numbers of *C. coccodes* microsclerotia. Real-time PCR will be a useful tool in developing effective management strategies for *V. dahliae* and *C. coccodes*. Data obtained to date suggest that the number of potato crops planted in a field directly influences the soil population of *C. coccodes*. As the number of times potato is grown on a field increases, so does the population of the black dot fungus. Fields with excessively high populations can be avoided and planted to a rotational crop. The rate of metham sodium can be adjusted based on the level of *V. dahliae* in the soil optimizing its use and minimizing the overuse of this agrochemical.

VCGs are known to exist for both *V. dahliae* and *C. coccodes* and represent a significant amount of diversity among populations of these two pathogens (Joaquim and Rowe, 1991; Strausbaugh, 1993; Heilmann *et al.*, 2006; Nitzan *et al.*, 2006*a*). Similar to *V. dahliae*, differences exist in aggressiveness among VCGs of *C. coccodes*, and it is possible to differentiate among these groups using amplified fragment length polymorphism (AFLP) analyses (Heilmann *et al.*, 2006; Nitzan *et al.*, 2006*a*). In AFLP analysis, once specific bands are converted to sequence-characterized amplified regions or SCAR markers, it will be a useful tool in further studies involving the biology and management of black dot (Heilmann *et al.*, 2006).

(ii) *Symptoms*

A typical visual symptom of potato early dying is loss of plant vigour during mid- to late summer followed by senescence and death of the crop before normal maturity. Initially the disease causes uneven chlorosis of the lower leaves on a few plants. Areas between the leaf veins turn yellow and later brown. Later wilting of the lower leaflets of the plants may occur, and infection proceeds up the infected stems, which often remain erect (Fig. 14.1). The disease causes vascular discolouration of stems and their tuber ends and plants may die without reaching maturity. It is not uncommon for one or two stems of the plant to die and the other portions of the plant to remain healthy (Rangahau, 2003). The symptoms often occur on one side of the plant or on individual leaves (Powelson and Rowe, 1993). A light brown vascular discolouration in basal stem tissues is usually present in symptomatic plants. In severe cases, plants across an entire field will die over a period of several weeks (Rowe and Powelson, 2002).

Fig. 14.1: Potato plants showing wilting symptoms due to early dying disease complex

A light brown vascular discolouration is often visible at the stem base when sliced (Fig. 14.2). Advanced symptoms usually do not occur until after flowering and may consist of a decline of isolated plants or in severe cases, the entire crop may mature early (Davis, 1981; Krikun and Orion, 1979; Mace *et al.*, 1981; Rowe, 1983). Infected potato tubers with Verticillium show symptoms of light brown discolouration of the vascular ring (Fig. 14.3).

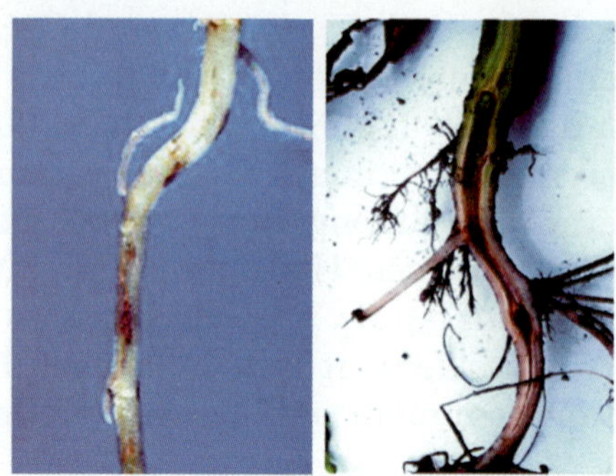

Fig. 14.2: Left - Root lesions, Right - Sliced potato root showing vascular discolouration

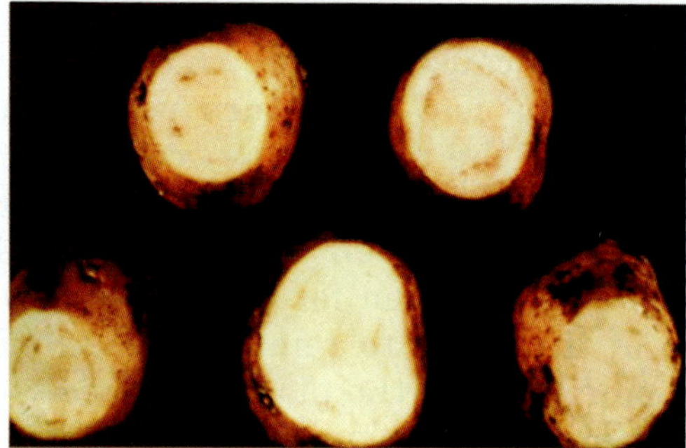

Fig. 14.3: Potato tubers infected with *Verticillium* showing symptoms of light brown discolouration of the vascular ring

(iii) *Etiology*

A major contributor to this syndrome is the soil-borne fungus *Verticillium*. Two species of the fungus, *Verticillium dahliae* and *V. albo-atrum*, cause the disease. *V. dahliae* grows better at slightly higher temperatures (25-28°C), whereas *V. albo-atrum* may be dominant pathogen in cooler production areas (below 25°C) (Powelson and Rowe, 1993; Rowe *et al.*, 1987). Microsclerotia are formed by *V. dahliae* and melanized hyphae are formed by *V. albo-atrum* (Powelson *et al.*, 1993).

Although the disease is primarily caused by the vascular wilt pathogen *V. dahliae*, coinfection of potato by the root-lesion nematode *Pratylenchus penetrans* can increase both the severity of the disease and tuber loss (Fig. 14.4) (LaMondia, 1999; Back *et al.*, 2002). Nematodes are thought to provide entry points for the fungus while feeding on potato roots, or they may reduce the natural host defense mechanism of the potato plant, allowing the fungus to attack roots and infect the host (Rangahau, 2003). Mitchell (1985) found that the effects of *P. penetrans* in combination with *V. dahliae* on symptom expression were additive.

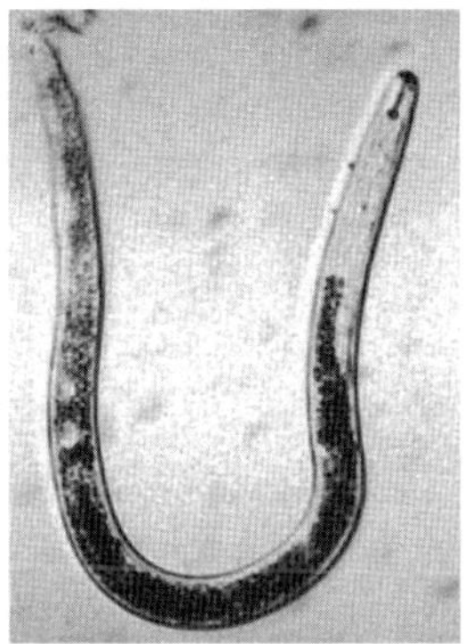

Fig. 14.4: The lesion nematode, *Pratylenchus penetrans*

An important interrelationship exists between *Pratylenchus* spp. and *Verticillium* spp. in potato (Morsink and Rich, 1968). In this situation, either pathogen alone may cause disease but damage is greater when nematodes and fungus occur together. Potatoes infected by *V. albo-atrum* also support greater reproduction of *P. penetrans* than fungus-free plants. Most evidence indicates strongly that *Verticillium* wilt generally promotes increases in reproduction of *Pratylenchus* spp.

A synergistic interaction of *P. penetrans* and *V. dahliae* has been demonstrated to reduce yields and tuber quality of potato, resulting in potato early-dying disease. Wounded roots may stimulate the germination of the dormant microsclerotia of the fungus due to increased root exudation or may facilitate access for the fungus to the vascular tissues. Some have also suggested that the nematode suppresses the production of phytoalexins. There was a strong reduction of photosynthesis rate following joint infection by both pathogens as compared to infection by either of these pathogens alone.

In early potato dying disease, *P. penetrans* predisposes plants to *V. dahliae* infection. Tests with three population levels of *P. penetrans* and two levels of *V. dahliae*, alone or in all combinations, showed that disease occurred when both pathogens interacted synergistically at population levels that individually had little or no effect. Yield reductions of 25-50% often occurred in the presence of both pathogens. Yield losses were most pronounced in seasons when high temperature stress occurred during tuberization (Rowe *et al.*, 1985).

A major contributor to this syndrome is the wilt fungus *V. dahliae* (MacGuidwin and Rouse, 1990). The fungus is favoured by moderate to high temperatures and is restricted to temperate climates (Joaquim and Rowe, 1990). *V. dahliae* survives from season to season as microsclerotia in the soil, either free or embedded in the plant debris, or as mycelium in the vascular ring of the tuber (Rowe and Powelson, 2002). These microsclerotia are stimulated to germinate in response to root exudates (Mace *et al.*, 1981). Kotcon *et al.* (1985) found the effects of root lesion nematode in combination with *V. dahliae* on symptom expression were additive. Severity of potato early dying and associated yield loss are also influenced by seasonal temperature as well as management factors such as irrigation and fertilization (Cappaert *et al.*, 1992). Many factors have been reported to influence the yield loss associated with the disease *e.g.*, climatic conditions (Martin *et al.*, 1982; Stevenson *et al.*, 1976), soil type (Martin *et al.*, 1981) potato cultivar (Bernard and Laughlin, 1976; Bird, 1981; Burpee and Bloom, 1978) and soil fertility (Bird, 1981).

(iv) *Biology and ecology*

Although the biology and ecology of *V. dahliae* is relatively well understood, there are aspects of the pathogen that required further definition to improve disease management. Soil pathogens can be affected by several soil physical properties (Rothrock, 1992), and tillage is known to affect several of these properties and exert an influence on disease development (Peters *et al.*, 2003, 2004). With *Verticillium* spp., soil inoculum levels are known to vary spatially and temporally, but much of that work was interpolated and quite old. Wilhelm (1950) demonstrated that the vertical distribution of *V. albo-atrum* varied considerably, with the greatest proportion residing in the upper 30 cm of soil, nearest the surface. More recently, *V. dahliae* was found to be concentrated primarily in the upper 20 cm soil strata (Ben-Yephet and Szmulewich, 1985; Hamm *et al.*, 2003). It has been determined that the inocula of *V. dahliae* is concentrated in the upper 10 cm (Taylor *et al.*, 2005). It is believed that changes in tillage practices, specifically those directed at soil conservation, are directly responsible for the concentration of inoculum in the upper 10 cm soil strata. It has been demonstrated that tillage practices influence survival of both pathogens as well as the efficacy of metham sodium (Taylor *et al.*, 2005).

(v) *Spread*

Contamination of clean fields with *Verticillium* can occur by wind, irrigation water or on contaminated equipment, but a primary means of infection is introduction of the fungus to the surface or within tissues of seed tubers. Australian studies indicate that up to 22% of certified seed may be infected with *Verticillium* even though the tolerance level in certified seed is 0%. In New Zealand, *Verticillium* may also occur naturally in some areas in the roots of native vegetation. Hence, the causal organism can exist as a part of the natural flora of the soil and may only require the introduction of a susceptible host crop to bring about a serious plant disease problem. Once established in the soil, *Verticillium* can survive for many years in a dormant state as tiny black structures called microsclerotia or as melanised hyphae. Microsclerotia are formed in dead or dying stems of infected plants and are the source of future infections in the field. The microsclerotia germinate in the presence of a susceptible host and under favourable environmental conditions. The fungus invades the roots through root hairs, usually behind the root tip, and establishes itself in the xylem, spreading upwards into stems, petioles and leaflets.

(vi) *Environmental factors*

The development of potato early dying is affected by external abiotic factors, including temperature and moisture. Temperature seems to be the most influencing factor in the development of *Verticillium* species. *V. albo-atrum* is the primary cause of PED in cooler production areas where summer temperatures do not normally exceed 21-24°C while *V. dahliae* is more widespread where average temperatures are often much higher. This reflects the optimum ranges for growth of the two species: the optimum for *V. dahliae* is 21-27°C and 21-24°C for *V. albo-atrum*. A wet season is more conducive to the development of *Verticillium* species than a dry season. Disease severity is greater with excessive soil moisture. An increase in soil water was related to an increase in the disease severity (Capper *et al.*, 1992). Vascular diseases in general are favoured by soil moisture and evapotranspiration conditions that result in rapid unrestricted flow of water - and pathogen propagules - through stems and leaves of the plant.

(vii) *Management methods*

The severity of PED that develops in a given crop is a function of the numbers of *Verticillium* and associated pathogens in the soil at planting, susceptibility of cultivar, and the extent to which the environment favours disease development. Selection of control strategies for potato early dying must be based on an understanding of the biology of the pathogens and the effects of various agricultural practices (Rowe *et al.*, 1987). Management practices are aimed at reducing the population of these pathogens in soil or changing the susceptibility of the host. No single practice will control potato early dying. Therefore, integrated management systems are emphasized in which several decisions are implemented throughout multiple-year rotational cropping programs. Potential components of integrated systems are cultural methods like crop rotation, green manures, fertilization, irrigation and vine removal along with host resistance, pesticides (primarily soil fumigation) and biological controls (Rangahau, 2003). Management practices that modify the rate of stem colonization by *V. dahliae* or increase the host's tolerance to colonization, may also be useful control measures for this disease (Rowe, 1987).

(a) *Cultural methods*: Severity of *V. dahliae* is compounded if fertilizer is applied through gravity furrow irrigation, whereas fertigation through a sprinkler system lessens the severity (Davis and Everson, 1986).

Conservation tillage: Conservation tillage is a commonly used practice in crops planted in rotation with potato. Conservation tillage, known variously as reduced tillage, no tillage or minimum tillage, are practices intended to preserve crop residue near the soil surface to enhance soil organic matter content, protect soil from erosion and several other benefits. Chisel ploughing, as opposed to the traditional tillage practice of mould board ploughing, is the most frequently used method of conservation tillage. Mechanical harvesting of potatoes leaves the crop residue at the soil surface where *V. dahliae* resides. The equipment currently utilized for reduced tillage practices do little to incorporate or mix the soil layers (Taylor *et al.*, 2005), resulting in >70% of the *V. dahliae* inoculum residing in the upper 10 cm of soil. In contrast, mould board ploughing inverts the soil strata thereby causing a significant amount of the *V. dahliae* inocula to reside in soil at depths of 20 cm. It is felt that older studies, and those studies conducted in potato production areas where mould board ploughing is still common, explain the contrasting results between the present work and that found in the literature. Since metham sodium is applied as a liquid but must volatilize into a gas to permeate through the soil profile to act on the microsclerotia of *V. dahliae*, the concentration of soil inoculum at the surface provides a significant challenge in successful soil fumigation. Our studies concluded that much of the metham sodium gas is lost to the atmosphere in the upper 10 cm resulting in very poor mortality of the *V. dahliae* microsclerotia and the concomitant loss of disease control (Taylor *et al.*, 2005). Simple inversion of the soil strata using mould board ploughing significantly improved disease control of early dying. Not only does mould board ploughing improve the efficacy of metham sodium fumigation (Taylor *et al.*, 2005) but this tillage practice increases the mortality of *V. dahliae* propagules in the soil and also significantly reduces black dot, an observation that supports previous studies (Denner *et al.*, 2000). These results are in direct contrast with an earlier study on *Rhizoctonia solani*, in which case disease severity was reduced with conventional tillage practices compared to mould board ploughing (Gudmestad *et al.*, 1978).

Crop rotation: Crop rotation has long been considered an important part of the disease management strategies employed for potato early dying. Although, 2 to 3 year rotation practices do not result in effective control of the disease, they are useful in the total management scheme. In Idaho, the minimum time to effectively reduce inoculum with grain rotation in moderately infested land is 5-10 years (Davis 1981; Davis and McDole, 1979). In Ohio, potato-wheat rotation has been found to be useful as part of an overall control strategy (Joaquim *et al.*, 1988). Potato yield increases were observed with barley-corn rotation (Davis and McDole, 1979).

The influence of rotation and green manure crops on early dying of potato was investigated in 120 field microplots infested with both pathogens. A single season of a Polynema marigold rotation crop significantly reduced *P. penetrans* recovery from roots of rye grown after the potato crop and increased marketable tuber yields. Nematodes were recovered from soil by sugar centrifugation four weeks after incorporation of the rotation crops. Saia oat, Rudbeckia and canola had greater numbers of saprophytic nematodes than potato, and tuber yield was correlated with increased recovery of saprophytes. In the second potato crop after a single year of rotation tuber yields were higher after Polynema marigold. Canola, while maintaining high lesion nematode numbers, resulted in high tuber yields. Rotation with Rudbeckia and canola resulted in the highest tuber yields.

In fields infested with root-knot nematodes and *V. dahliae,* Scholte (1990) found that a five-year rotation of maize-sugar beet-barley-barley-potato had significantly higher potato yields than continuous potato.

The length of rotation also affects the management of early dying, primarily by affecting the soil populations of *V. dahliae*. Soil populations of *V. dahliae* can be 2- to 4-fold higher in soils with a 2-year potato rotation compared to a 3-year potato rotation (one potato crop in three years). Higher soil populations make effective control via soil fumigation with metham sodium much more difficult. For example, if we accept an economic threshold of 8 vppg (Nicot and Rouse, 1987) and an efficacy of metham sodium of no more than 70% (Taylor *et al.*, 2005), then it is easy to understand that soil populations of *V. dahliae* of >25 vppg before fumigation will likely leave a resident population above threshold. For this reason, potato rotation durations should not exceed one potato crop in three years. Additionally, seasonal variations of *V. dahliae* populations clearly indicate that it takes one and a half to two years before the decomposition of potato crop debris will be sufficient enough to obtain effective fumigation (Joaquim *et al.*, 1988; Taylor *et al.*, 2005). When *V. dahliae* soil inoculum is protected by crop debris it cannot be killed by metham sodium, resulting in poor efficacy (Taylor *et al.*, 2005). The data generated is important in reducing soil populations through cultural practices such as mould board ploughing and by improving efficacy of metham sodium applications.

Soil fertility: As nitrogen availability increases, colonization of plant tissue by *V. dahliae* decreases. Early dying complex in 'Russet Burbank' has been found to be most severe when nitrogen is deficient. However, this inverse relationship between available nitrogen and severity of the disease has not been found with 'Nor gold Russet'. Phosphorus levels have significant effects on disease suppression. As phosphorus concentration in the soil increased to the optimum, the colonization of *V. dahliae* in potato stem-tissue decreased. Adequate nitrogen and phosphorus fertilization reduces disease incidence in 'Russet Burbank' and minimizes yield losses that occur with the intensive cropping of potato.

Researchers in the US demonstrated that adding nitrogen and phosphorus into soil to attain plants that were optimally fertilized, rather than nitrogen and phosphorus deficient suppressed potato early dying (wilt incidence and colonization) and also reduced the rate of increase of soil borne inoculum of *V. dahliae* (Rangahau 2003).

Davis *et al.* (1996) reported that in Idaho, the disease was controlled after green manure treatments of either Sudan grass or corn and an increase in yield was seen. Spent mushroom compost may be a means of reducing the effects of the disease on superior varieties of potato and increasing the yield when one or both pathogens are present (LaMondia, 1999).

Rangahau (2003) reported that incorporating organic material such as soymeal into soil a few months before planting potatoes to encourage naturally occurring antagonists in the individual paddock will reduce potato early dying disease.

The burning of potato vines may be of value as a first line of defense by delaying the build up of soil-borne inoculum and improving control of the disease (Mpofu and Hall, 2002). Also, *Verticillium*-free seed may reduce the rate of inoculum increase.

Irrigation: Maintaining plant vigour with careful irrigation and fertility practices is the best management strategy for controlling this disease. Amount and timing of irrigation have been used as a management tool to suppress the disease. Where potatoes are grown under irrigation, the amount of applied water can be reduced early in the season (before tuber initiation) to minimize water use by the crop without sacrificing tuber yield or quality (Rangahau, 2003). Severity of potato early dying was greater throughout each season when plants grown in infested soil were irrigated excessively compared to those receiving deficit irrigation (Cappaert *et al.*, 1992).

In 'Russet Burbank' potato, early dying can be suppressed if water is reduced before tuber initiation. In Wisconsin, tuber yield was significantly lower in deficit than in moderate or excessive pre-tuber initiation irrigation regimes (Cappaert *et al.*, 1994).

(b) *Chemical methods*: Methylisothiocyanate has been successfully used for soil fumigation to reduce the populations of soil-borne fungus, *V. dahliae* and alleviate the effects of disease complex involving wilt-inducing fungus (*V. dahliae*) and the lesion nematode (*P. penetrans*) on potato (Rowe *et al.*, 1987).

(c) *Host resistance*: Planting resistant cultivars is the most efficient, effective economical and environmentally sound means of managing potato early dying disease. Planting of resistant cultivars for five seasons can reduce pathogen levels by 60-70% compared with growing only susceptible cultivars (Selvanathan, 2003). Unfortunately, many commercial cultivars do not possess any resistance to potato early dying and even the most resistant cultivars will develop symptoms of the disease under favourable environmental conditions and when the soil inoculum levels are very high (Ranghau, 2003). Some commercial potato cultivars do exhibit varying degrees of tolerance to early dying, which can lessen long term pathogen survival. Times change, as do the challenges faced and the strategies required to manage potato early dying both effectively and economically. Successful management will involve five key areas: suppression of soil-borne inoculum, new plant protection chemicals, clean planting stock, enhanced host resistance and improved pathogen monitoring tools (Rowe and Powelson, 2002).

14.1.2. Cyst Nematode, *Globodera* spp. and Wilt, *Verticillium dahliae* Disease Complex

The association of fungus like *V. dahliae* with *G. rostochiensis* has been reported on potato indicating that the cyst nematode is involved in the disease complex.

Storey and Evans (1987) found that the pathology of the wilt fungus *V. dahliae* is dependent on the timing of invasion by the cyst nematode *G. pallida*, and also on the potato cultivar. The fungus was found to enter and use the invasion channels of juvenile *G. pallida,* if it was introduced at the same time as the nematode on cvs Pentland Javelin, Maris Peer and Maris Anchor. On both Maris Peer and Maris Anchor, lignified tissue developed following the invasion of the nematodes. If the fungus was introduced 8 days after the nematode, wilt symptoms caused by *V. dahliae* were less severe than on plants treated with the fungus alone. When sections of root tissue were examined, it could be seen that the lignified tissue had sealed the nematode penetration sites. In fact, the living tissue was partially protected by areas of woody tissue where this response had taken place. This response was not produced by cv. Pentland Javelin after nematode invasion, and *V. dahliae* was still able to colonize roots to a larger extent than on plants where it was introduced alone. In addition to the cavities caused during nematode invasion, nematodes produce other forms of mechanical damage to plant roots that are open to exploitation by soil borne fungi.

G. rostochiensis interacts with *V. dahliae* enhancing the severity of wilt caused by the fungus. The leaf symptoms appeared earlier and the tuber yields were depressed in the presence of the nematode (Jones, 1968).

(i) *Management methods*

(a) *Chemical methods*: Soil fumigation with methyl bromide resulted in yield increases of 68% for potato fields infested with *Globodera* spp. and *V. dahliae* (Hide and Corbett, 1974).

In *V. dahliae-G. rostochiensis* disease complex on potato, a mixture of methyl bromide and chloropicrin was found to have comparatively more fungicidal and nematicidal efficacy (Hide and Corbett, 1974).

14.1.3. Cyst Nematode, *Globodera* spp. and Root Rot, *Rhizoctonia solani* Disease Complex

Rhizoctonia solani (AG3) is a soil-borne fungus that is widespread in potato growing land in the UK. It causes stem and stolon canker diseases (Figs. 14.5 and 14.6) and black scurf on the tubers. Potato cyst nematodes (PCN), *Globodera pallida* and *G. rostochiensis* are the most damaging pests of the potato crop in the UK, causing losses in annual potato production with an estimated value in excess of £50 million. Potato cyst nematodes, *G. rostochiensis* and *G. pallida* in association with the fungus *R. solani* caused root rot complex in potatoes. In the glasshouse experiment the combined effects of *G. pallida* with *R. solani* or *G. rostochiensis* with *R. solani* showed greater *R. solani* diseases of potatoes compared with only *R. solani*. Grainger and Clark (1963) found that potato cyst nematode and black-scurf fungus *R. solani* together caused much greater reduction in yield than did the organisms separately.

Fig. 14.5: Germinating sprouts and seedlings killed by *R. solani* before they emerge from the soil

Fig. 14.6: Brown, sunken lesions on underground stems and stolons are caused by *R. solani*

The field experiment revealed a clear positive relationship between initial population densities of *G. pallida* and the incidence of stolons infected by *R. solani*, subsequent stolon pruning and stem canker. In both experiments a clear positive relationship was found between densities of nematodes within the potato roots and the incidence of infected stolons, stolon pruning and stem canker. Nematicide applied together with *R. solani* in the field experiment reduced the effect of *G. pallida* on the incidence and severity of diseases caused by *R. solani*.

14.1.4. Lesion Nematode, *Pratylenchus penetrans* and Root Rot, *Rhizoctonia solani* Disease Complex

P. penetrans increased the infection levels of *R. solani* on Russet Burbank potato roots (Kotcon *et al.*, 1985).

14.1.5. Root-knot Nematode, *Meloidogyne* spp. and Bacterial Wilt, *Ralstonia solanacearum* Disease Complex

The influence of root-knot nematodes on the development and severity of bacterial wilt has been elucidated (Akiew *et al.*, 1991). Plants infected with both *Meloidogyne* spp. and *R. solanacearum* have been found to express wilt symptoms earlier with greater severity than those infected with *R. solanacearum* alone (Johnson and Powell, 1969; Akiew *et al.*, 1991). Sitaramaiah and Sinha (1985) have suggested that the nematode induced stress is more important as a wilt-triggering factor than as a wounding agent. Infection by *Meloidogyne* spp. is known to predispose potato plants to *R. solanacearum* (Fig. 14.7) (Ateka *et al.*, 2001).

Fig. 14.7: Root-knot nematode and bacterial wilt disease complex in potato

Differences in bacterial wilt severity were significant ($P \leq 0.05$) among the treatments for all the three cultivars used in this test (Table 14.6). Bacterial wilt severity was higher in plants inoculated with *R. solanacearum* and *M. incognita* than those inoculated with *R. solanacearum* alone (Ateka *et al.*, 2001).

Analysis of variance revealed significant ($P \leq 0.05$) differences in bacterial wilt severity among the three cultivars. Plants whose roots were severed and inoculated with the two pathogens had the highest (3.4) overall bacterial wilt severity. Wilt severity was lower (3.2) but not significantly ($P > 0.05$) different in plants inoculated with both pathogens without root severing. Wilt severity in plants inoculated with *R. solanacearum* without root-wounding was significantly ($P = 0.05$) lower (1.1) than in plants inoculated with *R. solanacearum* and whose roots were severed (2.6). Disease progress in the three potato cultivars is depicted in Table 14.7. For all the cultivars, significant ($P = 0.05$) differences in bacterial wilt severity were observed

Table 14.6: Wilt severity in three potato cultivars with varying levels of resistance seven weeks after inoculation with *Ralstonia solanacearum* alone and together with *Meloidogyne incognita* (Mean of 18 replicates)

Inoculum	***Wilt severity in potato cultivars***		
	Asnate (susceptible)	***B 53 (intermediate)***	***Kenya Dhamana (resistant)***
R. solanacearum	1.7	1.5	1.0
R. solanacearum + Root wounding	3.6	1.8	1.4
R. solanacearum + *M. incognita* + Root wounding	4.5	3.7	2.0
R. solanacearum + *M. incognita*	4.1	3.5	1.9
Mean	3.4	2.6	1.6
CD (P = 0.05)	**0.84**	**0.86**	**0.62**

Table 14.7: Severity of bacterial wilt in three potato cultivars from second to seventh week after inoculation with *Ralstonia solanacearum* alone and together with *Meloidogyne incognita*

Cultivar/Inoculum	***Weeks after inoculation***					
	2	***3***	***4***	***5***	***6***	***7***
Asnate (Highly susceptible)						
R. solanacearum	0.4	1.5	2.0	2.1	2.1	2.2
R. solanacearum + Root wounding	1.4	3.2	3.7	4.2	4.3	4.4
R. solanacearum + *M. incognita* + Root wounding	3.3	3.6	4.6	4.7	4.7	5.0
R. solanacearum + *M. incognita*	2.7	4.2	4.1	4.2	4.8	4.0
B 53 (Intermediate)						
R. solanacearum	0.0	0.3	1.0	2.2	2.4	2.6
R. solanacearum + Root wounding	0.0	0.4	2.1	2.4	2.7	2.8
R. solanacearum + *M. incognita* + Root wounding	1.4	2.9	3.6	4.2	4.7	4.9
R. solanacearum + *M. incognita*	1.2	2.4	3.7	4.4	4.5	4.6
Kenya Dhamana (Resistant)						
R. solanacearum	0.0	0.1	0.3	1.3	1.9	2.6
R. solanacearum + Root wounding	0.0	1.0	1.2	1.3	2.3	2.6
R. solanacearum + *M. incognita* + Root wounding	0.0	0.8	2.0	2.6	3.1	3.1
R. solanacearum + *M. incognita*	0.0	0.6	1.6	2.6	3.0	3.4
CD (P = 0.05)	**0.58**	**0.77**	**1.01**	**0.99**	**1.02**	**1.07**

among potato plants inoculated with *R. solanacearum* alone without root wounding and those inoculated with *R. solanacearum* and *M. incognita* throughout the seven weeks. Bacterial wilt severity was consistently higher in plants whose roots were severed. When plants inoculated with *R. solanacearum* alone were compared with those inoculated with *R. solanacearum* in combination with *M. incognita*, significant ($P \leq 0.05$), differences in wilt severity were observed in each of the three cultivars throughout the assessment period. Bacterial wilt development was faster and the severity consistently higher in plants inoculated with the two pathogens. Similarly, comparisons between plants inoculated with the two pathogens simultaneously with severed roots and those treated the same way but with unsevered roots indicated non-significant ($P=0.05$) differences in each cultivar and for every week (Ateka *et al.*, 2001).

Bacterial wilt severity was higher and development faster in wounded plants inoculated with *R. solancearum* in combination with *M. incognita*. Likewise bacterial wilt severity was higher in potato plants inoculated with *R. solanacearum* with root-wounding than in plants inoculated with both *R. solanacearum* and *M. incognita* without root wounding (Ateka *et al.*, 2001).

Greenhouse studies revealed that infection by *R. solanacearum* in the presence of root-knot nematodes resulted in a higher severity and a faster development of bacterial wilt. Similar findings were reported by Ravichandra *et al.* (1990) and Akiew *et al.* (1991). The fact that bacterial wilt development was faster and severity consistently higher in plants inoculated with *R. solanacearum* in combination with *M. incognita* than in plants inoculated with *R. solanacearum* alone suggests that the resistance mechanism(s) of potato plants may have been weakened in the presence of *Meloidogyne* (Ateka *et al.*, 2001).

Nagesh *et al.* (1997) suggested a breakdown of resistance in potato plants inoculated with *R. solanacearum* together with *M. incognita*. Bacterial wilt severity was lower in potato cv. B53 plants inoculated with *R. solanacearum* with root-wounding than in plants inoculated with *R. solanacearum* and *M. incognita* without root wounding. This shows that the role of *M. incognita* in bacterial wilt development is more than just providing avenues for the entry of bacterium.

According to Napiere and Quimio (1980) and Samuel and Mathew (1983), the synergism between the two pathogens was mostly associated with wounds caused during penetration of larvae into roots. However, findings by Sitaramaiah and Sinha (1985) suggested that nematode induced stress was more important as a wilt-triggering factor than the wounding. Biochemical and physiological changes resulting from infection by root-knot nematodes may be responsible for the enhancement of wilt. According to Trudgill (1991), the biochemistry of a plant is considerably altered following nematode infection. These changes may either weaken the chemical resistance mechanisms of the plant or modify conditions in the infected potato tissue making the plant more suitable for bacterial colonization (Khan, 1993). This is in harmony with the suggestion by Jatala (1975) that infection by one pathogen may alter the response of a host to subsequent infection by another.

(i) Management methods

(a) *Host resistance*

Nematode infection caused reduction of plant height in all the potato cultivars screened (Table 14.8). Cultivar Rutuku CIP-720097 recorded the lowest (12.4%) reduction while the highest (50.6%) reduction was observed in cultivar Dutch Robjin. Differences in plant height between infected and uninfected plants were significant (P=0.05) in cultivars Desiree, K. Pink, K. Dhamana, Mauritius, Tigoni and Nyayo (Ateka *et al.*, 2001).

Table 14.8: Galling and egg mass indices and plant heights of sixteen potato cultivars 60 days after soil infestation with 6000 eggs/juveniles of *Meloidogyne incognita* (Mean of 8 replicates)

Cultivar	*Gall index*	*Egg mass index*	*Plant height*		
			Infected	*Uninfected*	*% change*
Tomato Cal J	7.6	5.0			
KP9373926	5.5	5.0	30.6	37.2	17.7
BS3	5.4	5.0	24.2	36.4	33.5
Dasrtee CIP-800048	5.1	50	22.0	38.4	42.7
KP92	5.0	5.0	4.4	40.4	14.9
K Pink	3.9	5.0	32.4	46.0	29.6
Kenya Dhamana	3.6	4.9	38.0	58.4	34.9
Cnua 720118	3.8	4.8	44.0	52.0	15.4
MaontiuS	3.3	4.5	28.6	36.4	21.4
Dutch Robjin	3.0	4.5	32.2	65.2	50.6
Asanie CIP38138120	3.0	4.8	39.0	59.0	33.9
H Tana	2.6	4.9	32.8	44.8	26.8
Rutuku OP-720097	2.6	4.5	50.0	57.0	12.3
Fur aha	2.5	4.4	27.2	32.6	16.6
Tigcni	2.5	4.4	42.8	59.6	28.2
Nyayo	1.8	4.5	47.0	58.4	19.5
CD (P = 0.05)	**0.9**	**0.4**			

The result of this study reinforces the need to control root-knot in fields heavily infested with *R. solanacearum* in order to gain maximum benefit from the use of wilt resistant potato cultivars. Use of potato cultivars resistant to *Meloidogyne* are recommended as a component of integrated bacterial wilt disease management.

14.1.6. Tobacco Rattle Virus/Corky Ringspot of Potato (Spraing)

Tobacco rattle virus is reported in many areas of the world and more than 400 plant species in 50 plant families are susceptible to infection (Brunt *et al.*, 1996; Dallwitz, 1980; Dallwitz *et al.*, 1993; Hooker, 1981; Ploeg *et al.*, 1989).

Spraing is the term given to a range of internal and external symptoms in potato tuber flesh consisting of brown, necrotic rings, arcs and flecks. Tobacco rattle virus (TRV) cause these symptoms. In some cases external concentric rings may be visible on the surface of the tuber. The severity of symptoms in the tuber can range from mild (a few flecks or faint brown lines) to severe (heavy, dark chocolate coloured arcs with a corky texture). A few symptomatic tubers may be found or the stock could be heavily affected (>50%). Spraing symptoms may not be evident at harvest. Fluctuations in temperature during storage can exacerbate development of the internal tuber symptoms.

Corky ringspot disease of potato is caused by tobacco rattle virus (Tobravirus), which is vectored by stubby-root nematodes (*Paratrichodorus* spp. and *Trichodorus* spp.). The condition is sometimes referred to as spraing, a Scottish word meaning a bright streak or stripe (de Bokx, 1972).

(a) *Symptoms*

Foliar symptoms: Generally TRV spraing infected tubers produce healthy plants. However, in some cases several healthy and symptomatic, stunted stems will be produced, which is termed as stem mottle. Symptoms may present as pinching and "scorching" of the leaf tip, mottles, distorted/wavy leaf margins and occasionally finely etched lines and rings or chevron type markings reminiscent of PMTV (Fig. 14.8).

Fig. 14.8: Left – Yellow ring spots induced by TRV. Right – Leaf twisting with necrosis induced by TRV

Corky ringspot symptoms vary depending on virus strain, potato cultivar, and time of infection. Symptoms often include brown necrotic rings, arcs, and diffuse spots which are considered quality defects and may result in after-harvest devaluation or rejection of either table or processed potatoes.

Corky ringspot symptoms in potato include necrotic rings and pits on the tuber surface and range from diffuse brown spots to concentric rings or arcs of brown, necrotic tissue to dark-brown necrotic tissue which extends through tuber flesh (Mojtahedi *et al.*, 2001). The virus is usually detectable when symptoms are seen but may also be present in asymptomatic tissue (Charlton, 2006).

Tubers from soil with a history of tobacco rattle virus serve as a reservoir. The virus may spread to daughter tubers when infected tubers are used as seed (Crosslin *et al.*, 1999). Newly formed potato tubers are quite vulnerable to tobacco rattle virus infection and tubers as small as 3-cm in diameter had corky ringspot blemishes in tobacco-rattle infested fields in Florida (Fig. 14.9) (Weingartner *et al.*, 1975).

Fig.14.9: External and internal symptoms of corky ringspot virus on potato tubers

Several species of *Trichodorus* and *Paratrichodorus* transmit M-type isolates of tobacco rattle virus (Harrison and Robinson, 1986). Stubby-root nematodes are migratory ectoparasites that are mobile during each stage of their life cycle (Stark and Love, 2003) and feed primarily on meristematic cells or root tips which hinders root elongation (Crow, 2005). Damaged root tips may swell and lateral roots may emerge behind them, resulting in root proliferation.

(b) *Transmission*

Transmission of TRV is via free-living nematodes of the genera *Trichodorus* and *Paratrichodorus*. As a consequence of the biology of the vector TRV tends to be a greater issue on light, sandy soils. Nematodes favour wet conditions and heavy rainfall or over irrigation can exacerbate incidence of the disease in affected areas. The virus has been shown to have a very broad host range and can infect over 400 plant species. These include crops such as beetroot (*Beta vulgaris*) and spinach (*Spinacia oleracea*). In addition to these species the virus can also infect many common field weeds including fathen (*Chenopodium album*), chickweed (*Stellaria media*), knotweed (*Polygonum aviculare*), bindweed (*Polygonum convolvulus*), groundsel (*Senecio vulgaris*) and dandelion (*Taraxacum officinale*).

TRV affected tubers will rarely be systemically infected and will generally grow into healthy plants. Whilst spraing symptoms are still the most visible and economically damaging characteristic of this disease, some commonly grown varieties will become systemically infected. These varieties can act as a reservoir of virus for non-infective nematodes which feed on the infected tubers and acquire the virus. It is important therefore to know whether land contains viruliferous nematodes and the varietal susceptibility to TRV.

(c) *Biology*

The life cycle of stubby-root nematodes is not well studied. Eggs of *Paratrichodorus minor*, hatch at 53.3 ± 7.3 degree-days using a basal threshold of 10°C (DD10C) after egg deposition. Second stage population peaks occurred at 28 DD10C, third stage at 67 DD10C, fourth stage at 109 DD10C and adults at 143 DD10C. Population densities increased in the presence of a suitable host but declined to 33% of the initial level after 300 DD10C in the absence of a host (Schneider and Ferris, 1987). The optimum temperature range for development and reproduction of *P. allius* is 21 to 24°C (Ayala *et al.*, 1970).

In general, Trichodoridae nematodes prefer coarser textured soils and avoid desiccation stress by migrating vertically to moister soil layers. Vertical movement of Trichodoridae nematodes appears greatest when soil pores are half-full of water and least in waterlogged or dry soil conditions (Decraemer, 1995). Since soil moisture is often maintained below field capacity for optimum yield and quality in potato production, vertical migration may explain the difficulty of consistently finding stubby-root nematodes during routine soil sampling procedures to 30-cm depth in Washington and Oregon (Charlton, 2006).

Stubby-root feeding injury is not important economically and is rarely visible in potato, however, corky ringspot virus can be transmitted at very low population densities of the nematode (Charlton, 2006).

(d) *Management Guidelines for Corky Ringspot Virus*

(i) *Cultural methods*

The wide host range for both virus and vectors make eradication of corky ringspot in potato fields difficult. Alfalfa (*Medicago sativa*), a rotation crop in Oregon and Washington, and spearmint (*Mentha cardiaca*) do not appear to serve as hosts of tobacco rattle virus (Mojtahedi *et al.*, 2002). Since the virus is lost each time the nematode molts, and since it does not pass through the nematode egg stage, *P. allius* are cleansed of tobacco rattle virus after being reared on these plants for a period of 1 to 3 months (Boydston *et al.*, 2004). Therefore, it may be possible to eliminate tobacco rattle virus from populations of *P. allius* rotation crops that do not harbour the virus. Of course, the crops must be kept free of weed hosts for tobacco rattle for the strategy to be effective (Charlton, 2006).

The main method of control for virus is through field intelligence and effective cultural control. By ensuring that seed tubers are free from spraing diseases the opportunity for these diseases to spread from field to field can be minimized. The soil bait test offered as a diagnostic service by the Plant Clinic can help growers make informed decisions on which fields to use by identifying fields infected with both the vectors and their respective viral pathogens simultaneously.

(ii) *Chemical control of the nematode vector*

Spraing disease of potato was found to be checked by methomyl used at 9 kg/ha and dazomet at 170 kg/ha only for one year.

Aldicarb (Temik 15G), a systemic carbamate pesticide, suppresses corky ringspot damage in potato by controlling stubby-root nematodes (Weingartner and Shumaker, 1990; Rykbost *et al.*, 1992).

1,3-dichloropropene (Telone II) was effective in reducing corky ringspot in Washington but not in Florida or the Klamath Basin of Oregon (Ingham *et al.*, 2000; Weingartner *et al.*, 1990; Rykbost *et al.*, 1995). The fumigation failure in Florida may be attributed to the ability of *P. minor* to escape the toxicant in deep soil and then migrate to upper levels soil profile treated areas once the active ingredient had dissipated. D-D at 23 kg/ha checked the spread of the virus for two years (Cooper and Thomas, 1971; Mass, 1974).

Metham sodium (Vapam HL) delivered through chemigation (water-run) is not effective against corky ringspot of potatoes when used alone (Ingham *et al.*, 2000).

14.2. TOMATO, *LYCOPERSICON ESCULENTUM*

Tomato (*Lycopersicon esculentum*) is the most important and remunerative vegetable crop in India. It is being cultivated in 0.572 million hectares producing 10.261 million tonnes of tomatoes with an average yield of 18 tonnes per hectare (Bijay Kumar, 2009) (Tables 14.9 and 14.10). Andhra Pradesh, Orissa, Karnataka, West Bengal, Bihar, Gujarat, Maharashtra, Chhattisgarh, Jharkhand and Madhya Pradesh are the major tomato-growing states in India. A rich source of minerals, vitamins and organic acids, tomato fruits provides 3-4% total solids, 15-30 mg/ 100 g ascorbic acid, 7.5-10.0 mg/100 ml titratable acidity and 20-50 mg/100 g fruit weight of lycopene. Tomato fruits contain 0.4% protein, 2.5% carbohydrates, 1.5 mg iron and 2 mg vitamin C in 100 g of fresh weight. Fruits are good for people suffering from constipation, jaundice and indigestion.

14.2.1. Root-knot Nematode, *Meloidogyne incognita* and Wilt, *Fusarium oxysporum* f. sp. *lycopersici*/*F. solani* Disease Complex

The root-knot nematode and *Fusarium* wilt are frequently associated with tomato resulting in considerable damage to the crop. Jenkins and Coursen (1957) induced

Table 14.9: Major Tomato Producing Countries of the World (2007-08)

Country	***Area (in '000 ha)***	***Production (in '000 MT)***	***Productivity (in MT/ha)***	***% Share of world production***
China	1455	33645	23	26
USA	175	11500	66	9
India	572	10261	18	8
Turkey	270	9920	37	8
Egypt	194	7550	39	6
Italy	118	6026	51	5
Iran	140	5000	36	4
Spain	56	3615	65	3
Brazil	57	3364	59	3
Mexico	130	2900	22	2
Others	1552	34140	22	26
Total	**4719**	**127920**	**39.0**	

Source: FAO (except Indian data) (Indian data source: Indian Horticulture Database, 2006)

Table 14.10: Area, production and productivity of leading tomato growing states in India (2007-08)

State	***Area (in'000 ha)***	***Production (in '000 MT)***	***Productivity (MT/ha)***	***% Share of India production***
Andhra Pradesh	83.1	1579.0	19.0	32.2
Orissa	100.7	1344.2	13.3	28.7
Karnataka	47.2	1285.1	27.2	17.5
West Bengal	51.1	956.7	18.7	4.3
Bihar	46.2	921.9	19.9	4.3
Gujarat	30.8	739.6	24.0	1.9
Maharashtra	32.2	715.3	22.2	1.5
Chhattisgarh	37.7	404.1	10.7	1.4
Jharkhand	17.5	350.2	20.0	1.2
Madhya Pradesh	22.7	340.5	15.0	1.1
Others	102.5	1624.0	15.8	6.0
Total	**571.7**	**10260.6**	**17.9**	

(*Source*: Indian Horticulture Database, 2007-08)

wilting in *Fusarium* wilt-resistant tomato variety 'Chesapeake' only when root-knot nematodes were present along with fungal inoculum. Furthermore, when *M. hapla* was combined with the fungus, only 60% of the plants wilted, whereas *M. incognita acrita* promoted wilt in 100% of the plants.

Bhagawati and Goswami (2000) reported the interaction of *M. incognita* and *F. oxysporum* f. sp. *lycopersici* on tomato. It was found that when either or both pathogens were inoculated simultaneously or nematode was inoculated 10 days prior to inoculation of the fungus, the symptoms were visible by 20 days as against no such symptoms in plants where fungus was inoculated 10 days prior to nematodes. The intensity of wilt after 40-60 days of inoculation was significantly higher when simultaneously inoculated or prior inoculation of nematodes compared to prior inoculation of fungus. The number of galls and egg masses and nematode population in soil was significantly reduced when simultaneously inoculated or prior inoculation of nematodes.

The interaction among soil-borne pathogens is well documented in tomatoes on old production land when *Fusarium* wilt disease and root-knot nematode (*Meloidogyne* spp.) are both present (Fig. 14.10). Young plants are very susceptible to the combination of pests, collapsing prior to harvest. The nematode, by impairing water and nutrient availability, disrupts root function and plant growth processes. These effects combined with the vascular blocking due to the wilt fungi can be particularly severe, and if widespread, result in total crop failure. The root-knot nematode, by causing the development of root galls, provides a nutrient-rich food source which the fungi colonize rapidly. Root-knot nematodes can thus significantly enhance disease development and yield loss, elevating primary or secondary pathogens to major pest status even though population levels or pathogenic potential of the fungi were initially very low and yield losses would have been minimal in the absence of the nematode.

Fig. 14.10: The simultaneous occurrence of both root-knot nematode (*Meloidognyne* spp.) and *Fusarium* wilt disease (*Fusarium oxysporum*) causing enhanced disease development and tomato yield loss

Goode and McGuire (1967) have pointed to another possibly important aspect of the Fusarium-root-knot interrelationship. They found that nematode infection enables certain races of *F. oxysporum* f. sp. *lycopersici* to attack tomato varieties ordinarily resistant to those races and suggested that the fungus mutates within the host.

Inoculation of *M. incognita* juveniles prior to *F. oxysporum* f. sp. *lycopersici* inoculation advanced the wilt of tomato by one week, while nematode inoculation after fungus did not increase wilting (Table 14.11) (Nagesh *et al.*, 2006).

Table 14.11: Progress of *Fusarium* wilt in the presence of *M. incognita*

Treatment	***Wilt index at weekly intervals starting 10 DAP***					
	1	***2***	***3***	***4***	***5***	***6***
M. incognita (Mi)	0	0	0	0	0	0
F. oxysporum f. sp. *lycopersici (Fo)*	0	0	+	++	++	-
Mi + Fo (Mi inoculated 7 days prior to *Fo)*	0	+	+	++	++	-
Fo + Mi (Fo inoculated 7 days prior to *Mi)*	0	0	+	++	++	-

+ = Wilting of young leaves, ++ = Wilting of whole plant, - = Dead plant

Wang and Bergeson (1974) suggested that changes in total sugar concentration in the xylem sap which reach maximum concentration four weeks after nematode inoculation may contribute to the enhancement of *Fusarium* wilt in tomato infected with *M. incognita.*

Management methods

(i) *Biomanagement*

(a) *Antagonistic bacteria*: Root infection caused by *M. javanica* and the soil-borne root-infecting fungi *F. oxysporum* and *F. solani* was effectively suppressed following application of *P. aeruginosa*. Biocontrol and growth promoting potential of the bacterium was enhanced when soil was kept at 50% or 75% moisture holding capacity. An inoculum level 2.5 × 10^8 cfu/ml of *P. aeruginosa* was optimal for the enhancement of plant growth (Siddiqui and Ehteshamul-Haque, 2001).

(b) *Antagonistic fungi*: Soil application of *T. harzianum* at 50 kg/ha (2 × 10^8 cfu/g) was effective in improving the plant growth parameters and fruit weight followed by *P. fluorescens*, carbofuran, *A. niger*, *T. virens* and neem seed powder. Carbofuran was found highly effective against *M. incognita*, topsin-M against *F. solani* and *T. harzianum* was effective against both the pathogens (Table 14.12) (Haseeb *et al.*, 2006).

Paenibacillus polymyxa GBR-462; GBR-508 and *P. lentimorbus* GBR-158 strains showed the strongest antifungal and nematicidal activities. These three strains used in pot experiment reduced the symptom development of the disease complex caused by *M. incognita* and *F. oxysporum* f. sp. *lycopersici* on tomato (wilting and plant death), and increased plant growth. The control effects were estimated to be 90–98%, and also reduced root gall formation by 64–88% compared to the untreated control (Son *et al.*, 2009).

Table 14.12: Effect of bioagents, botanicals and chemicals for the management of disease complex in tomato

Treatment	*Fruit weight (g)*	*Reproduction factor*	*Root-knot index*	*% wilt incidence*
Uninoculated control	230	–	–	–
Inoculated control	103	8.8	2.6	50
Neem seed powder	180	2.8	1.0	15
Neem leaves	155	4.5	1.5	23
Murraya leaves	150	5.5	1.7	23
P. fluorescens	198	4.4	4.5	9
T. harzianum	203	3.3	1.2	6
T. virens	185	5.1	1.6	12
P. lilacinus	172	1.4	0.7	18
A. niger	195	4.3	1.5	11
Farm yard manure	130	8.4	2.5	43
Mint manure	142	6.9	2.0	30
Carbofuran	197	1.1	0.6	10
Topsin-M	173	8.0	2.3	5
CD (P = 0.05)	**8.639**	**0.22**	**0.12**	**1.03**

(ii) *Chemical methods*

Application of metham sodium through drip irrigation was found to have an edge over D-D for root-knot and *Fusarium* sp. complex of tomato (Roberts *et al.*, 1988).

(iii) *Host resistance*

Abawi and Barker (1984) found that the monogenic resistance in tomato against *F. oxysporum* f. sp. *lycopersici* race 1 (Florida, MH-1, Manapal) remained unaffected in the presence of *M. incognita*.

(iv) *Integrated management*

(a) Bioagents and botanicals: *T. harzianum* controlled wilt in the presence of *M. incognita* for six weeks after transplanting. Tomato roots that received *P. lilacinus* along with *T. harzianum* and neem cake were free from root-knot nematodes (*M. incognita*) and did not wilt due to *F. o.* f. sp. *lycopersici* till harvest. The roots were also free from *Fusarium* infection. The above treatment also reduced the % of wilt (10% compared to 90% in control), root-knot index (1.8 compared to 4.4 in control) and increased root colonization with bioagents (74% compared to 0% in control), parasitization of egg masses (56% compared to 0% in control) and eggs/ egg mass (44% compared to 0% in control) (Nagesh *et al.*, 2006 (Table 14.13).

(b) *Bioagents, cultural methods and host resistance*: Deep ploughing and exposing soil to hot sun in summer, removal and burning of crop debris, soil application

Table 14.13: Effect of integration of neem cake and bioagents for the management of *Meloidogyne incognita* infecting tomato

Treatment	***Healthy plants (%)***	***Root-knot index***	***Colonization by bioagents***		
			Root colonization (%)	***Colonization of egg masses/eggs (%)***	
				Egg masses	***Eggs***
Control	10	4.4	–	–	–
P. lilacinus	30	3.6	48	48	46
T. harzianum	40	4.0	50	44	32
Neem cake	10	4.2	–	–	–
P. lilacinus + Neem cake	60	2.8	62	62	58
T. harzianum + Neem cake	70	3.8	68	36	32
P. lilacinus + *T. harzianum* + Neem cake	90	1.8	74	56	44
CD (P = 0.05)	**8.87**	**1.78**	**3.66**	**4.11**	**3.22**

of *T. viride* and *P. lilacinus*, use of wilt resistant varieties like Utkal Pallavi, Utkal Deepti, Utkal Kumari, Utkal Urbasi, etc. help in controlling the disease complex.

(c) *Arbuscular mycorrhizal fungi and botanicals*: Bhagawati *et al.* (2000) demonstrated that although the mustard cake and AM fungus, *Glomus etunicatum* were effective in reducing the damage caused by *M. incognita* and *F. o.* f. sp. *lycopersici* on tomato, the performance of concomitant application of both (bioagent and botanical) was much better than their individual application.

(d) *Two bioagents*: Efficacy of *Pseudomonas aeruginosa* alone or in combination with *Paecilomyces lilacinus* was evaluated in the control of root-knot nematode and root-infecting fungi under field conditions. The biocontrol fungus and the bacterium significantly suppressed soil-borne root-infecting fungi such as *Macrophomina phaseolina, Fusarium oxysporum, Fusarium solani, Rhizoctonia solani* and *Meloidogyne javanica*, the root-knot nematode. *P. lilacinus* parasitized eggs and female of *M. javanica* and this parasitism was not significantly influenced in the presence of *P. aeruginosa* (Siddiqui *et al.*, 2000).

14.2.2. Root-knot Nematode, *Meloidogyne* spp. and Root Rot, *Rhizoctonia solani* Disease Complex

Abu-Elamayem *et al.* (1978) observed that damping-off of tomato was more severe in soil infested with both *M. javanica* and *R. solani* than with the fungus alone. *M. javanica* increased the extent of damage by pre- and post-emergence phases of damping off caused initially by *R. solani* in tomato (Nath *et al.*, 1984).

Plants in untreated field soil or in sterilized soil inoculated with both organisms, developed a root rot in about 42 days. If the nematode preceded the fungus by three weeks, the root rot was more severe and appeared within 14-21 days. The fungus penetrated either directly or through ruptures in the root created by the mature female nematode. *R. solani* colonized nematode giant cells and root xylem cells. Vascular discolouration occurred both in roots and stem, however, no fungus was isolated from stems.

M. incognita predisposed roots to *R. solani* which resulted in severe root rot and subsequent plant death. Tomato plants inoculated with either *R. solani* or *M. incognita* alone were free of root decay for the entire period of six-week study. Three weeks after nematode and fungus inoculation, black sclerotia of *R. solani* were visible on nematode- induced galls, while on non-galled portions on the same root systems were free of sclerotia. Prior to root rot development, *R. solani* demonstrated marked preference for root galls on nematode infected roots. It is hypothesized that the leakage of nutrients from the root was responsible for attracting the fungus to the galls and for initiating sclerotial formation.

Five weeks after inoculation, distinct brown lesions were observed only on the galls of plants inoculated with both *M. incognita* and *R. solani*. Lower leaves of plants were chlorotic and suffered premature leaf drop.

M. incognita and *R. solani* are also frequently associated with tomato causing considerably greater damage to this crop (Haseeb, 2003). Highest reduction in fruit yield, plant height, fresh and dry weight was observed in plants inoculated with nematode-fungus simultaneously followed by nematode 7 days prior to fungus (Kumar and Haseeb, 2009) (Table 14.14). The fact that prior inoculation of nematode

Table 14.14: Effect of *Meloidogyne incognita* and *Rhizoctonia solani* on plant growth and fruit yield of tomato cv. K-25.

Treatment	*Fruit yield (kg/plant)*	*Plant height (cm)*	*Fresh plant weight (g)*	*Dry plant weight (g)*
Untreated uninoculated	2.530	67.5	200.5	38.8
M. incognita (4000 J_2/4 kg soil)	1.280	50.8	155.0	29.2
R. solani (10 g mycelial mat/4 kg soil)	1.440	53.7	168.0	31.7
M. incognita + *R. solani* simultaneously	0.487	26.7	94.8	17.6
M. incognita 7 days prior to *R. solani*	0.658	31.8	112.0	21.0
R. solani 7 days prior to *M. incognita*	0.821	40.0	124.7	23.4
CD (P = 0.05)	–	**2.41**	**7.68**	**1.46**

caused more damage indicates that the roots are predisposed by *M. incognita* for subsequent damage by *R. solani*.

The highest reproduction rate (14.9) and root-knot index (3.5) was observed in plants inoculated with nematode alone followed by nematode inoculation 7 days prior to the fungus (Table 14.15). The highest root infection by the fungus was observed in simultaneous inoculation followed by nematode inoculation 7 days prior to the fungus (Kumar and Haseeb, 2009).

Table 14.15: Effect of *Meloidogyne incognita* and *Rhizoctonia solani* on nematode multiplication, root galling and fungus infection on tomato cv. K-25.

Treatment	***Final nema popn. (soil + roots)***	***Reprodu -ction factor***	***Root-knot index***	***% root infection by R. solani***
Untreated uninoculated	–	–	–	–
M. incognita (4000 J_2/4 kg soil)	59,500	14.9	3.50	–
R. solani (10g mycelial mat/4 kg soil)	–	–	–	30.0
M. incognita + *R. solani* simultaneously	38,700	9.7	2.00	63.5
M. incognita 7 days prior to *R. solani*	45,550	11.4	2.50	55.0
R. solani 7 days prior to *M. incognita*	35,850	9.0	1.35	48.5
CD (P = 0.05)	**2808.55**	**0.64**	**0.13**	**2.65**

Management methods

(i) *Chemical methods*

Soil fumigation usually decreased disease severity due to *R. solani* and *M. incognita* disease complex in tomato (Golden and Van Gundy, 1975).

Abu-Elamayem *et al.* (1978) reported that the combined use of CGA 12223 and benomyl provided significant reduction in the disease intensity.

(ii) *Biomanagement*

(a) *Antagonistic bacteria*: Root infection caused by *M. javanica* and the soil-borne root-infecting fungus *R. solani* was effectively suppressed following application of *Pseudomonas aeruginosa*. Biocontrol and growth promoting potential of the bacterium was enhanced when soil was kept at 50% or 75% moisture holding capacity. An inoculum level 2.5×10^8 cfu/ml of *P. aeruginosa* was optimal for the enhancement of plant growth (Siddiqui and Ehteshamul-Haque, 2001).

(b) *Antagonistic fungi*: *Trichothecium roseum* was found most effective in mitigating the adverse effect of *R. solani* and *M. incognita* followed by *Trichoderma viride*.

The effect of *T. roseum* was more pronounced when the plants were inoculated both at seedling emergence and transplantation stages (Arya and Saxena, 2009).

(iii) *Integrated management*

(a) *Two bioagents*: Efficacy of *Pseudomonas aeruginosa* alone or in combination with *Paecilomyces lilacinus* was evaluated in the control of root-knot nematode and root-infecting fungi under field conditions. The biocontrol fungus and the bacterium significantly suppressed soil-borne root-infecting fungi such as *Macrophomina phaseolina, F. oxysporum, F. solani, R. solani* and *M. javanica*, the root-knot nematode. *P. lilacinus* parasitized eggs and female of *M. javanica* and this parasitism was not significantly influenced in the presence of *P. aeruginosa* (Siddiqui *et al.*, 2000).

P. aeruginosa-Bacillus subtilis treatment was the most effective in the suppression of root-rot disease complex with enhancement of plant growth (Siddiqui and Ehteshamul-Haque, 2001).

14.2.3. Lesion Nematode, *Pratylenchus* spp. and Wilt, *Verticillium* spp. Disease Complex

Verticillium wilt seems to render plants more suitable for colonization by lesion nematodes, *Pratylenchus* spp. When the fungus is added to soil already infested with the nematodes, reproduction of the latter is increased in roots of tomato (Mountain and McKeen, 1962).

14.2.4. Cyst Nematode, *Globodera rostochiensis* and Root Rot, *Rhizoctonia solani* Disease Complex

G. rostochiensis has been reported to interact with both *R. solani* and *Colletotrichum atramentarium* (the causal fungus of brown root rot) in tomato (Dunn, 1968; Dunn and Hughes, 1964).

14.2.5. Root-knot Nematode, *Meloidogyne incognita;* Reniform Nematode, *Rotylenchulus reniformis* and Wilt, *Fusarium oxysporum* f. sp. *lycopersici* Disease Complex

Inoculation of *M. incognita* prior to *R. reniformis* and *F. oxysporum* f. sp. *lycopersici*, the plants were synergistically affected (biomass-18 g) and increased the severity of wilt disease complex. While prior establishment of *R. reniformis* not only negatively affected the *M. incognita* population but also reduced the *Fusarium* wilt severity. The population density of both the nematodes was observed to be mostly competitive in nature, one establishes at the cost of the other. In *R. reniformis* dominated plants not only a significant decrease in final population (18%) of *M. incognita* was reduced but also recorded a good recovery in biomass (40%) in comparison to the plants where *M. incognita* dominated *R. reniformis* and given prior to *F. oxysporum* f. sp. *lycopersici* (Singh *et al.*, 2007).

14.2.6. Root-knot Nematode, *Meloidogyne incognita*, Reniform Nematode, *Rotylenchulus reniformis* and Bacterial Wilt, *Ralstonia solancearum* Disease Complex

Pani and Das (1972) have reported the association of root-knot nematode with bacterial wilt of tomato.

Haider *et al*. (1987) reported the significant reduction in the root-knot index and larval development of *M. incognita* in soil where *R. solanacearum* was present. *R. solanacearum* and *M. incognita* alone as well as in different combinations reduced plant growth and yield significantly with the nematode followed by the bacterium combination showing the maximum reduction in growth.

Disease severity was greatly increased when wounding was done just before inoculation with *R. solanacearum* and decreased when wounding was carried out six days before inoculation (Abd-El-Ghafar and Abd-El-Kader, 1997).

Napiere (1980) and Napiere and Quinio (1980) found that wilt disease development occurred earlier and with a higher mortality rate in both wilt-resistant and susceptible tomato cultivars grown in *R. solanacearum* and *M. incognita* – infested soil (Fig. 14.11).

Fig. 14.11: Root-knot nematode and bacterial wilt disease complex

Management methods

(i) *Integrated methods*

(a) *Bioagents and botanicals*: Treatment of nursery bed with *Pseudomonas fluorescens* (10^9 cfu/g) and *Trichoderma harzianum* (10^6 cfu/g) each at the rate of 20 g/m^2 and subsequent application of 5 MT of farm yard manure enriched with 5 kg each of *P. fluorescens* (10^9 cfu/g) and *P. lilacinus* (10^6 cfu/g) per hectare, significantly reduced *R. reniformis* and *M. incognita* in tomato roots by 74% and 70%, respectively; reduced the incidence of bacterial wilt; and increased the yield by 24.2%. Cost: benefit ratio (calculated for the additional cost of the bio-pesticides and additional returns accrued by the application of the bio-pesticide) was 1: 4.4 (Rao *et al*., 2009).

14.2.7. Stubby Root Nematode, *Trichodorus christiei* and Wilt, *Verticillium albo-atrum* Disease Complex

The effect of the stubby root nematode *T. christiei* on the incidence of infection of tomato cultivar Bonny Best by *V. albo-atrum* was determined. The infection incidence increased at most fungal inoculum densities when *T. christiei* was also present. The infection incidence by *V. albo-atrum* also increased as the inoculation density of *T. christiei* was increased. Infection by the fungus did not enhance reproduction of either nematode in or on the host roots; however, an increase in root weight occurred when both the fungus and *T. christiei* were present (Conroy and Green, 1974) (Tables 14.16 and 14.17).

Table 14.16: Infection % of tomato cv. Bonny Best by *Verticillium albo-atrum* alone and in combination with *Trichodorus christiei* at controlled inoculum densities

Verticillium albo-atrum		*Verticillium albo-atrum + Trichodorus christiei*		
Inoculum density	*Infection (%)*	*Inoculum densities of T. christiei*		
		5,000	10,000	15,000
0	0.0	0.0	0.0	0.0
25	26.6	33.3	33.3	40.0
50	40.0	60.0	53.3	73.3
100	73.3	80.6	93.3	100.0
200	100.0	100.0	100.0	100.0

Table 14.17: The effect of infection by *Verticillium albo-atrum* on final soil population of *Trichodorus christiei* and weight and surface of roots of tomato cv. Bonny Best

Inoculum density	*Microsclerotia/g fungus*	*Nematodes/kg soil*		*Root weightg/ replicate**	*Root surface area/replicate***
		Initial	*Final*		
0	0	0	0	1.53	43.8
25	0	10,000	29,416	2.43	32.1
50	50	10,000	40,271	2.56	32.6
100	100	10,000	31,285	2.90	32.8
200	200	10,000	36,585	2.86	38.0

* Mean root weight, g/replicate (3 plants) air-dried. Increases in root weights in treatments 3, 4 and 5 are statistically significant (P = 0.05) from treatment 1. Difference between treatments 1 and 2 is not statistically significant.

** Expressed as ml of titrated base. No significant differences between treatments.

14.2.8. Tomato Black Ring Transmitted by *Longidorus elongatus*

Management methods

(i) *Chemical methods*

Pentachloronitrobenzene (PCNB) has been found to effectively check the population increase of nematodes like *Longidorus elongatus* and *Xiphinema diversicaudatum*.

This efficacy would eventually influence the spread of tomato blacking virus (Murant and Taylor, 1965; Taylor and Gordon, 1970).

14.3. BRINJAL, *SOLANUM MELONGENA*

Brinjal (*Solanum melongena*) is a widely grown vegetable crop in Asian countries. It is being cultivated in 0.566 million hectares producing 9.596 million tonnes of brinjal with an average yield of 17 tonnes per hectare (Bijay Kumar, 2007) (Tables 14.18 and 14.19). In India, it is adapted to a wide range of climatic conditions

Table 14.18: Major Brinjal Producing Countries of the World (2007-08)

Country	*Area (in '000 ha)*	*Production (in '000 MT)*	*Productivity (in MT/ha)*	*% Share of world production*
China	1202	18033	15	54
India	566	9596	17	29
Egypt	43	1000	23	3
Turkey	30	791	26	2
Indonesia	53	390	7	1
Iraq	22	380	17	1
Japan	12	375	31	1
Italy	12	271	23	1
Sudan	12	230	19	1
Philippines	21	198	9	1
Others	124	1954	16	6
Total	**2097**	**33219**	**18.45**	

Source: FAO (except Indian data) (Indian data *source*: Indian Horticulture Database, 2007-08)

Table 14.19: Area, production and productivity of leading brinjal growing states in India (2007-08)

State	*Area (in'000 ha)*	*Production (in '000 MT)*	*Productivity (MT/ha)*	*% Share of India production*
West Bengal	153.9	2734.9	17.8	28.5
Orissa	129.0	1932.5	15.0	20.1
Bihar	54.6	1158.2	21.2	12.1
Gujarat	55.8	987.7	17.7	10.3
Maharashtra	29.4	478.7	16.3	5.0
Andhra Pradesh	30.6	458.8	15.0	4.8
Jharkhand	19.0	379.8	20.0	4.0
Karnataka	15.6	352.6	22.6	3.7
Chhattisgarh	22.3	327.5	14.7	3.4
Madhya Pradesh	14.9	223.3	15.0	2.3
Others	41.0	561.3	13.7	5.8
Total	566.1	9595.8	16.9	

(*Source*: Indian Horticulture Database, 2007-08)

from north to south and east to west. In hilly regions, it is grown only in summer. Brinjal is used in a variety of culinary preparations. Pickles and industrially processed foods are also produced. It is rich in vitamin A and B and good for diabetic patients. The major brinjal producing states include West Bengal, Orissa, Bihar, Gujarat, Maharashtra, Andhra Pradesh, Jarkhand, Karnataka, Chhattisgarh and Madhya Pradesh.

14.3.1. Root-knot Nematode, *Meloidogyne incognita* and Wilt, *Fusarium* spp. Disease Complex

Brinjal was not susceptible to *F. oxysporum* unless *M. incognita* was also present (Smits and Noguera, 1982) (Fig. 14.12).

Fig. 14.12: Root-knot *Fusarium* wilt complex in brinjal

Management methods

(i) *Integrated methods*

Treatment of seedlings with carbofuran at 1 kg *a.i.*/ha + *T. harzianum* at 50 kg/ha (having 10^8 cfu/g) was found highly effective in increasing the number and fresh weight of brinjal seedlings/bed and maximum reduction in root-knot index due to *M. incognita*. However, minimum percent root infection by *F. solani* was observed in carbofuran at 1 kg *a.i.*/ha + bavistin at 1 kg *a.i.*/ha treated seedlings (Kumar *et al.,* 2009).

Studies were undertaken to determine the effect of bio-control agents *viz.*, *Aspergillus niger*, *Trichoderma harzianum*, *Paecilomyces lilacinus*, and *Pseudomonas fluorescens* alone at 100 kg/ha each having 10^8 cfu/g culture, pesticides *viz.*, carbofuran and bavistin each at 2 kg *a.i.*/ha and organic amendment *viz.*, neem seed powder at 500 kg/ha alone and in combination (at half dose of each treatment material) against root-knot-wilt disease complex caused by *M. incognita* and *F. solani* on brinjal. Results indicated that all the treatments alone and in combination significantly improved the number and fresh weight of the seedlings and reduced the root-knot index and percent root infection. Carbofuran + *T. harzianum* was found highly effective in increasing the number and fresh weight of seedlings/bed. Maximum reduction in root-knot index was also observed in carbofuran + *T. harzianum* treated seedlings. However, minimum percent root infection by *F. solani* was observed in carbofuran + bavistin treated seedlings (Kumar *et al.*, 2009).

14.3.2. Root-knot Nematode, *Meloidogyne incognita* and Root Rot, *Ozonium texanum* var. *parasiticum* Disease Complex

M. incognita and *O. texanum* var. *parasiticum* together acted synergistically by reducing the germination of brinjal to 28 per cent (Nath *et al.*, 1976).

14.3.3. Root-knot Nematode, *Meloidogyne incognita* and Root Rot, *Colletotrichum atramentarium* Disease Complex

Synergistic interactions were reported between *M. incognita* and *C. atramentarium* in brinjal.

14.3.4. Lesion Nematode, *Pratylenchus penetrans*, and Wilt, *Verticillium albo-atrum* Disease Complex

Mountain and McKeen (1962) demonstrated a synergistic relationship between *P. penetrans*, and *V. albo-atrum*. At low and intermediate levels of *Verticillium* inoculum, the nematodes increased wilt and larger number of nematodes occurred in eggplant roots. The combined effect of these organisms significantly retarded growth of shoots and roots.

Verticillium wilt seems to render plants more suitable for colonization by lesion nematodes, *Pratylenchus* spp. When the fungus is added to soil already infested with the nematodes, reproduction of the latter is increased in roots of eggplant (Fig. 14.13).

Fig. 14.13: Lesion nematode and *Verticillium* wilt disease complex in brinjal

14.3.5. Root-knot Nematode, *Meloidogyne incognita* and Root Rot, *Macrophomina phaseolina* Disease Complex

Studies were undertaken to determine the effect of sequential, simultaneous and single inoculation of *M. incognita* (2,500 J_2/kg soil) and *M. phaseolina* (5.0 g mycelium including sorghum grains/kg soil) on growth and yield of *Solanum melongena* cv. Nageena. Highest reduction in shoot-root fresh and dry weights, number of fruits and fruit weight was observed in plants inoculated with nematode and fungus simultaneously. Highest reproduction rate (13.4) of *M. incognita* and root-knot index (3.5) were observed in plants inoculated with the nematode alone, whereas, highest root colonization (19.2%) by fungus was observed in plants inoculated with nematode and fungus simultaneously (Haseeb and Archana, 2009).

14.3.6. Root-knot Nematode, *Meloidogyne incognita* and Bacterial Wilt, *Ralstonia solanacearum* Disease Complex

Eggplant is prone to many soil borne diseases among which the bacterial wilt (*R. solanacearum*) in combination with root-knot nematode (*M. incognita*) takes heavy toll every year all over the world (Naik, 2003). Combined inoculation of nematode and the bacterium had a greater pathogenic effect than pathogens inoculated individually (Fig. 14.14) (Ravichandra *et al*., 1990; Savitha and Lingaraju, 1996).

Fig. 14.14: Root-knot nematode and bacterial wilt disease complex in brinjal

The root-knot nematode, *M. incognita*, present along with *R. solanacearum*, greatly increases the incidence and severity of bacterial wilt of brinjal (Table 14.20). The root-knot nematode is responsible for breaking bacterial wilt resistance in "Pusa Purple Cluster" cultivar of brinjal (Reddy *et al*., 1979). *M. incognita* interacts with *R. solanacearum* in increasing the speed of development and severity of wilt disease of eggplant. Maximum disease occurred when the bacterium and nematode were inoculated simultaneously. Inoculation with *R. solanacearum* alone resulted in less severe disease. It was concluded that root-knot nematodes modify the plant tissues facilitating bacterial colonization (Nayar *et al*., 1988).

Management methods

(i) *Biomanagement*

Biocontrol agents *Pseudomonas fluorescens* at 5 g/kg soil and *T. harzianum* at 3 g/kg soil significantly reduced the final soil nematode population and wilt disease incidence in brinjal. The highest reduction in bacterial population in soil was observed in the treatment *P. fluorescens* (Barua and Bora, 2008) (Table

Table 14.20: Effect of *M. incognita* and *R. solanacearum* alone and in combination on bacterial wilt incidence in brinjal (Husain and Bora, 2009).

Treatment	*No. of galls/root system*	*% wilt incidence*
M. incognita at 1000 J_2/kg soil	108.30	0.00
M. incognita at 2000 J_2/kg soil	110.00	0.00
R. solanacearum at 25 ml/kg soil	0.00	25.00
R. solanacearum at 50 ml/kg soil	0.00	25.00
M. incognita at 1000 J_2/kg soil + *R. solanacearum* at 50 ml/kg soil	86.10	100.00
M. incognita at 2000 J_2/kg soil + *R. solanacearum* at 50 ml/kg soil	43.30	100.00
M. incognita at 1000 J_2/kg soil + *R. solanacearum* at 25 ml/kg soil	89.00	75.00
M. incognita at 2000 J_2/kg soil + *R. solanacearum* at 25 ml/kg soil	91.20	75.00
Control	0.00	0.00

14.21). Further, combined application of wheat bran formulations of *P. fluorescens* and *T. harzianum* significantly reduced root galling, egg mass production and final population of *M. incognita* and *R. solanacearum* in soil and increased population of the antagonists in soil (Barua and Bora, 2009).

Table 14.21: Effect of different treatments on root galling, egg mass production and population of nematodes and bacteria in brinjal

Treatment/ Dose	*No. of galls/plant*	*Egg masses /plant*	*Nema popn. in soil*	*Nema popn. in root*	*Bact. popn (10^8cfu/g soil)*
T. harzianum – 1g/kg soil	69	44	355	59	21
T. harzianum – 2g/kg soil	60	36	298	52	27
T. harzianum – 3g/kg soil	52	31	282	46	15
P. fluorescens – 5g/kg soil	64	40	309	54	13
P. fluorescens – 10g/kg soil	56	34	290	49	7
P. fluorescens – 15g/kg soil	44	28	235	40	4
Carbofuran – 1.5 kg *a.i.*/ha	26	16	176	24	21
Streptocycline – 500 ppm	83	38	452	70	11
Neem cake – 15 g/kg soil	39	24	226	96.8	11
Control	92	53	484	416	42
CD (P = 0.05)	**0.093**	**0.129**	**80.00**	**11.1**	**2.75**

Root dip of brinjal seedlings in *P. fluorescens* (5 g/litre of water for 30 min) gave maximum suppression of root-knot nematode population in soil followed by *P. chlamydosporia* (soil application at 10 g/plant + root dip at 5 g/litre of water for 30 min), neem cake, *P. fluorescens* (soil application at 10 g/plant + root dip at 5 g/litre of water for 30 min), *P. chlamydosporia* as soil application (10 g/plant), *P. fluorescens* as soil application (10 g/plant) and *P. chlamydosporia* as root dip (5 g/litre of water for 30 min) (Dhawan *et al*., 2008).

Nursery bed treatment with *P. chlamydosporia* at 100 g/m^2 gave reduced root galling; and increased seedling height, seedling weight and root colonization by *P. chlamydosporia* (Table 14.22) (Naik, 2003). The seedlings were highly vigorous and colonized by *P. chlamydosporia* and reached main field when transplanted.

Table 14.22: Effect of *P. chlamydosporia* on plant growth and management of root-knot nematodes infecting brinjal under nursery conditions

Treatment	*Plant height (cm)*	*Seedling weight (g)*	*P. chlamydosporia root colonization (cfu/g)*
P. chlamydosporia at 25 g/m^2	18.8	2.1	13,560
P. chlamydosporia at 50 g/m^2	21.6	2.4	19,780
P. chlamydosporia at 100 g/m^2	23.7	2.7	21,470
Control	15.2	1.6	–
CD (P = 0.05)	**3.34**	**0.48**	**785.79**

The seedlings raised in nursery beds treated with *P. chlamydosporia* at 100 g/m^2 when transplanted in the main field gave reduced root galling, egg mass production and egg parasitization; and increased fruit yield and root colonization by *P. chlamydosporia* (Table 14.23) (Naik, 2003).

Table 14.23: Effect of *P. chlamydosporia* on plant growth and management of root-knot nematodes infecting brinjal under field conditions

*Treatment**	*Root-knot index*	*No. of egg masses /5 g root*	*% egg parasitization by bioagent*	*Fruit yield (kg/6m^2)*	*P. chlamydosporia root colonization (cfu/g)*
T_1	6.5	25	42.59	7.4	13,560
T_2	5.2	20	59.35	8.2	19,780
T_3	5.1	18	65.37	8.5	21,470
T_4	7.8	49	0.00	5.2	–
CD (P = 0.05)	**0.37**	**6.84**	**9.23**	**0.86**	**785.79**

* T_1 - *P. chlamydosporia* at 25 g/m^2, T_2 - *P. chlamydosporia* at 50 g/m^2, T_3 - *P. chlamydosporia* at 100 g/m^2, T_4 - Control.

(ii) *Integrated methods*

(a) *Two bioagents*: The nursery seedling stand was good in seed treatment with *P. fluorescens* at 50 g/kg + nursery bed treatment with *P. fluorescens* at 50 g/m^2 followed by nursery bed treatment with *T. harzianum* + *P. fluorescens*, and neem based *P. fluorescens* (Naik, 2003) (Table 14.24).

Table 14.24: Effect of integration of bioagents on nursery stand of brinjal seedlings

Treatment	***Nursery stand of seedlings****
Seed treatment with talc based *P. fluorescens*	4.0
Nursery bed treatment with neem based *P. fluorescens*	3.6
Seed treatment with talc based *P. fluorescens* + Nursery bed treatment with neem based *P. fluorescens*	4.6
Nursery bed treatment with *T. harzianum* + *P. fluorescens*	4.0
Control	3.0

* 3 – Poor, 4 – Good, 5 – Very good

Integration of seedling dip for 15 min in talc based formulation of *P. fluorescens* + *P. chlamydosporia* (5 + 5 g/litre) with soil application of the above bioagents gave maximum survival of brinjal plants under field conditions (53.33%) compared to 33.50% survival in control (Naik, 2003) (Table 14.25).

Table 14.25: Effect of integration of bioagents on survival of brinjal plants under field conditions

Treatment	***Survival of plants (%)***
Seedling dip treatment for 15 min in talc based formulation of *P. fluorescens* at 10 g/litre	43.33 (41.15)
Seedling dip treatment for 30 min in talc based formulation of *P. fluorescens* at 10 g/litre	43.33 (41.15)
Seedling dip treatment for 15 min in talc based formulation of *P. fluorescens* at 10 g/litre + soil drenching	46.66 (43.05)
Seedling dip treatment for 15 min in talc based formulation of *P. fluorescens* + *T. harzianum* at 5 + 5 g/litre + soil application	53.33 (46.89)
Control	33.50 (35.37)
CD (P = 0.05)	**4.92**

Seed treatment with talc based formulation of *P. fluorescens* at 50 g/kg gave least mortality of brinjal plants (13.18%) followed by seed treatment with neem based formulation of *P. fluorescens* at 50 g/kg (16.66%) compared to 45.29% mortality in control plants (Table 14.26) (Naik, 2003).

Table 14.26: Effect of bioagents on mortality of brinjal plants

Treatment	***Mortality of plants (%)***
Seed treatment with talc based formulation of *P. fluorescens* at 50 g/kg	13.18 (21.30)
Seed treatment with neem based formulation of *P. fluorescens* at 50 g/kg	16. 66 (24.12)
Soil application of talc based formulation of *P. fluorescens* + *T. harzianum* at 25 + 25 g/bed	25.45 (30.33)
Soil application of neem based formulation of *P. fluorescens* + *T. harzianum* at 25 + 25 g/bed	22.13 (28.04)
Control	45.29 (44.60)
CD (P = 0.05)	**4.212**

Results related to root-knot and wilt disease complex management in eggplant under field conditions revealed that the combination formulation of *T. harzianum* and *P. chlamydosporia* performed very well. The yield was 2.024 kg/3m^2 followed by combination of *P. fluorescens* + *P. chlamydosporia* where the yield was 1.782 kg

Table 14.27: Effect of integration of bioagents on root galling, wilt disease incidence and fruit yield of brinjal

Treatment	***No. of fruits/ plant***	***Fruit yield (kg/3m^2)***	***Root-knot index***	***Wilt disease incidence (%)***
Neem + wheat bran formulation (8:2) of *T. harzianum*	36.64	1.614	2.23	33.25 (35.24)
Liquid broth formulation of *P. fluorescens*	38.10	1.657	1.12	31.23 (33.96)
Neem + pongamia + wheat bran formulation (4:4:2) of *P. chlamydosporia*	39.47	1.733	2.00	30.68 (33.65)
T. harzianum + *P. fluorescens*	39.00	1.680	1.93	27.28 (31.50)
T. harzianum + *P. chlamydosporia*	42.97	2.024	1.80	22.18 (28.11)
P. fluorescens + *P. chlamydosporia*	40.64	1.782	1.81	25.13 (30.07)
Neem + wheat bran	34.70	1.571	3.10	46.21 (42.82)
Neem + pongamia + wheat bran	37.60	1.471	3.28	48.30 (44.03)
Control	28.30	1.133	4.20	98.00 (81.87)
CD (P = 0.05)	**3.514**	**0.245**	**0.15**	**3.21**

and *T. harzianum* + *P. fluorescens* (1.680 kg). Root gall index (RGI) and wilt disease incidence due to *R. solanacearum* (WDI) was minimum in combination treatment of *T. harzianum* + *P. chlamydosporia* followed by *P. fluorescens* + *P. chlamydosporia* and *T. harzianum* + *P. fluorescens* where the RGI was 1.80, 1.81 and 1.93, respectively and that of WDI was 22.18%, 25.13% and 27.28%, respectively (Table 14.27) (Naik, 2003).

The highest reduction of *M. incognita* and *R. solanacearum* population in soil was observed in combined application of *T. harzianum* and *P. fluorescens*. *P. fluorescens* was proved to be more promising followed by *T. harzianum* in suppressing the population of *R. solanacearum*, while *T. harzianum* proved more effective than *P. fluorescens* in reducing root galling. The population of both the bioagents increased in the soil (Table 14.28) (Barua and Bora, 2009).

Table 14.28: Effect of *T. harzianum* and *P. fluorescens* alone and in combination for the management of disease complex

Treatment	***No. of galls/root system***	***Popn. of R. solanacearum (x 10^8cfu/ml)***	***Popn. of T. harzianum (x 10^7cfu/ml)***	***Popn. of P. fluorescens (x 10^8cfu/ml)***
T. harzianum	36.00	18.20	42.00	0.00
P. fluorescens	45.60	6.60	0.00	40.00
T. harzianum + *P. fluorescens*	29.40	2.40	23.00	23.00
Inoculated control	93.80	22.60	0.00	0.00
Uninoculated control	0.00	0.00	0.00	0.00
CD (P = 0.05)	**0.43**	**1.00**	**0.75**	**0.72**

(b) *bioagents, botanicals, chemicals and cultural methods*: Integration of two summer ploughing during May-June, half recommended dose each of carbofuran (0.75 kg *a.i.*/ha), neem cake (7.5 g/spot), streptocycline (250 ppm at 30 ml/spot) and full dose of *T. harzianum* (150 g/spot) improved plant growth parameters and fruit yield with corresponding decrease in the nematode reproduction rate (*M. incognita*), bacterial population in the soil and bacterial wilt incidence (*R. solanacearum*) in brinjal (Hussain and Bora, 2008) (Table 14.29).

14.3.7. Root-knot Nematode, *Meloidogyne* spp. and Mosaic Virus Disease Complex

Brinjal plants having multiple infections of mosaic virus and root-knot nematodes (*M. incognita* / *M. javanica*) are frequently encountered. In such situation, brinjal crop appear to suffer more seriously.

Combined infection of both the pathogens considerably lowered the plant growth parameters compared with control or infection with either pathogen. Simultaneous

Table 14.29: Effect of integration of different practices for the management of disease complex in brinjal

Treatment	*No. of galls/ root system*	*No. of egg masses/root system*	*% wilt incidence*	*Yield (kg/4m²)*
2 Summer ploughings + full dose of carbofuran (1.5 kg *a.i.*/ha)	65	59	27.77	3.21
2 Summer ploughings + full dose of neem cake (15 g/spot)	71	69	27.77	3.10
2 Summer ploughings + full dose of streptocycline (500 ppm at 30 ml/spot)	128	112	38.88	2.43
2 Summer ploughings + full dose of *T. harzianum* (150 g/spot)	67	63	27.77	3.10
2 Summer ploughings + ½ dose carbofuran + full dose *T. harzianum*	35	26	5.53	3.90
2 Summer ploughings + ½ dose neem cake + full dose *T. harzianum*	57	47	11.10	3.30
2 Summer ploughings + ½ dose streptocycline + full dose *T. harzianum*	59	55	24.99	3.20
2 Summer ploughings + ½ dose carbofuran + ½ dose neem cake + ½ dose streptocycline	28	20	5.53	4.20
2 Summer ploughings + ½ dose carbofuran + ½ dose neem cake + ½ dose streptocycline + full dose *T. harzianum*	24	17	5.53	4.90
Control	362	207	66.66	1.6
CD (P = 0.05)	**6.69**	**6.75**	**11.15**	**0.83**

infection of both pathogens exhibited more severe impact on growth parameters as compared to the single infections indicating synergistic effect on growth of brinjal. Root-knot index, number of galls and root population of the nematode were higher in the treatments with nematode alone or nematode prior to the virus or in simultaneous inoculations of both the pathogens (Table 14.30) (Chovatiya *et al.*, 2005).

14.4. CHILLI, *CAPSICUM ANNUUM*

Chilli (*Capsicum annuum*) is valued for its diverse commercial uses. India is a major producer, exporter and consumer of chilli. Indian chillies reach over 90 countries in the world. Bangladesh, Bahrain, Canada, Sri Lanka, Saudi Arabia, USA and UAE are the leading importing countries. In India, chilli is grown in almost all states. It is being cultivated in 0.809 million hectares producing 1.325

Table 14.30: Effect of interaction between brinjal mosaic virus and root-knot nematode on brinjal

Treatment	*Plant height (cm)*	*Fresh shoot weight (g)*	*No. of galls/ groot*	*No. of eggs & larvae/g root*
Virus	32.1	60.1	–	–
Virus (prior) + Nematode (later)	29.7	29.5	103	19000
Nematode	30.2	45.0	167	24220
Nematode (prior) + Virus (later)	30.0	41.9	137	31240
Virus + Nematode (simultaneously)	26.9	31.3	131	23840
Uninoculated control	40.4	95.2	–	–
CD (P = 0.05)	**3.1**	**33.7**	**30.5**	**7474.0**

million tonnes of chillies with an average yield of 1.63 tonnes per hectare (Table 14.31). The chilli export for the year 2007-08 was 209,000 MT which was worth Rs.10975 million. Andhra Pradesh has been the largest chilli-growing state followed by Karnataka and Maharashtra. Productivity of chilli is highest in Andhra Pradesh, followed by Arunachal Pradesh and Punjab. The productivity of chilli in Andhra Pradesh and Punjab is higher since the crop is raised under irrigated conditions compared with Maharashtra and Karnataka. Chilli is rich in vitamins A and C.

Table 14.31: State-wise area and production of chilli (2006-07)

State	*Area (ha)*	*Production (MT)*	*Productivity (MT/ha)*
Andhra Pradesh	214000	766000	3.579
Karnataka	137850	164300	1.191
Orissa	75120	63930	0.851
West Bengal	52250	63596	1.216
Maharashtra	98030	47080	0.480
Gujarat	33320	46730	1.402
Tamil Nadu	61420	42720	0.695
Madhya Pradesh	47630	40810	0.856
Punjab	10110	16350	1.617
Rajasthan	18370	16180	0.880
Uttar Pradesh	14990	14580	0.972
Assam	15450	9980	0.645
Jammu & Kashmir	560	5760	10.285
Uttaranchal	1650	5410	3.229
Manipur	8910	5340	0.600
Bihar	2878	3017	1.048

Table 14.31: *(Contd...)*

Table 14.31: *(Contd...)*

State	*Area (ha)*	*Production (MT)*	*Productivity (MT/ha)*
Arunachal Pradesh	1900	3000	1.578
Chhattisgarh	6250	2140	0.342
Tripura	1840	2085	1.133
Meghalaya	1870	1380	0.738
Haryana	1380	1090	0.789
Mizoram	990	990	1.000
Nagaland	860	976	1.135
Andaman & Nicobar	360	870	2.417
Kerala	820	780	0.951
Himachal Pradesh	620	160	0.258
Pondicherry	9	19	2.111
TOTAL	**809437**	**1325273**	**1.603**

14.4.1. Root-knot Nematode, *Meloidogyne javanica* and Wilt, *Fusarium oxysporum/F. solani* Disease Complex

The disease complex involving *Meloidogyne javanica* and *Fusarium oxysporum / F. solani* inflict severe losses to chilli crop (Fig. 14.15) (Haseeb, 2003).

Fig. 14.15: Front - Chilli plants infected with root-knot and *Fusarium* wilt disease complex. Back – Healthy plants

Management methods

(i) *Integrated methods*

(a) *Bioagents and botanicals*: Integration of *T. harzianum* at 50 kg/ha with neem seed powder at 250 kg/ha in nursery beds was found to be highly effective in increasing the number of germinated seedlings (87/0.5 m^2 compared to 39/0.5 m^2 in control) and fresh weight of seedlings/bed (146.9g compared to 61g in control). The above treatment was also effective in reducing root galling due to

M. incognita (0.7 RKI compared to 3.5 RKI in control) and percent root infection by *F. solani* (5% compared to 41.5% in control) in nursery beds (Table 14.32) (Kumar *et al.*, 2009).

Table 14.32: Effect of biocontrol agents and organic amendments against *M. incognita-F. solani* disease complex in chilli

Treatment	***No. of seedlings emerged/bed***	***Fresh total wt. of seedlings/bed (g)***	***Root-knot index***	***% root infection***
Untreated control	39	61.0	3.5	41.5
Trichoderma harzianum	67	110.3	1.5	15.0
Aspergillus niger	62	100.0	2.3	21.5
Paecilomyces lilacinus	60	97.5	1.5	15.0
Pseudomonas fluorescens	65	105.7	1.9	17.3
Neem seed powder	64	105.0	1.3	15.5
Farm yard manure	45	71.9	3.3	35.5
T. harzianum + Neem seed powder	87	146.9	0.7	5.0
A. niger + Neem seed powder	77	128.7	1.0	13.0
P. lilacinus + Neem seed powder	74	123.0	0.9	13.5
P. fluorescens + Neem seed powder	81	136.7	0.7	5.5
T. harzianum + Farmyard manure	80	134.5	1.2	7.7
A. niger + Farmyard manure	71	115.3	2.0	15.5
P. lilacinus + Farmyard manure	70	113.0	1.3	23.5
P. fluorescens + Farmyard manure	75	124.9	1.5	13.0
Carbofuran	72	117.0	0.6	25.5
CD (P = 0.05)	**3.53**	**7.91**	**0.06**	**0.65**

(b) *Two bioagents*: *Pseudomonas aeruginosa* and *Paecilomyces lilacinus* when used together significantly reduced infection of the disease complex on chilli (Perveen *et al.*, 1998). Use of *P. aeruginosa* and *P. lilacinus* significantly ($P<0.05$) increased plant height of chilli. *P. aeruginosa* and *P. lilacinus* used alone or together significantly ($P<0.05$) reduced infection of root knot nematode *M. javanica* and root infecting fungi *viz.*, *Macrophomina phaseolina, Rhizoctonia solani, F. solani* and *F. oxysporum* on chilli. *P. aeruginosa* was more effective than *P. lilacinus* in reducing the *M. javanica* root knot nematode infection (Table 14.33).

Table 14.33: Effect of *P. aeruginosa* and *P. lilacinus* on plant height and control of root rot disease complex in chilli

Treatment	*Plant height (cm)*	*Root-knot index*	*Infection %*			
			M. phaseolina	*R. solani*	*F. solani*	*F. oxysporum*
Control	10.5	3.3	31	19	75	37
P. lilacinus (PL)	14.5	2.9	19	0	56	44
P. aeruginosa (PA)	16.5	2.5	12	6	75	21
PL + PA	14.7	2.1	12	25	69	12
CD (P = 0.05)	**2.2**	**0.34**	**6.1**	**6.1**	**6.1**	**6.1**

Bare root-dip treatment with *Pseudomonas aeruginosa* along with or without *T. harzianum, T. koningii* and *T. hamatum* significantly controlled infection of roots by *F. solani* and *M. javanica* on chilli. Combined use of *T. harzianum* along with *P. aeruginosa* caused the greatest reduction in root galling by *M. javanica* (Siddiqui *et al.*, 1999).

14.4.2. Root-knot Nematode, *Meloidogyne incognita* and Damping-Off, *Pythium aphanidermatum* Disease Complex

P. aphanidermatum and *R. solani* were both found to interact with *M. incognita* on chilli, causing some loss of nematode resistance in two cultivars tested (Hasan, 1985), whereas the interaction of *P. debaryanum* with this nematode appeared to be due to physiological response of the plants to nematode infection, making the roots more susceptible to invasion by the fungus (Brodie and Cooper, 1964).

Management methods

(i) *Integrated methods*

(a) *Bioagents and botanicals*: Biological control of *P. aphanidermatum-M. incognita* disease complex in chilli with organic amendments (FYM and neem cake); antagonistic organisms *T. viride* and *T. harzianum* (against *P. aphanidermatum*); and *Paecilomyces lilacinus* (against *M. incognita*) was reported by Karthikeyan *et al.* (1999).

14.4.3. Root-knot Nematode, *Meloidogyne* sp. and Root Rot, *Phytophthora* sp. Disease Complex

Kim *et al.* (1989) reported that that the disease incidence was raised to 54% when hot pepper was inoculated with *Meloidogyne* sp. and *Phytophthora* sp.

14.5. BELL PEPPER, *CAPSICUM ANNUUM*

14.5.1. Root-knot Nematode, *Meloidogyne incognita* and Bacterial Wilt, *Ralstonia solanacearum* Disease Complex

Capsicum is prone to many soil borne diseases among which the bacterial wilt

(*R. solanacearum*) in combination with root-knot nematode (*M. incognita*) takes heavy toll every year all over the world (Fig. 14.16) (Naik, 2003).

Fig. 14.16: Root-knot nematode-bacterial wilt disease complex in sweet pepper

Management methods

(i) *Integrated methods*

(a) *Two bioagents*: The seedling stand was good when the nursery beds were combinedly treated with neem based formulations of *T. harzianum* + *P. fluorescens* followed by seed treatment with talc based *P. fluorescens* + nursery bed treatment with neem based *P. fluorescens* (Naik, 2003) (Table 14.34).

Table 14.34: Effect of integration of bioagents on nursery stand of capsicum seedlings

Treatment	*Nursery stand of seedlings**
Seed treatment with talc based *P. fluorescens*	3.4
Nursery bed treatment with neem based *P. fluorescens*	3.2
Seed treatment with talc based *P. fluorescens* + Nursery bed treatment with neem based *P. fluorescens*	4.2
Nursery bed treatment with *T. harzianum* + *P. fluorescens*	4.4
Control	3.0

* 3 – Poor, 4 – Good, 5 – Very good

Combined application of neem based formulations of *P. fluorescens* and *P. chlamydosporia* / *T. harzianum* at 40 g/m^2 in nursery beds and transplanting these seedlings in the main field resulted in significant reduction in disease index and root-knot index in capsicum to the tune of 70% and increased the crop yield by 37% (Rao *et al*., 2002) (Table 14.35). Combinations of the bioagents did not affect the colonization of the individual bioagents on the roots and hence the transplants carried the bioagents to the main field. This has resulted in the effective management of the pathogens involved in the disease complex.

Table 14.35: Effect of integration of neem-based bioagents on plant growth and management of root-knot and bacterial wilt disease complex and yield of capsicum.

Treatment	*Seedling weight (g)*	*Root-knot index (1-10)*	*Disease index (1-9)*	*Yield in kg/4 m²*
Seed treat. with *P. fluorescens*	421	5.6	6.4	4.3
Seed treat. with neem-based *P. fluorescens*	428	5.2	6.7	4.7
Nursery treat. with *P. fluorescens*	435	4.6	5.4	4.8
Nursery treat. with *P. chlamydosporia*	364	4.4	7.3	3.2
Nursery treat. with *T. harzianum*	374	4.8	7.0	3.4
Nursery treat. with *P. fluorescens* + *P. chlamydosporia*	463	4.1	5.2	4.0
Nursery treat. with *P. fluorescens* + *T. harzianum*	493	3.8	3.5	5.1
Control	340	8.7	8.2	2.6
CD (P = 0.05)	**27.20**	**0.49**	**0.38**	**0.25**

Integration of seed treatment with *P. fluorescens* and nursery bed treatment with *P. chlamydosporia* at 25 or 50 g/m^2 increased plant height, weight of seedlings and root colonization with bioagents; and reduced root galling in capsicum (Table 14.36) (Naik, 2003). The seedlings were highly vigorous which were colonized by the bioagents and reached the main field when transplanted.

Table 14.36: Effect of integration of bioagents on plant growth, root galling and root colonization by bioagents in capsicum under nursery conditions

*Treatment**	*Plant height (cm)*	*Seedling weight (g)*	*No. of galls/10 seedlings*	*P. chlamydosporia root colonization (cfu/g)*	*P. fluorescens root colonization (cfu/g)*
T_1	12.58	3.6	62	15,896	–
T_2	14.63	3.8	59	17,459	–
T_3	15.87	4.0	55	–	12,563
T_4	18.54	4.5	51	16,256	12,349
T_5	17.42	4.4	53	16,789	11,897
T_6	11.23	3.2	82	–	–
CD (P = 0.05)	**1.67**	**0.34**	**7.96**	**959.72**	**729.65**

* T_1 _ Nursery bed treatment *P. chlamydosporia* at 25 g/m^2, T_2 _ Nursery bed treatment *P. chlamydosporia* at 50 g/m^2, T_3 _ Seed treatment with *P. fluorescens*, T_4 _ T_1 + T_3, T_5 _ T_2 + T_3, T_6 _ Control.

The seedlings raised in nursery beds treated with *P. chlamydosporia* at 25 or 50 g/m^2 + seed treatment with *P. fluorescens* when transplanted in the main field gave least root galling and increased fruit yield, colonization of roots with the bioagents, and parasitization of eggs by *P. chlamydosporia* (Table 14.37) (Naik, 2003).

Table 14.37: Effect of integration of bioagents on fruit yield, root galling and root colonization by bioagents in capsicum under field conditions

Treatment*	***Root-knot index***	***Fruit yield (kg/6m^2)***	***P. chlamydosporia root colonization (cfu/g)***	***P. fluorescens root colonization (cfu/g)***	***% eggs parasitized by P. chlamydosporia***
T_1	6.3	4.4	22,456	–	41.00
T_2	5.6	4.7	25,789	–	45.00
T_3	6.0	5.1	–	22,568	–
T_4	4.8	5.3	21,679	21,789	40.00
T_5	4.4	5.5	24,587	21,567	42.67
T_6	8.1	4.0	–	–	–
CD (P = 0.05)	**0.45**	**0.28**	**1246.76**	**1089.86**	**3.78**

* T_1 _ Nursery bed treatment *P. chlamydosporia* at 25 g/m^2, T_2 _ Nursery bed treatment *P. chlamydosporia* at 50 g/m^2, T_3 _ Seed treatment with *P. fluorescens*, T_4 _ T_1 + T_3, T_5 _ T_2 + T_3, T_6 _ Control.

Naik (2003) reported that combination of *T. harzianum* along with *P. fluorescens* increased the yield (3.320 kg/3m^2 plot) followed by combination of *T. harzianum* and *P. chlamydosporia* (2.960 kg/3m^2). The least gall index was present where the combination of *T. harzianum* and *P. fluorescens* was used (1.5) followed by *T. harzianum* + *P. chlamydosporia* (1.7) and *P. fluorescens* + *P. chlamydosporia* (1.7). But among combination treatment of bio-agents, all the treatments were on par (*P. fluorescens* + *T. harzianum*, *P. fluorescens* + *P. chlamydosporia*, *T. harzianum* + *P. chlamydosporia*). Results related to the percent bacterial wilt disease (*R. solanacearum)* incidence also showed similar trend in which *T. harzianum* + *P. fluorescens* performed very well and the percent incidence was 16.90 followed by *T. harzianum* + *P. chlamydosporia* (17.20) (Table 14.38).

Naik *et al.* (2003) reported that soil application of organically developed *P. lilacinus* (50 g/m^2 of formulated product containing 10^6 spores/g) along with pure culture of *Bacillus pumilis* (10^8 cfu/ml) to the nursery beds of bell pepper increased plant growth, root galling and root colonization by the bioagents. The above seedlings when transplanted in field significantly reduced root galling and wilt incidence, and increased root colonization, propagule density in soil, egg parasitization and fruit yield in the main field (Tables 14.39 and 14.40).

Table 14.38: Effect of integration of bioagents on root galling, wilt disease incidence and fruit yield of capsicum

Treatment	*No. of fruits/ plant*	*Fruit yield (kg/3m²)*	*Root-knot index*	*Wilt disease incidence (%)*
Neem + wheat bran formulation (8:2) of *T. harzianum*	31.2	2.30	1.90	23.8 (29.20)
Liquid broth formulation of *P. fluorescens*	34.2	2.40	1.75	21.4 (27.56)
Neem + pongamia + wheat bran formulation (4:4:2) of *P. chlamydosporia*	32.8	2.04	1.92	26.7 (31.11)
T. harzianum + *P. fluorescens*	46.2	3.32	1.50	16.9 (24.20)
T. harzianum + *P. chlamydosporia*	40.6	2.96	1.70	17.2 (24.50)
P. fluorescens + *P. chlamydosporia*	40.0	2.48	1.70	17.8 (24.95)
Neem + wheat bran	27.2	1.93	2.25	30.5 (33.52)
Neem + pongamia + wheat bran	29.8	2.00	2.00	36.4 (37.11)
Control	26.2	1.78	3.80	96.0 (78.46)
CD (P = 0.05)	**7.38**	**0.41**	**0.32**	**3.28**

Table 14.39: Effect of integration of bioagents on plant growth and management of disease complex in bell pepper in nursery beds

*Treatment**	*Seedling length (cm)*	*Seedling weight (g)*	*No. of galls/ 10 plants*	*Root colonization (cfu/g)*	
				P. lilacinus	*B. pumilis*
T1	12.63	3.8	52	13,342	–
T2	15.53	4.0	50	16,539	–
T3	15.17	4.2	55	–	14,431
T4	18.43	4.4	44	12,984	13,964
T5	19.25	4.6	40	16,839	14,291
T6	10.53	3.3	76	–	–
CD (P = 0.05)	**1.29**	**0.37**	**6.28**	**1087.95**	**976.75**

* T1 – Nursery bed treatment with *P. lilacinus* (25 g/ m^2), T2 – Nursery bed treatment with *P. lilacinus* (50 g/ m^2), T3 – Nursery bed treatment with *B. pumilis*, T4 – T1+ T3, T5 – T2 + T3, T6 – Control.

14.5.2. Root-knot Nematode, *Meloidogyne incognita* and Foot and Root Rot, *Phytophthora capsici* Disease Complex

Capsicum seed treatment with consortia of *P. fluorescens* (3 strains) at the rate of 10 g/kg and subsequent application of consortia formulation of *P. fluorescens* at the rate of 15 g/m^2, significantly reduced *M. incognita* in capsicum roots

Table 14.40: Effect of integration of bioagents on root galling, wilt incidence, root colonization, propagule density and yield of bell pepper under field conditions

*Treatment**	*Root-knot index*	*Mortality%*	*Yield kg/ 6m²*	*Root colonization (cfu/g)*		*Propagule density (cfu/g soil)*	
				P. lilacinus	*B. pumilis*	*P. lilacinus*	*B. pumilis*
T1	6.5	68.90	4.8	30,653	–	21,346	–
T2	5.3	52.45	5.4	34,289	–	23,467	–
T3	7.0	34.62	7.5	–	25,875	–	17,321
T4	4.3	25.86	8.2	31,764	21,459	20,457	15,347
T5	4.1	17.69	8.5	34,762	20,764	22,689	14,569
T6	8.4	91.45	0.5	–	–	–	–
CD(P=0.05)	**0.58**	**8.94**	**0.35**	**1176.52**	**1087.54**	**1234.87**	**897.34**

* T1 – Nursery bed treatment with *P. lilacinus* (25 g/ m^2), T2 - Nursery bed treatment with *P. lilacinus* (50 g/ m^2), T3 - Nursery bed treatment with *B. pumilis*, T4 – T1 + T3, T5 – T2 + T3, T6 – Control.

by 74%; reduced the incidence of *Phytophthora* foot and root rot and increased the yield by 26%. Cost benefit ratio (calculated for the additional cost of the bio-pesticides and additional returns accrued by the application of the bio-pesticide) was 1:3.8 (Rao *et al.*, 2009).

14.6. PEA, *PISUM SATIVUM*

Pea (*Pisum sativum*) is an important vegetable crop widely grown throughout the world. As a cool season crop, it is extensively grown in temperate zone; but restricted to cooler altitudes in the tropics and winter season in the sub-tropics. A rich source of proteins (25%), amino acids and sugars (12%), green peas are an all time favourite vegetables. It is rich in vitamins A, B, C and minerals. It is being cultivated in 0.314 million hectares producing 2.560 million tonnes of peas with an average yield of 8.2 tonnes per hectare (Bijay Kumar, 2009) (Tables 14.41 and 14.42). Garden pea is grown in Uttar Pradesh, Madhya Pradesh, Himachal Pradesh, West Bengal, Punjab, Andhra Pradesh, Uttarakhand, Haryana, Bihar and Chhattisgarh.

14.6.1. Root-knot Nematode, *Meloidogyne incognita* and Wilt, *Fusarium oxysporum* f. sp. *pisi* Disease Complex

Pathogenic level of *M. incognita* was calculated to be 2 J_2/g soil as indicated by the reproduction of the nematode. On the basis of wilt index on pea, 1 g of mycelium mat of *F. oxysporum* f. sp. *pisi* could produce significantly high index value (Fig. 14.17). The presence of root knot nematode irrespective of the presence of other organism showed reduction in shoot length to a significant manner.

Table 14.41: Major Peas producing countries of the world (2007-08)

Country	*Area (in '000 ha)*	*Production (in '000 MT)*	*Productivity (in MT/ha)*	*% Share of world production*
India	314	2560	8	31
China	251	2509	10	30
USA	87	875	10	10
France	31	355	12	4
UK	33	330	10	4
Egypt	27	280	10	3
Morocco	18	110	6	1
Turkey	15	101	7	1
Hungary	17	92	6	1
Italy	13	90	7	1
Others	315	1230	4	14
Total	**1119**	**8532**	**8.18**	

Source: FAO (except Indian data) (Indian data source: Indian Horticulture Database, 2007-08)

Table 14.42: Area, production and productivity of leading peas growing states in India (2007-08)

State	*Area (in'000 ha)*	*Production (in '000 MT)*	*Productivity (MT/ha)*	*% Share of India production*
Uttar Pradesh	158.1	1398.6	8.8	54.6
Madhya Pradesh	21.5	236.4	11.0	9.2
Himachal Pradesh	17.4	203.4	11.7	7.9
West Bengal	20.9	123.8	5.9	4.8
Punjab	18.5	111.0	6.0	4.3
Andhra Pradesh	2.1	74.9	35.7	2.9
Uttarakhand	11.1	72.5	6.5	2.8
Haryana	10.0	70.9	7.1	2.8
Bihar	9.0	57.6	6.4	2.3
Chhattisgarh	9.9	44.8	4.5	1.8
Others	35.0	166.1	4.7	6.5
Total	**313.5**	**2560.0**	**8.2**	

(*Source*: Indian Horticulture Database, 2007-08)

However, maximum reduction in shoot length was observed in nematode inoculated one week prior to fungus. Both the organisms suppressed the plant growth characters in general and shoot length in particular. Maximum nematode population was found to be in nematode alone and nematode followed by fungus inoculation. The wilt index on pea after 50 days was recorded to be 3.5, 4.0 and 4.2 in F + N, F→N and N→F, respectively.

Fig. 14.17: Root-knot nematode and Fusarium wilt complex in pea

14.6.2. Reniform Nematode, *Rotylenchulus reniformis* and Wilt, *F. oxysporum* f. sp. *pisi* Disease Complex

The interaction of *Rotylenchulus reniformis* and *F. oxysporum* f. sp. *pisi* race 3 and its importance in "early yellowing" disease of peas was investigated by Labruyere *et al.* (1959). Neither pest alone causes severe root rot in peas but together they cause an extensive decay of the root cortex which is the chief symptom in roots of peas with "early yellowing".

14.6.3. Lance Nematode, *Hoplolaimus uniformis* and Wilt, *F. oxysporum* f. sp. *pisi* Disease Complex

A nematode-fungus relationship occurs in peas. The chief symptoms of the "early yellowing" disease, root rot, is dependent upon the presence of both *H. uniformis* and *F. oxysporum* f. sp. *pisi* race 3 (Labruyere *et al.*, 1959).

14.6.4. Stunt Nematode, *Tylenchorhynchus claytoni* and Wilt, *Fusarium oxysporum* Disease Complex

T. claytoni increases more rapidly on pea roots infected with *F. oxysporum* f. sp. *pisi* race 1 than in roots free from the fungus (Davis and Jenkins, 1963).

14.6.5. Root-knot Nematode, *Meloidogyne incognita* and Root Rot, *Rhizoctonia solani* Disease Complex

Peas were damaged by *M. incognita* or *R. solani* but plant growth was depressed even further when plants were inoculated with both organisms, with the maximum effect occurring when the two were inoculated simultaneously (Husain *et al.*, 1985).

14.6.6. Lesion Nematode, *Pratylenchus crenatus* and Root Rot, *Thielaviopsis basicola* Disease Complex

The minor pathogen, *Thielaviopsis basicola*, was assisted by *P. crenauns* in penetrating pea roots (Green *et al.*, 1983).

14.7. FRENCH BEAN, *PHASEOLUS VULGARIS*

14.7.1. Root-knot Nematode, *Meloidogyne incognita* and Root Rot, *Macrophomina phaseolina* Disease Complex

(i) Management methods

(a) *Cultural methods*: The beneficial effects of neem cake were observed not only against *M. incognita* and the fungus *M. phaseolina* individually, but also when both the pathogens formed a disease complex on French bean.

14.7.2. Root-knot Nematode, *Meloidogyne incognita* and Root Rot, *Fusarium solani* Disease Complex

M. incognita in association with soil-borne fungus such as *F. solani* was responsible for root rot disease complex in French bean (Fig. 14.18). In the presence of nematode, the root rot due to *F. solani* increased considerably (Singh *et al.*, 1981).

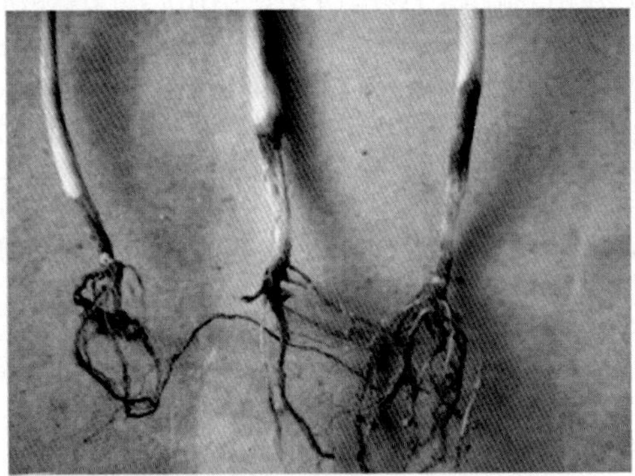

Fig. 14.18: Root-knot and *Fusarium* wilt complex in French bean

14.7.3. Root-knot Nematode, *Meloidogyne incognita* and Root Rot, *Rhizoctonia solani* Disease Complex

M. incognita in association with soil-borne fungus such as *R. solani* was responsible for root-rot disease complex in French bean. In the presence of nematode, the root rot due to *R. solani* increased considerably (Reddy *et al.*, 1979).

14.7.4. Root-knot Nematode, *Meloidogyne incognita* and *Rhizobium leguminosarum* bv. *phaseoli* Interaction

Infection by root-knot nematodes is known to suppress nodulation and hence nitrogen fixation in beans. *Meloidogyne* infection caused significant ($P \leq 0.05$) reductions in nodulation. In a second pot experiment, bean cv. GLP-24 was inoculated with *R. leguminosarum* bv. *phaseoli* alone and in various combinations with *M. incognita*. Both nodulation and the dinitrogen fixation processes were adversely affected especially in plants where nematode inoculation preceded rhizobial inoculation (Kimenju *et al.,* 1999).

Effect of *Meloidogyne* Infection on Nodulation of Different Bean Genotypes

Numbers of nodules on bean plants infected with *M. incognita* were significantly ($P \leq 0.05$) lower than in control, with the exception of lines E1 and M28 (Table 14.43). Reduction in nodulation ranged from 35.4% in line NOB to 100% in line E4. Numbers of nodules in plants inoculated with *M. incognita* were significantly ($P \leq 0.05$) different among the bean genotypes. For instance, no nodules were observed on the roots of bean line E4 inoculated with nematodes as compared to more than 111 nodules on roots of bean line NOB. *Meloidogyne* infection caused significant ($P \leq 0.05$) reduction in proportions of effective nodules in the bean genotypes tested, with the exception of bean cv. GLP-24 (Kimenju *et al.,* 1999).

Table 14.43: Numbers of nodules and percentage of effective nodules in *Meloidogyne*-infected and non-infected bean plants

Bean genotype	*Nodule numbers in:*		*Effective nodules (%) in*	
	Inoculated[1]	*Control*[2]	*Inoculated*	*Control*
GLP-24	90.0	166.8	70.0	98.0*
GLP-1004	20.0	58.6	14.0	50.0
NOB	111.2	172.2	72.0	90.0
KK8	70.2	132.2	36.0	78.0
KK14	88.6	161.8	38.0	88.0
KK15	77.4	157.6	40.0	80.0
KK22	38.0	124.6	32.0	62.0
E1	42.6	92.8	20.0	74.0
E3	19.4	121.4	8.0	40.0
E4	0	54.0	0	76.0
M14	1.4	30.2	0	68.0
M24	45.4	99.8	22.0	62.0
M26	41.6	92.4	10.0	58.0
M28	25.4	39.2*	20.0	70.0
M29	27.6	59.4	46.0	74.0
M30	9.2	28.0	8.0	38.0
L31	32.2	116.6	36.0	70.0
L32	32.4	107.8	34.0	66.0
L40	4.6	40.0	12.0	44.0
L45	19.0	46.2	18.0	50.0
CD ($P \leq 0.05$)	**19.9**	**24.7**	**15.2**	**12.6**

[1]Inoculated = Nematode-infected, [2]Control = Nematode-free; Data are means of five replications

Effect of Time of Inoculation with *Meloidogyne incognita* on Bean Nodulation

The number of nodules was significantly ($P \leq 0.05$) higher in bean cv. GLP-24 plants inoculated with *rhizobia* alone than in plants inoculated with combinations of *rhizobia* and *M. incognita*. Plants inoculated with nematodes at emergence and with *rhizobia* 10 days after emergence (DAE) had the lowest number of nodules on their roots. The percentage of effective nodules was significantly ($P \leq 0.05$) higher in bean plants inoculated with *rhizobia* alone than those inoculated with *rhizobia* and *M. incognita* (Table 14.44). Plants inoculated with nematodes at emergence had lower ($P \leq 0.05$) proportions of functional nodules than those inoculated with nematodes 10 DAE (Kimenju *et al.*, 1999).

Table 14.44: Effect of *Meloidogyne* infection on bean nodulation and functioning of nodules in bean cv. GLP-24 plants[1]

Treatments	*Nodule number/plant*	*Effective nodules (%)*
Rhizobium alone at emergence	96.5[a2]	83.0[a]
Nematode + Rhizobium (both at emergence)	74.4[b]	48.8[c]
Nematode (at emergence) + Rhizobium (10 DAE)	54.2[c]	47.4[c]
Rhizobium (at emergence) + Nematodes (10 DAE)	78.2[b]	66.0[b]
DAE = Days after emergence		

[1]Data are means of 10 replications. [2]Means followed by the same letter along the columns are not significantly ($P \leq 0.05$) different by Least significant difference test

14.8. COWPEA, *VIGNA UNGUICULATA*

14.8.1. Root-knot Nematodes, *Meloidogyne* spp. and Root Rot, *Rhizoctonia solani* Disease Complex

Synergistic interactions were reported between *M. javanica* and *R. solani* in cowpea. Inoculation of cowpea with *M. incognita* and *R. solani* led to breakdown of resistance to both organisms (Khan and Husain, 1989), and the greatest decrease in plant dry weight occurred if the nematode was inoculated two weeks before the fungus (Varshney *et al.*, 1987).

Management methods

(i) *Biomanagement*

(a) *Antagonistic fungi*: Khan and Husain (1988, 1990) reported that application of *P. lilacinus* at 2 g mycelium/pot reduced damage to cowpea caused by *M. incognita, R. reniformis* and *R. solani* disease complex in cowpea.

The root-knot nematode reproduction parameters such as number of galls, egg masses and final soil nematode population under *M. incognita* alone and *M. incognita* + *R. solani* complex, were also reduced maximally and significantly in *Trichoderma viride* and *Gliocladium virens* application at 10 g/kg soil. Similarly, under complex of both the pathogens, *T. viride* and *G. virens* enhanced all the plant growth parameters of cowpea significantly (Verma *et al.*, 2010).

14.8.2. Root-knot Nematode, *Meloidogyne incognita* and Root Rot, *Macrophomina phaseolina* Disease Complex

Management methods

(i) *Integrated methods*

(a) *AMF and botanicals*: Devi and Goswami (1992) demonstrated that *G. fasciculatum* in cowpea together with mustard cake helped in reducing the disease severity caused by *M. incognita* and *M. phaseolina*. They observed that the pre-establishment of AMF checked the entry of *M. incognita* larvae as also colonization of the pathogenic fungus.

14.8.3. Root-knot Nematode, *Meloidogyne incognita* and Wilt, *Fusarium oxysporum* Disease Complex

The wilt fungus, *F. oxysporum* and root-knot nematode, *M. incognita* co-infect cowpea. Histopathological studies revealed that in nematode + fungus inoculated cowpea roots, conidia of *F. oxysporum* could be observed in the cortex as well as xylem vessels adjacent to the giant cells but not inside the giant cells induced by *M. incognita* (Singh *et al.*, 2007).

Management methods

(i) *Integrated methods*

(a) *Chemicals and botanicals*: The minimum gall diameter index was reported in reduced dose of both neem cake and carbofuran (Singh *et al.*, 2007).

14.8.4. Cyst Nematode, *Heterodera cajani* and Root Rot, *Rhizoctonia solani* Disease Complex

If *R. Solani* is added to cowpea before *H. cajani*, nematode reproduction was severely depressed (Walia and Gupta, 1986a, 1986b) and only when the fungus was added two weeks after the nematode there was an adverse effect on plant growth (Walia and Gupta, 1986a).

14.9. PIGEONPEA, *CAJANUS CAJAN*

14.9.1. Root-knot Nematode, *Meloidogyne* spp. and Wilt, *Fusarium udum* Disease Complex

The wilt disease complex caused by *F. udum* in association with *Heterodera cajani, M. incognita* and *R. reniformis* has been reported as the most severe constraint in the cultivation of pigeonpea. Among plant-parasitic nematodes, *M. incognita* is considered most serious threat to its cultivation. Similarly, among fungi, *Fusarium* wilt is one of the most destructive diseases of pigeon pea. When both these pathogens attack the crop, intensity of damage is increased by several folds. Inoculation with *F. udum* and *H. cajani* together significantly increased wilt severity in pigeon pea seedlings compared with inoculation of the fungus alone.

Simultaneous or sequential inoculation of *M. incognita* and *F. udum* increased the severity of the disease. Highest reduction in plant height, fresh/dry weight was observed in plants inoculated with *M. incognita* and *F. udum* simultaneously followed by *M. incognita* prior and *F. udum* 7 days later and *F. udum* prior and *M. incognita* 7 days later, respectively. Reproduction of *M. incognita* was enhanced in the presence of *F. udum* but per cent root colonization by *F. udum* was suppressed in the presence of *M. incognita*. Highest final nematode population and gall index of *M. incognita* were observed in simultaneous inoculation of *M. incognita* and *F. udum* and lowest in *F. udum* prior and *M. incognita* 7 days later, while highest per cent root colonization was found in *F. udum* prior and *M. incognita* 7 days later followed by *M. incognita* and *F. udum* simultaneously and *M. incognita* prior and *F. udum* 7 days later, respectively (Table 14.45) (Perveen *et al.,* 1988).

Table 14.45: Effect of *Meloidogyne incognita* and *Fusarium udum* on plant growth, root galling and disease index in pigeonpea

Treatment	***Plant height (cm)***	***Dry plant wt. (g)***	***Gall index***	***Disease index***
Control	33.9	1.03	–	–
M. incognita	32.6	0.73	1.75	0.00
F. udum	32.6	0.81	–	45.32
M. incognita + *F. udum* simultaneouly	23.6	0.38	2.15	38.21
M. incognita (pre) + *F. udum* (post)	27.3	0.57	2.75	32.14
F. udum (pre) + *M. incognita* (post)	30.6	0.65	1.25	42.33
CD (P = 0.05)	**1.3**	**0.031**	**0.143**	**0.273**

Effect of Disease Complex on Wilt Resistant Accessions

Use of varieties resistant to Fusarium wilt is one of the best options to avoid the losses caused by the disease complex. However, sometimes the varieties developed as resistant to Fusarium wilt when grown under farmers' field conditions become susceptible to wilt due to the presence of root-knot nematodes.

Among 10 pigeon pea accessions (identified as resistant to Fusarium wilt) (Vishwadhar and Chaudhary, 2000) evaluated against combined infection of *M. javanica* and *F. udum* under pot culture conditions, increase in wilting was observed in five accessions namely, ICP 8859, AWR 74/15, KPL 44, ICPL 89049 and ICPL 12745. In these accessions wilting started 30 days after inoculation of fungus. Maximum wilting was observed in AWR 74/15 (60%) followed by ICP 8859 (50%) and ICPL 89049 (50%). Wilting increased from 8 to 33% in KPL 44, 15 to 60% in AWR 74/15, 25 to 50% in ICP 8859 and ICPL 89049, and 15 to 50% in ICP 12745 when *M. javanica* was present with *F. udum*. Whereas in other 5 accessions (KPL

43, PI 397430, BWR 370, GPS 33 and ICPL 89048), wilting was not influenced much in the presence of *M. javanica* (Singh *et al*., 2004).

The plant height, fresh and dry shoot and root weights were significantly lower in combined inoculation compared to either nematode or fungus alone. The root-knot index varied from 3.0 to 4.5 in these accessions in both treatments having nematode alone or nematode and fungus together. The lowest root-knot index was observed in KPL 43 (1.5) and GPS 33 (1.75) (Table 14.46) (Singh *et al*., 2004).

Table 14.46: Effect of combined inoculation of *M. javanica* and *F. udum* on plant growth and disease complex in different accessions of pigeon pea

Pigeonpea accessions	*Plant height (cm)*	*Fresh shoot weight (g)*	*Fresh root weight (g)*	*Gall index*	*% Wilt incidence*
KPL 43	25.9	4.94	5.12	2.45	0
KPL 44	17.4	2.50	3.67	1.97	33
AWR 74/15	18.3	1.92	2.55	1.56	60
ICP 8859	14.7	1.52	2.33	3.39	50
ICPL 89049	24.3	2.97	4.23	2.17	50
PI 397430	24.8	3.73	3.57	3.06	0
BWR 370	28.4	4.18	4.92	3.17	0
GPS 33	32.0	3.93	5.75	2.25	0
ICPL 89048	28.3	3.65	4.17	3.25	0
ICP 12745	19.4	2.67	3.19	2.50	50
CD (P = 0.05)	**2.40**	**0.43**	**NS**	**1.60**	-

Management methods

(i) *Biomanagement*

(a) *Antagonistic fungi*: In a farmers' field trial conducted at Arania, Bulandshahr district of Uttar Pradesh during 2001; wilting of plants was not seen in plots treated with carbofuran, neem seed powder, neem jivan, *T. harzianum* and *Aspergillus niger*, 1-5% wilting was observed in plots treated with *P. lilacinus* (Haseeb and Shukla, 2005) (Table 14.47).

Pigeon pea seed treatment with *Trichoderma harzianum* and *Pochonia chlamydosporia* gave maximum control of the disease complex in which wilt severity decreased by 79-81%, and root galling by 44-49%, and egg mass production by 33-35% (Khan *et al*., 2010).

(ii) *Host resistance*

Five accessions of pigeonpea which were resistant to *F. udum* (KPL 43, PI 397430, BWR 370, GPS 33 and ICPL 89048) were also found resistant to combined inoculation of *M. javanica* and *F. udum*, which indicated that wilt resistance in these accessions was not influenced by the presence of *M. javanica* (Singh *et al*., 2004).

Table 14.47: Effect of seed treatment with biopesticides, chemicals and botanicals on parasitic nematodes infecting pigeonpea under field conditions during 2001.

Treatment	*Grain yield (kg)/100 m²*	*% wilting of plants*	*% root infection by F. udum*	*Nematodes/ g of root*	*Cysts & larvae/100 ml soil*
Carbofuran	27.57	0.0	10.0	15	20
Neem seed powder (soil application)	22.66	0.0	5.0	18	22
Neem seed powder (seed treatment)	19.66	0.0	5.0	17	26
T. harzianum	18.66	0.0	5.0	20	27
Neem jivan	17.33	0.0	10.0	24	27
A. niger	16.49	0.0	10.0	26	30
P. lilacinus	15.83	2.0	15.0	30	36
Dimethoate	15.83	5.0	20.0	31	32
Latex of *Calotropis*	12.66	5.0	25.0	30	33
Control	8.83	33.0	65.0	44	37
CD (P = 0.05)	**1.58**	**1.8**	**2.7**	**2.9**	**2.1**

(iii) *Chemical methods*

Aldicarb, carbofuran and phorate reduced disease severity both in plants inoculated with both organisms and in plants inoculated with fungus alone. Greatest reduction occurred at the highest dose (20 mg *a.i.*/kg soil) (Hasan, 1989).

(iv) *Integrated methods*

(a) *Bioagents, botanicals and chemicals*: In a trial conducted in the field during 2002, no wilting was observed in plots treated with neem seed powder + dimethoate/*T. harzianum*/*P. lilacinus*/latex, and neem seed powder and *T. harzianum* alone. Neem seed powder + *T. harzianum* was found to be the most effective treatment in increasing the yield and suppressing the pathogens, followed by carbofuran, neem seed powder + *P. lilacinus*, neem seed powder + dimethoate and neem seed powder + latex. Data regarding *F. udum* infection in roots gathered at the pre-flowering stage (90 DAS) indicated that all the treatments maintained significant protection of the roots as compared to control (Haseeb and Shukla, 2005) (Table 14.48).

(b) *Bioagents and AMF*: The detrimental effects of a disease complex on pigeonpea involving the sedentary endoparasite *H. cajani* and the fungus *F. udum* were reduced following application of the fungi *P. lilacinus* and *P. chlamydosporia* together with the arbuscular mycorrhizal fungus *Gigaspora margarita* (Siddiqui and Mahmood, 1995).

The treatment constituting FYM, karanj oilseed cake and arbuscular mycorrhizal fungus, *Glomus fasciculatum* reduced the disease incidence caused by

Table 14.48: Effect of seed treatment with bioagents, chemicals and botanicals on wilt and parasitic nematodes infecting pigeon pea under field conditions during 2002

Treatment	*Grain yield (kg)/100 m²*	*% wilting of plants*	*% root infection by F. udum*	*Nematodes/ g of root*	*Cysts & larvae/100 ml soil*
Carbofuran	23.47	5.0	35	9	27
Neem seed powder (soil application)	19.55	0.0	20	23	45
Neem seed powder (seed treatment)	18.33	0.0	20	30	43
T. harzianum	15.46	0.0	25	31	44
P. lilacinus	12.24	5.0	35	36	55
Dimethoate	11.75	10.0	50	39	58
Latex of *Calotropis*	11.34	10.0	50	39	59
NSP (soil) + Dimethoate	21.05	0.0	15	16	36
NSP (soil) + *T. harzianum*	23.35	0.0	15	10	28
NSP (soil) + *P. lilacinus*	21.85	0.0	20	14	34
NSP (soil) + latex	20.75	0.0	15	18	43
Control	5.35	45.0	75	51	67
CD (P = 0.05)	**1.22**	**2.8**	**3.9**	**0.68**	**3.1**

root knot nematode, *M. incognita* and root wilt fungus, *F. udum* on pigeonpea to a great extent with the most promising improvement in plant growth parameters (Goswami *et al.*, 2007).

14.10. GREEN GRAM, *VIGNA RADIATA*

14.10.1. Root-knot Nematode, *Meloidogyne* spp. and Wilt, *Fusarium oxysporum* Disease Complex

Nematode-fungus disease complex particularly of *Meloidogyne incognita* and *Fusarium oxysporum* poses a great problem to the cultivation of green gram (*Vigna radiata*) by inflicting severe yield losses. The present study was carried out to determine the comparative efficacy of chemicals, neem seed powder and microbial antagonists against *M. incognita* - *F. oxysporum* disease complex on green gram cv ML-1108 (Haseeb *et al.*, 2005).

Data (Table 14.49) show that the effect of various treatments on all the growth and yield parameters *viz.* shoot length, fresh weights of shoot and root, seed weights were mostly significant ($P \geq 0.05$) as compared to untreated inoculated plants. However, no significant differences were found in shoot height of plants treated with bavistin (24.1 cm) and *P. fluorescens* (22.6 cm). Maximum increase in all the growth and yield parameters was found in carbofuran treated plants followed by *A. indica* seed powder, *T. harzianum*, bavistin and *P. fluorescens* treated plants as compared to untreated inoculated plants.

Table 14.49: Effect of carbofuran, bavistin, neem seed powder, *Trichoderma harzianum* and *Pseudomonas fluorescens* against *Meloidogyne incognita* and *Fusarium oxysporum* disease complex on *Vigna radiata* cv. ML-1108

Treatments	***Fresh weight (g)***		***Shoot height (cm)***	***Seed weight (g)***
	Root	***Shoot***		
Nematode + fungus (Control)	22.1 (28.3)	4.8 (36.2)	18.6 (32.3)	5.34 (37.1)
Nematode + fungus + carbofuran	29.2 (5.2)	7.0 (6.8)	25.9 (5.8)	7.88 (7.2)
Nematode + fungus + neem seed powder	27.8 (9.7)	6.7 (10.2)	24.8 (9.8)	7.62 (10.4)
Nematode + fungus + *T. harzianum*	26.3 (14.6)	6.4 (15.1)	22.7 (17.3)	7.12 (16.2)
Nematode + fungus + bavistin	24.1 (21.8)	6.0 (20.1)	22.1 (19.7)	6.77 (20.4)
Nematode + fungus + *P. fluorescens*	23.8 (22.6)	5.4 (22.8)	21.9 (20.5)	6.66 (21.7)
CD (P = 0.05)	**2.5**	**0.6**	**2.0**	**0.70**

Each value is an average of three replicates; Values in parenthesis represent percent reduction over uninoculated control.

Data presented in (Table 14.50) show that the greatest suppression of nematode reproduction *i.e.* reproduction factor (Rf) and root-knot index was achieved by application of carbofuran (*Rf* =0.46, *RKI* =0.25) followed by neem seed powder (*Rf* =0.99, *RKI* =0.50), *T. harzianum* (*Rf* =1.27, *RKI* =1.50), *P. fluorescens* (*Rf* =1.38, *RKI* =1.75) and bavistin (*Rf* =1.50, *RKI* =2.25), respectively as compared to controls (*Rf* =1.74, *RKI* =3.00). Bavistin was highly effective in suppression of tap root colonization by fungus (8% root colonized) followed by neem seed powder (30%), *T. harzianum* (35%), *P. fluorescens* (65%) and carbofuran (75%), respectively.

14.11. CLUSTER BEAN, *CYAMOPSIS TETRAGONOLOBA*

14.11.1. Root-knot Nematode, *Meloidogyne javanica* and Root Rot, *Macrophomina phaseolina, Rhizoctonia solani, Fusarium solani, F. oxysporum* Disease Complex

Management methods

Integrated methods

Pseudomonas aeruginosa and *Paecilomyces lilacinus* used alone or together significantly (P<0.05) reduced infection of root knot nematode *M. javanica* and

Table 14.50: Effect of carbofuran, bavistin, neem seed powder, *Trichoderma harzianum* and *Pseudomonas fluorescens* on root-knot development, of *Meloidogyne incognita* and infection of *Fusarium oxysporum* on *Vigna radiata* cv ML-1108

Treatments	***Final nematode population (Pf)***			***Reproduction factorRf = Pf / Pi***	***Root-knot index***	***Disease index (%)***
	Root	***Soil***	***Total***			
Nematode + fungus (Control)	1488	2000	3488	1.74	3.00	95
Nematode + fungus + carbofuran	420	500	920	0.46	0.25	75
Nematode + fungus + neem seed powder	1474	500	1974	0.99	0.50	30
Nematode + fungus + *T. harzianum*	1536	1000	2536	1.27	1.50	35
Nematode + fungus + bavistin	1508	1500	3008	1.50	2.25	8
Nematode + fungus + *P. fluorescens*	1512	1250	2762	1.38	1.75	65
CD (P = 0.05)	**144.1**	**121.9**	**258.4**	**0.11**	**0.14**	**6.0**

Each value is an average of three replicates; *Pi*: Initial population.

root infecting fungi *viz., M. phaseolina, R. solani, F. solani* and *F. oxysporum* on cluster bean. *P. aeruginosa was* more effective than *P. lilacinus* in reducing the *M. javanica* root knot nematode infection. Combined use of *P. lilacinus* and *P. aeruginosa* was more effective in reducing the infection of root knot nematode and *F. solani* on guar than either used alone. Use of *P. aeruginosa* and *P. lilacinus* significantly ($P<0.05$) increased plant height of guar (Perveen *et al.*, 1998) (Table 14.51).

Table 14.51: *Effect of P. aeruginosa* and *P. lilacinus* on plant height and control of root rot disease complex in cluster bean

Treatment	***Plant height (cm)***	***Root-knot index***	***Infection %***			
			M. phaseolina	***R. solani***	***F. solani***	***F. oxysporum***
Control	24.5	4.1	31	0	75	81
P. lilacinus (PL)	28.0	3.5	6	0	62	81
P. aeruginosa (PA)	27.7	2.0	19	0	44	50
PL + PA	32.0	1.4	6	0	37	44
CD (P = 0.05)	**2.2**	**0.34**	**6.1**	**6.1**	**6.1**	**6.1**

14.12. CABBAGE, *BRASSICA OLERACEA* VAR. *CAPITATA* AND CAULIFLOWER, *BRASSICA OLERACEA* VAR. *BOTRYTIS*

Cabbage (*Brassica oleracea* var. *capitata*) is an important winter vegetable of cole group. It is a rich source of vitamins A, B and C and also contains minerals. It is being cultivated in 3.121 million hectares producing 69.819 million tonnes of cabbages with an average yield of 30.72 tonnes per hectare (Bijay Kumar, 2009) (Table 14.52). It covers about 4% of total area under vegetables. India comes next to China in cabbage production. In India, it is being cultivated in 0.266 million hectares producing 5.888 million tonnes of cabbage with an average yield of 22.2 tonnes per hectare (Bijay Kumar, 2009) (Table 14.53). It is now grown almost throughout the year. West Bengal, Orissa, Bihar, Gujarat, Maharashtra, Uttar Pradesh, Chhattisgarh, Haryana, Karnataka and Madhya Pradesh are the major cabbage-growing states.

Table 14.52: Major cabbage producing countries of the world (2007-08)

Country	*Area (in '000 ha)*	*Production (in '000 MT)*	*Productivity (in MT/ha)*	*% share of world production*
China	1772	36355	21	52
India	266	5888	22	8
Russia	170	5054	24	6
Korea	48	3000	63	4
Japan	54	2390	44	3
Poland	36	1376	38	2
Ukraine	64	1300	20	2
Indonesia	57	1250	22	2
USA	30	1171	39	2
Romania	46	1120	24	2
Others	578	11935	21	17
Total	**3121**	**69819**	**30.72**	

Source: FAO (except Indian data) (Indian data *source*: Indian Horticulture Database, 2007-08)

Table 14.53: Area, production and productivity of leading cabbage growing states in India (2007-08)

State	*Area (in'000 ha)*	*Production (in '000 MT)*	*Productivity (MT/ha)*	*% Share of India production*
West Bengal	73.2	2016.1	27.5	34.3
Orissa	33.8	935.6	27.6	15.9
Bihar	37.4	638.1	17.0	10.8
Gujarat	23.0	396.2	17.3	6.7

Table 14.53: (*Contd...*)

Table 14.53: (*Contd*...)

State	*Area (in'000 ha)*	*Production (in '000 MT)*	*Productivity (MT/ha)*	*% Share of India production*
Maharashtra	15.1	383.0	25.4	6.5
Uttar Pradesh	11.4	183.0	16.0	3.1
Chhattisgarh	11.3	183.0	15.7	3.0
Haryana	11.3	152.6	13.4	2.6
Karnataka	6.8	145.2	21.2	2.5
Madhya Pradesh	2.5	141.2	56.2	2.4
Others	36.8	716.5	18.1	12.2
Total	**265.4**	**1884.8**	**22.2**	

(*Source*: Indian Horticulture Database, 2007-08)

Cauliflower (*Brassica oleracea* var. *botrytis*) is the most popular winter vegetable among cole crops. It is rich in vitamin C. It is being cultivated in 0.321 million hectares producing 5.797 million tonnes of cauliflowers with an average yield of 18.0 tonnes per hectare (Bijay Kumar, 2009) (Tables 14.54 and 14.55). West Bengal, Bihar, Orissa, Haryana, Jarkhand, Maharashtra, Gujarat, Chhattisgarh, Madhya Pradesh and Uttar Pradesh are the major cauliflower-growing states. With the development of new varieties, it is now being grown in non-traditional areas like Andhra Pradesh, Tamil Nadu and Kerala.

Table 14.54: Major cauliflower producing countries of the world (2007-08)

Country	*Area (in '000 ha)*	*Production (in '000 MT)*	*Productivity (in MT/ha)*	*% share of world production*
China	404	8585	21	44
India	321	5797	18	29
USA	72	1241	17	6
Spain	25	450	18	2
Italy	18	433	24	2
France	27	370	14	2
Mexico	22	305	14	2
Poland	16	277	18	1
Pakistan	11	209	18	1
UK	17	186	11	1
Others	134	2037	15	10
Total	**1065**	**19890**	16.9	

Source: FAO (except Indian data) (Indian data *source*: Indian Horticulture Database, 2007-08)

Table 14.55: Area, production and productivity of leading cauliflower growing states in India (2007-08)

State	*Area (in'000 ha)*	*Production (in '000 MT)*	*Productivity (MT/ha)*	*% Share of India production*
West Bengal	66.9	1682.1	25.1	29.0
Bihar	60.7	1023.9	16.9	17.7
Orissa	45.2	644.2	14.3	11.1
Haryana	24.2	349.9	14.4	6.0
Jharkhand	20.7	331.9	16.0	5.7
Maharashtra	13.1	330.7	25.3	5.7
Gujarat	17.4	314.2	18.1	5.4
Chhattisgarh	16.6	251.9	15.2	4.3
Madhya Pradesh	10.8	172.5	16.0	3.0
Uttar Pradesh	7.7	158.4	20.4	2.7
Others	37.4	536.8	14.3	9.3
Total	**320.6**	**5796.6**	**18.1**	

(*Source*: Indian Horticulture Database, 2007-08)

14.12.1. Root-knot Nematode, *Meloidogyne incognita* and Club Root, *Plasmodiophora brassicae* Disease Complex (Fig. 14.19)

Fig. 14.19: Root-knot and club root complex in cabbage

Management methods

(i) *Biological methods*

Among different PGPR strains tested, *P. fluorescens* – Pf1 effectively reduced the root-knot index. Combined application of carbendazim and carbofuran gave least club root index both in cabbage and cauliflower under greenhouse conditions (Tables 14.56 and 14.57) (Loganathan *et al.*, 2001).

Table 14.56: Efficacy of bioformulations against club root disease-root-knot complex in cabbage and cauliflower under greenhouse conditions.

Treatment	***Cabbage***		***Cauliflower***	
	Club root index (PDI)	***Root-knot index***	***Club root index (PDI)***	***Root-knot index***
T. viride – TvO	35.55	2.66	36.66	2.33
T. viride – TvOL	37.77	2.33	35.55	3.00
T. viride – TvMNT7	26.66	1.33	22.22	1.00
T. viride – TvF9	36.66	2.00	37.77	1.66
T. viride – TvV4	32.77	2.33	31.66	1.33
T. harzianum	30.32	2.66	32.77	1.66
P. fluorescens – PfO2	34.71	2.00	36.66	1.33
B. subtilis – Bs1	38.88	1.66	38.32	2.00
P. fluorescens – Pf1	28.88	1.33	24.44	1.33
P. fluorescens – PfCHAO	33.88	1.66	34.70	2.66
T. viride – TvK	48.88	3.33	46.66	3.00
T. virens – Tv1	47.77	3.66	48.66	4.00
P. fluorescens – PfO1	51.11	4.33	48.32	3.66
B. subtilis – Bs2	48.88	3.66	49.99	4.33
Carbendazim	22.22	4.00	17.77	4.33
Carbofuran	33.33	3.00	28.88	2.66
Cabendazim + Carbofuran	19.99	3.00	15.55	2.33
P. brassicae	60.00	2.66	52.22	2.00
M. incognita	17.80	4.66	26.66	4.33
Control	67.77	4.66	57.72	4.66

Table 14.57: Efficacy of bioformulations against club root disease-root-knot complex in cabbage under field conditions.

Treatment	***Club root index (PDI)****	***Decrease in nema. Popn. over initial (%)***	***Root-knot index***	***Yield (MT/ha)***
T. viride – TvO	50.0 (45.00)	36.2	2.66	39.99
T. viride – TvOL	40.0 (39.22)	35.7	2.66	44.44
T. viride – TvMNT7	20.0 (26.56)	38.5	2.00	53.33
T. viride – TvF9	38.0 (38.05)	36.5	2.33	45.77
T. viride – TvV4	36.5 (37.16)	33.3	2.66	46.66
T. harzianum	36.2 (36.97)	32.5	3.00	48.88
P. fluorescens – PfO2	46.0 (42.70)	37.4	2.00	44.44
B. subtilis	43.0 (40.97)	29.0	3.33	41.33
P. fluorescens – Pf1	23.0 (28.66)	46.2	1.33	59.10
P. fluorescens – PfCHAO	45.0 (42.13)	38.6	2.00	47.40
Carbendazim	18.0 (25.10)	26.4	3.66	51.11
Carbofuran	46.0 (42.70)	30.2	2.00	45.77
Cabendazim + Carbofuran	11.5 (19.81)	31.1	2.00	54.66
Control	60.0 (50.79)	+52.2	4.00	33.77

Values in parenthesis are transformed values.

(ii) *Integrated methods*

(a) *Bioagents and botanicals*: PGPR strains (*Pseudomonas fluorescens, Bacillus subtilis*) combined with fungal biocontrol agents (*Trichoderma viride, T. harzianum*) were found to be effective in reducing the nematode-fungal disease complex in cabbage (Loganathan *et al.*, 2001). The bioformulation mixture of *P. fluorescens, T. viride* and chitin effectively reduced the disease complex in cabbage under greenhouse conditions (Samiyappan, 2003) (Table 14.58).

Table 14.58: Efficacy of bioformulation mixtures against club root disease-root-knot nematode complex in cabbage under greenhouse conditions.

Treatment	*Club root index*	*Nematode incidence*	
		Population	*Root-knot index*
Trichoderma viride	25.99 (30.65)	129	2.66
Pseudomonas fluorescens	28.20 (32.07)	112	2.33
T. viride + *P. fluorescens*	25.33 (30.22)	114	2.33
T. viride + Chitin	25.44 (30.29)	108	2.33
P. fluorescens + Chitin	25.66 (30.43)	111	2.00
T. viride + *P. fluorescens* + Chitin	22.22 (28.12)	108	2.00
Chitin alone	31.70 (34.26)	139	3.00
Carbendazim	19.90 (26.49)	264	4.66
Carbofuran	39.90 (39.17)	106	1.66
Cabendazim + Carbofuran	15.00 (22.79)	103	1.66
Plasmodiophora brassicae alone	48.90 (44.37)	0.033	0.133
Meloidogyne incognita alone	0.03 (0.60)	280	5.00

14.12.2. Root-knot Nematode, *Meloidogyne incognita* and Wilt, *Fusarium oxysporum* f. sp. *conglutinans* Disease Complex

Pathak and Keshar (2004) reported that presence of *M. incognita* resulted in highest wilting due to *F. oxysporum* f. sp. *conglutinans* in cauliflower. They reported that the plants exposed to the wilt fungus in absence of nematodes recorded low severity, which suggested the delay in entry of pathogen due to predisposing agent. Maximum wilt score due to *F. oxysporum* f. sp. *conglutinans* (4.0) and wilt symptom index (80.0) was observed in cauliflower plants which received *M. incognita* 14 days before the inoculation of wilt fungus.

14.12.3. Stunt Nematode, *Tylenchorhynchus brassicae* and Root Rot, *Rhizoctonia solani* Disease Complex

T. brassicae alone did not affect the percentage emergence of cauliflower seedlings. *R. solani* was highly destructive by itself, but the adverse effect on the emergence of seedlings was greater when nematode infestation occurred along with fungal infection (Khan *et al.*, 1971).

14.13. PUMPKIN, *CUCURBITA PEPO*

14.13.1. Root-knot Nematode, *Meloidogyne javanica* and Root Rot/Wilt, *Macrophomina phaseolina, Fusarium oxysporum, F. solani* Disease Complex

Management methods

(i) *Biomanagement*

Chahal and Chahal (1998) reported beneficial effects of *T. harzianum* in reducing the number of galls and disease complexes formed by root-knot nematodes and root rot fungi on cucurbits.

(ii) *Integrated methods*

(a) *Two bioagents*: *Pseudomonas aeruginosa* and *Paecilomyces lilacinus* when used together significantly reduced infection of the disease complex on pumpkin (Perveen *et al.*, 1998). *P. aeruginosa* and *P. lilacinus* used alone or together significantly ($P<0.05$) reduced infection of root knot nematode *M. javanica* and root infecting fungi *viz.*, *M. phaseolina, R. solani, F. solani* and *F. oxysporum* on pumpkin. *P. aeruginosa* was more effective than *P. lilacinus* in reducing the *M. javanica* infection. Combined use of *P. lilacinus* and *P. aeruginosa* was more effective in reducing the infection of *M. phaseolina* and *F. oxysporum* on pumpkin than either used alone. Use of *P. aeruginosa* and *P. lilacinus* significantly ($P<0.05$) increased plant height and fresh shoot weight of pumpkin (Tables 14.59 and 14.60).

Table 14.59: Effect of *P. aeruginosa and P. lilacinus on growth of pumpkin.*

Treatment	*Germination%*	*Fresh weight of shoot (g)*	*Plant height (cm)*
Control	40	4.2	26.5
P. lilacinus (PL)	50	6.7	32.2
P. aeruginosa (PA)	55	7.2	30.0
PL + PA	50	5.9	28.2
CD (P = 0.05)	**11.1**	**0.55**	**2.2**

Table 14.60: Effect of *P. aeruginosa* and *P. lilacinus* in the control of root disease complex in pumpkin

Treatment	*Root-knot index*	*Infection %*			
		M. phaseolina	*R. solani*	*F. solani*	*F. oxysporum*
Control	3.1	87	25	56	75
P. lilacinus (PL)	1.2	94	0	44	62
P. aeruginosa (PA)	0.6	50	6	19	44
PL + PA	0.5	19	0	19	6
CD (P = 0.05)	**0.34**	**6.1**	**6.1**	**6.1**	**6.1**

14.14. CUCUMBER, *CUCUMIS SATIVUS*

14.14.1. Dagger Nematode, *Xiphinema americanum* and Tobacco Ringspot Virus Disease Complex

(i) Management methods

(a) *Chemical methods*: Benomyl application inhibited feeding of *X. americanum* on cucumber, which was reflected in a marked reduction of its transmission ability of tobacco ring spot virus (McGuire and Good, 1970; Rodriguez-Kabana and King, 1977).

14.15. WATERMELON, *CITRULLUS LANATUS*

14.15.1. Root-knot Nematode, *Meloidogyne javanica* and Wilt, *Fusarium solani* Disease Complex (Fig. 14.20)

Fig. 14.20: Root-knot nematode and *Fusarium* wilt disease complex

Management methods

(i) *Integrated methods*

(a) *Two bioagents*: *Pseudomonas aeruginosa* and *Paecilomyces lilacinus* when used together significantly reduced infection of *M. javanica* and *F. solani* on watermelon. *P. aeruginosa* and *P. lilacinus* used alone or together significantly ($P<0.05$) reduced infection of root knot nematode *M. javanica* and root infecting fungi viz., *M. phaseolina, R. solani, F. solani* and *F. oxysporum* on watermelon. *P. aeruginosa* was more effective than *P. lilacinus* in reducing the *M. javanica* root knot nematode infection (Perveen *et al.*, 1998). Combined use of *P. lilacinus* and *P. aeruginosa* was more effective in reducing the infection of *F. solani* on watermelon than either used alone. Use of *P. aeruginosa* and *P. lilacinus* significantly ($P<0.05$) increased plant height and fresh shoot weight on watermelon (Tables 14.61 and 14.62).

Table 14.61: Effect of *P. aeruginosa and P. lilacinus* on growth of watermelon.

Treatment	*Germination %*	*Fresh weight of shoot (g)*	*Plant height (cm)*
Control	60	2.5	35.1
P. lilacinus (PL)	70	3.0	37.5
P. aeruginosa (PA)	80	3.7	35.0
PL + PA	75	3.2	38.5
CD (P = 0.05)	**11.1**	**0.55**	**2.2**

Table 14.62: Effect of *P. aeruginosa* and *P. lilacinus* in the control of root disease complex in watermelon

Treatment	*Root-knot index*	*Infection %*			
		M. phaseolina	*R. solani*	*F. solani*	*F. oxysporum*
Control	3.8	62	44	69	12
P. lilacinus (PL)	2.8	69	12	81	0
P. aeruginosa (PA)	1.5	25	12	69	0
PL + PA	1.5	37	0	44	0
CD (P = 0.05)	**0.34**	**6.1**	**6.1**	**6.1**	**6.1**

14.16. OKRA, *ABELMOSCHUS ESCULENTUS*

Okra (*Abelmoschus esculentus*) is an annual vegetable crop grown in tropical and sub-tropical regions. It is being cultivated in 0.409 million hectares producing 4.192 million tonnes of okra with an average yield of 10.3 tonnes per hectare (Bijay Kumar, 2009) (Table 14.63). West Bengal, Bihar, Orissa, Gujarat, Andhra Pradesh,

Table 14.63: Area, production and productivity of leading okra growing states in India (2007-08)

State	*Area (in'000 ha)*	*Production (in '000 MT)*	*Productivity (MT/ha)*	*% share of India production*
West Bengal	71.5	815.3	11.4	19.4
Bihar	57.2	707.3	12.4	16.9
Orissa	71.5	621.1	8.7	14.8
Gujarat	41.5	365.9	8.8	8.7
Andhra Pradesh	30.3	454.6	15.0	10.8
Chhattisgarh	23.0	202.4	8.8	4.8
Maharashtra	26.3	165.4	6.3	3.9
Jharkhand	24.5	343.2	14.0	8.2
Uttar Pradesh	9.2	102.1	11.1	2.4
Haryana	14.9	117.8	7.9	2.8
Others	39.0	297.6	7.6	7.1
Total	**409.0**	**4192.8**	**10.3**	

(*Source*: Indian Horticulture Database, 2007-08)

Chhattisgarh, Maharashtra, Jharkhand, Uttar Pradesh and Haryana are the major okra-growing states. Tender green fruits are cooked in curry and soup. The root and stem are used for clearing cane juice in preparation of 'gur'. High iodine content of fruits helps to control goitre while leaves are used in inflammation and dysentery. It is a good source of vitamins A, B and C. The fruits also help in cases of renal colic, leucorrhoea and general weakness. The dry seed contains 13-22% good edible oil and 20-24% protein. The oil is used in soap, cosmetic industry and as vanaspati while protein is used for fortified preparations. The crushed seed is fed to cattle for more milk production and the fibre is utilized in jute, textile and paper industry.

14.16.1. Root-knot Nematode, *Meloidogyne incognita* and Wilt, *Fusarium oxysporum* f. sp. *vasinfectum* Disease Complex

Meloidogyne-Fusarium disease complex has been considered important on many crops including okra leading to reduction in its productivity.

Management methods

(i) *Biomanagement*

(a) *Antagonistic bacteria and fungi*: Soil application of 25 g/m^2 of *Pseudomonas fluorescens* (2 × 10^6 cfu/g) or *Pochonia chlamydosporia* (2 × 10^6 cfu/g) enriched deoiled neem cake has proved to be an effective treatment in combating the damage caused by *M. incognita* and *F. oxysporum* f. sp. *vasinfectum* to the tune of 68 and 57%, respectively. This treatment also increased the yield of okra fruits by 24% under field conditions (Chaya *et al*., 2010).

(b) *Antagonistic fungi*: Among chemicals, botanicals and biocontrol agents evaluated, carbofuran was the most effective treatment in improving plant growth followed by *Trichoderma harzianum*, neem seed powder, neem cake, *Aspergillus niger* and bavistin respectively. Highest suppression of root colonization of the fungus (*F.*

Table 14.64: Effect of chemicals, botanicals and bioagents on plant growth and disease complex in okra

Treatment	*Plant height (cm)*	*Fresh weight of plant (g)*	*Root-knot index*	*Disease index (%)*
Uninoculated control	71.0	112.7	–	–
Inoculated control (N + F)	48.7	62.4	3.75	42.5
Carbofuran	67.7	107.2	0.25	20.0
Bavistin	55.9	93.5	2.75	5.0
Neem seed powder	62.9	101.9	0.90	7.5
Neem cake	60.6	98.0	1.90	12.5
T. harzianum	65.1	104.0	0.75	17.5
A. niger	57.5	95.6	2.10	15.0
CD (P = 0.05)	**0.61**	**1.38**	**0.36**	**1.45**

o. f. sp. *vasinfectum*) was achieved in plants treated with bavistin (5.0) followed by neem seed powder (7.5), neem cake (12.5), *A. niger* (15.0), *T. harzianum* (17.5) and carbofuran (20.0). The greatest suppression of root-knot development was achieved by carbofuran (0.25) followed by *T. harzianum* (0.75), neem seed powder (0.90), neem cake (1.90), *A. niger* (2.20) and bavistin (2.75). In general, carbofuran and *T. harzianum* were highly effective against the nematodes, bavistin against the fungus and neem seed powder and neem cake against both the pathogens (Table 14.64) (Abuzar and Haseeb, 2006).

(ii) *Chemical methods*

Good (1964) found that among several soil fumigants tested at 15 and 30 cm depth against root-knot-*Fusarium* wilt of okra, only methyl bromide significantly reduced the disease complex. D-D, on other hand, could slightly check the nematode-fungal complex but only when applied at 15-cm depth, whereas DBCP was much more effective than D-D at the same depth.

14.16.2. Root-knot Nematode, *Meloidogyne incognita* and Root Rot, *Rhizoctonia solani* Disease Complex

Plants in untreated field soil or in sterilized soil inoculated with both organisms, developed a root rot in about 42 days. If the nematode preceded the fungus by three weeks, the root rot was more severe and appeared within 14-21 days. The fungus penetrated either directly or through ruptures in the root created by the mature female nematode. *R. solani* colonized nematode giant cells and root xylem cells. Vascular discoloration occurred both in roots and stem, however no fungus was isolated from stems.

M. incognita predisposed roots to *R. solani* which resulted in severe root rot and subsequent plant death. Okra plants inoculated with either *R. solani* or *M. incognita* alone were free of root decay for the entire period of six-week study.

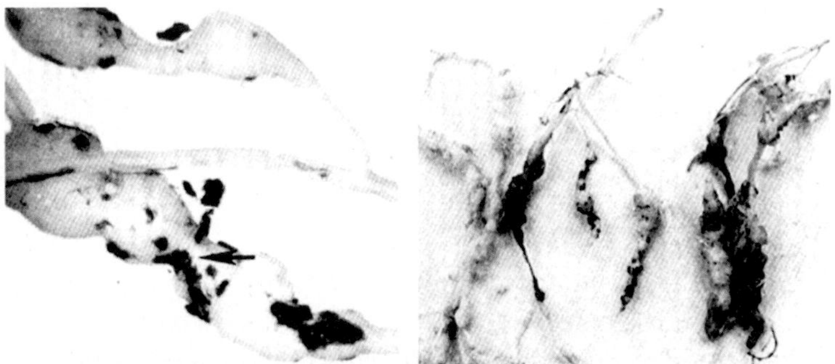

Fig. 14.21: Okra roots infected with *Meloidogyne incognita* and *Rhizoctonia solani*. Left - Five-week-old root galls with massive sclerotia (Arrow) and egg masses on root surface; Right - *M. incognita* and *R. solani* infected okra roots six weeks after nematode infection showing severe root rot.

Three weeks after nematode and fungus inoculation, black sclerotia of *R. solani* were visible on nematode-induced galls (Fig. 14.21), while on non-galled portions on the same root system were free of sclerotia. Prior to root rot development, *R. solani* demonstrated marked preference for root galls on nematode infected roots. It is hypothesized that the leakage of nutrients from the root was responsible for attracting the fungus to the galls and for initiating sclerotial formation.

Five weeks after inoculation, distinct brown lesions were observed only on the galls of plants inoculated with both *M. incognita* and *R. solani*. Lower leaves of plants were chlorotic and suffered premature leaf drop.

M. incognita and *R. bataticola* or *Sclerotium rolfsii* when inoculated simultaneously in soil, reduce the germination of seeds in okra (Chhabra and Sharma, 1981). Combined attack of both these pathogens cause significantly greater damage to the crop than that of the damage caused by either pathogen alone (Bhagawati *et al.*, 2007).

Management methods

(i) *Chemical methods*

Soil fumigation usually decreased disease severity due to *R. solani* and *M. incognita* disease complex in okra (Golden and Van Gundy, 1975).

(ii) *Biomanagement*

(a) *Antagonistic fungi*: *Paecilomyces lilacinus* can be effectively used against *R. solani – M. incognita* complex of okra (Shahzad and Ghaffar, 1989).

(iii) *Integrated methods*

(a) *Bioagents and botanicals*: Chaitali *et al.* (2003) observed that *T. viride* combined with neem cake controlled the disease complex better than *T. viride* combined with groundnut cake in okra.

Integration of *P. lilacinus* and *T. viride* with mustard cake gave significant reduction in root galling, egg mass production, fecundity, reproduction factor and nematode population both in soil and roots coupled with increase in plant growth parameters (Tripathi and Singh, 2006).

(b) *Bioagents and chemicals*: The lowest pre-emergence (5.01%) and post-emergence (7.73%) damping-off were observed in the treatment where both carbosulfan 25 SD and carbendazim 50 WP were applied together as seed treatment (T_{10}) which was at par with the treatment with dual application of *T. harzianum* (6.26%) and *P. fluorescens* (7.73%) as seed treatment (T_4). As evidenced from the results, the seed treatment was found to be significantly superior to soil application of bioagents (Bhagawati *et al.*, 2009) (Table 14.65).

Table 14.65: Effect of bioagents and chemicals for the management of disease complex caused by *Meloidogyne incognita* and *Rhizoctonia solani*

*Treatment**	*Pre-emergence damping -off (%)*	*% decrease over inoculated control*	*Post-emergence damping -off (%)*	*% decrease over inoculated control*
T_1	32.98	–	34.34	–
T_2	11.05	66.49	13.99	59.26
T_3	10.86	67.07	12.52	63.54
T_4	6.26	81.02	7.73	77.45
T_5	20.45	37.99	24.95	27.34
T_6	18.69	43.33	25.14	26.79
T_7	18.88	42.75	23.38	31.92
T_8	20.26	38.57	24.94	27.37
T_9	18.78	43.06	24.76	27.88
T_{10}	5.01	84.81	7.73	77.45
T_{11}	0.00	–	0.00	–
CD (P = 0.05)	**3.94**	**–**	**4.25**	**–**

* T_1– *M. incognita* + *R. solani* (Inoculated & untreated control), T_2 - *M. incognita* + *R. solani* + *T. harzianum* (seed treatment), T_3 – *M. incognita* + *R. solani* + *P. fluorescens* (seed treatment), T_4 _ *M. incognita* + *R. solani* + *T. harzianum* + *P. fluorescens* (seed treatment), T_5 – *M. incognita* + *R. solani* + *T. harzianum* (soil application), T_6 – *M. incognita* + *R. solani* + *P. fluorescens* (soil application), T_7 _ *M. incognita* + *R. solani* + *T. harzianum* + *P. fluorescens* (soil application), T_8 _ *M. incognita* + *R. solani* + carbosulfan (seed treatment), T_9 _ *M. incognita* + *R. solani* + carbendazim (seed treatment), T_{10} – *M. incognita* + *R. solani* + carbosulfan + carbendazim 50 (seed treatment), T_{11} – Uninoculated & untreated control.

The maximum plant height, dry weight of shoot and root were recorded in the uninoculated and untreated control (T_{11}) followed by the treatment receiving *T. harzianum* and *P. fluorescens* as seed treatment (T_4) which were at par with the treatment receiving carbosulfan 25 SD and carbendazim 50 WP as seed treatment (T_{10}) (Bhagawati *et al.*, 2009) (Table 14.66).

Table 14.66: Effect of bioagents and chemicals on plant growth parameters

Treatment	*Plant height (cm)*	*% increase over T_1*	*Shoot weight (g)*		*Root weight (g)*	
			Dry	*% increase over T_1*	*Dry*	*% increase over T_1*
T_1	27.30	–	1.80	–	0.32	–
T_2	36.40	33.33	3.12	73.33	0.48	50.00
T_3	36.82	34.87	2.96	64.44	0.47	46.88
T_4	44.26	62.12	3.98	121.11	0.72	125.00
T_5	31.85	16.67	2.24	24.44	0.53	65.63
T_6	32.76	20.00	2.28	26.67	0.47	46.88

Table 14.66: (*Contd...*)

Table 14.66: (*Contd...*)

Treatment	*Plant height (cm)*	*% increase over T_1*	*Shoot weight (g)*		*Root weight (g)*	
			Dry	*% increase over T_1*	*Dry*	*% increase over T_1*
T_7	36.10	32.24	2.90	61.11	0.50	56.25
T_8	32.58	19.34	2.34	30.00	0.46	43.75
T_9	31.40	15.01	2.38	32.22	0.51	59.34
T_{10}	43.60	59.70	4.10	127.78	0.68	112.50
T_{11}	45.94	–	4.25	–	0.74	–
CD (P = 0.05)	**3.26**	**–**	**0.38**	**–**	**0.13**	–

* T_1– *M. incognita* + *R. solani* (Inoculated & untreated control), T_2 – *M. incognita* + *R. solani* + *T. harzianum* (seed treatment), T_3 – *M. incognita* + *R. solani* + *P. fluorescens* (seed treatment), T_4 _ *M. incognita* + *R. solani* + *T. harzianum* + *P. fluorescens* (seed treatment), T_5 – *M. incognita* + *R. solani* + *T. harzianum* (soil application), T_6 – *M. incognita* + *R. solani* + *P. fluorescens* (soil application), T_7 _ *M. incognita* + *R. solani* + *T. harzianum* + *P. fluorescens* (soil application), T_8 _ *M. incognita* + *R. solani* + carbosulfan (seed treatment), T_9 _ *M. incognita* + *R. solani* + carbendazim (seed treatment), T_{10} – *M. incognita* + *R. solani* + carbosulfan + carbendazim 50 (seed treatment), T_{11} – Uninoculated & untreated control.

The minimum number of galls and egg masses in roots as also nematode population in soil were recorded in the treatment receiving carbosulfan 25 SD and carbendazim 50 WP as seed treatment (T_{10}) which was on par with the treatment receiving *T. harzianum* and *P. fluorescens* as seed treatment (T_4). Further, the treatment with *T. harzianum* and *P. fluorescens* as seed treatment was found to be significantly better in reducing the host infection and nematode multiplication than soil application of both these bioagents (Bhagawati *et al.*, 2009) (Table 14.67).

14.16.3. Reniform Nematode, *Rotylenchulus reniformis* and Root Rot, *Rhizoctonia solani* Disease Complex

Glasshouse studies proved that when the reniform nematode, *R. remformis* and the fungus *R. solani* were present, okra plants succumbed to wilt at an early stage and when the nematode infection preceded the fungus, wilt appeared earlier than when the plants were infected first with the fungus. The fungus caused wilt 45–60 days after germination in the absence of the nematode, depending upon the level of inoculum and the age of the plant, while the initial injury by the nematode shortened the period to 34-43 days after germination. When both the pathogens were inoculated at high level 15 days after germination, wilt occurred 23 days after germination. A low initial/nematode inoculum preceding the fungal inoculum also caused early wilting irrespective of the density of the latter (Kumar and Sivakumar, 1981).

Table 14.67: Effect of bioagents and chemicals on root galling, egg mass production and nematode population in soil

Treatment	***No. of galls/root system***	***% decrease over control (T_1)***	***No. of egg masses/ root system***	***% decrease over control (T_1)***	***Nema popn./ 200 ml soil***	***% decrease over control (T_1)***
T_1	236.00	–	138.25	–	1846	–
T_2	70.50	70.13	44.68	67.68	595	67.77
T_3	74.75	68.33	46.25	66.55	614	66.74
T_4	54.25	77.01	26.25	81.01	348	81.14
T_5	102.25	56.67	58.75	57.50	815	55.85
T_6	107.50	54.45	60.25	56.42	792	57.10
T_7	72.50	69.28	42.50	69.26	831	54.98
T_8	76.00	67.80	40.26	70.88	820	55.58
T_9	146.00	38.14	62.30	54.94	830	55.04
T_{10}	51.75	78.07	30.75	77.76	336	81.80
T_{11}	0.00	–	0.00	–	0.00	–
CD (P = 0.05)	**0.71**	**–**	**0.67**	**–**	**2.59**	**–**

* T_1– *M. incognita* + *R. solani* (Inoculated & untreated control), T_2 – *M. incognita* + *R. solani* + *T. harzianum* (seed treatment), T_3 – *M. incognita* + *R. solani* + *P. fluorescens* (seed treatment), T_4 – *M. incognita* + *R. solani* + *T. harzianum* + *P. fluorescens* (seed treatment), T_5 – *M. incognita* + *R. solani* + *T. harzianum* (soil application), T_6 – *M. incognita* + *R. solani* + *P. fluorescens* (soil application), T_7 – *M. incognita* + *R. solani* + *T. harzianum* + *P. fluorescens* (soil application), T_8 – *M. incognita* + *R. solani* + carbosulfan (seed treatment), T_9 – *M. incognita* + *R. solani* + carbendazim (seed treatment), T_{10} – *M. incognita* + *R. solani* + carbosulfan + carbendazim 50 (seed treatment), T_{11} – Uninoculated & untreated control.

14.17. CARROT, *DACUS CAROTA*

14.17.1. Root-knot Nematode, *Meloidogyne* sp. and Root Rot, *Pythium ultimum* Disease Complex

Application of metham sodium through drip irrigation was found to have an edge over D-D for root-knot and *Fusarium* sp. disease complex of carrot (Roberts *et al.*, 1988).

14.18. RADISH, *RAPHANUS SATIVUS*

14.18.1. Root-knot Nematode, *Meloidogyne hapla* and Root Rot, *Rhizoctonia solani* Disease Complex

Khan and Muller (1982) used gnotobiotic culture to investigate interactions between *R. solani* and *M. hapla* on radish and showed that prior infection of roots by the nematode helped the fungus to colonize the roots.

14.8.2. Root-knot Nematode, *Meloidogyne incognita* and Wilt, *Erwinia carotovora* s. sp. *carotovora* Disease Complex

The combination treatment of *Pseudomonas fluorescens* and *Paecilomyces lilacinus* enriched in neem cake applied at 10 g/m^2 increased the root colonization of both the bioagents and reduced the incidence of *M. incognita* and *E. carotovora* s. sp. *carotovora* by 68 and 56%, respectively. There was also a significant increase in the yield of carrot to the tune of 23% (Sowmya *et al.*, 2010).

14.19. BEET ROOT, *BETA VULGARIS*

14.19.1. Cyst Nematode, *Heterodera schachtii* and Damping-off, *Rhizoctonia solani* Disease Complex

The invasion process of *H. schachtii* (beet-cyst nematode) was found to facilitate the infection of beet root (*Beta vulgaris*) by the damping-off fungus *R. solani* (Polychronopoulos *et al.*, 1969). During their investigation, beet root seedlings grown in either nematode-infested or nematode-free soil were exposed to *R. solani* before being examined microscopically over a series of 12 hour intervals for 3 days. On inspection of the seedlings 36 hour after inoculation, distinct differences could be seen between the two treatments. When in combination with *H. schachtii*, the hyphae of *R. solani* were found to grow vigorously through the epidermis and cortex. Closer examination showed that hyphal colonization frequently followed tracts made by invading nematode juveniles. On the epidermal surfaces of the seedlings, the pathogen was found to produce fewer infection cushions in the presence of nematodes than when it was present alone. The authors suggested that infection cushion synthesis could have been hindered in some way by the invading nematodes. However, nematode invasion sites may provide *R. solani* with the necessary portals for penetration and entry, consequently reducing the need for developing more sophisticated infection structures such as infection cushions.

14.19.2. Docking Disorder, *Longidorus* spp., *Trichodorus* spp.

Longidorus leptocephalus, L. attenuatus, L. elongatus, Trichodorus spp and viruses are involved in Docking disorder of beet root (named for Docking region in south of England). Symptoms include stunted growth in spring due to nematodes and virus symptoms on foliage. Effects are most pronounced in spring, heavy rainfall in May seems to increase the problem. In July the affected plants start to grow again and may achieve almost normal foliage, but a much reduced tap root. The severity of Docking disorder varies from year to year with the climate.

(i) *Identification*

Docking disorder of beet was named after a village in the north-west of Norfolk. Although recognized and documented for many years it was not until the early 1960's that free-living nematodes were identified as a prime causal agent of this condition.

The main free living nematodes species responsible for causing damage to beet root are the stubby root nematodes (*Trichodorus* spp. and *Paratrichodorus* spp.) and the needle nematodes (*Longidorus* spp.).

(ii) *Symptoms*

Hatched larvae are attracted to the emerging crop plants to feed on the roots tips by inserting their piercing mouth parts (hollow stylets) into cells of the root and the root hairs. Enzymes are injected into these cells and the resulting substrate is withdrawn into the gut of the nematode to provide nourishment.

Feeding on the tap root, at an early stage of plant development, can result in its growing tip being killed which encourages lateral roots to take over. The net result is short stubby stunted plants with numerous laterally growing side roots which gives rise to the condition described as 'fanging' (Fig. 14.22).

Affected plants can be stunted resulting in so called 'hen and chick symptoms' where a large healthy plant has a small plant as its neighbour (Fig. 14.23).

Fig. 14.22: Damaged leaves and roots of beet root

Fig. 14.23: Docking disorder affected field of beet root

(iii) *Life cycle*

The nematodes lay eggs near host plants and after a few days the larvae wriggle out in search of food. Several generations occur each season and populations comprise a mixture of both adults and juveniles. *Longidorus* spp can live in the soil for over a year.

(iv) *Importance*

Crops grown on light sandy soils in wet, cold conditions are most at risk. It has been estimated that yield losses of up to 17.5 MT/ha can occur.

(v) *Threshold*

The current threshold is 200 nematodes per litre of soil but with late drilled crops when soils are warm the threshold for *Trichodorus* spp. can probably be safely doubled to 400 nematodes per litre of soil. However, *Longidorus* nematodes are larger than most species and are consequently damaging at relatively low populations (50 per litre of soil).

(vi) *Management*

Management by rotation is difficult as several nematode species are involved, each with a differing host range. Also host ranges of these nematodes are not completely known. 1, 3-D nematicide applied in the plant row prior to planting reduces the nematode populations and increases yield, but only in well-drained alkaline soils. The treatment is not always effective at economically feasible rates.

14.20. CELERY

14.20.1. Root-knot Nematode, *Meloidogyne hapla* and Root Rot, *Pythium irregulare* Disease Complex

Pyroxychlor provided significant reduction in *M. hapla* - *P. irregulare* root rot disease complex in celery (Starr and Mai, 1976).

14.20.2. Lesion Nematode, *Pratylenchus penetrans* and Root Rot, *Trichoderma viride* Disease Complex

P. penetrans and *T. viride* caused more reduction in root and shoot growth in celery than either organisms alone (Edmunds and Mai, 1966).

14.21. LETTUCE, *LACTUCA SATIVA*

14.21.1. Nematodes, *Meloidogyne hapla, Pratylenchus penetrans* and Root Rot, *Pythium tracheiphilum* Disease Complex

The association of the fungus, *P. tracheiphilum* with two nematodes (*P. penetrans* and *M. hapla*) was studied. The plants inoculated two weeks after seeding had significant and consistent differences on lettuce growth. In the relationship between *P. tracheiphilum* with *P. penetrans*, a negative interaction seemed to exist; since no significant increase of the damage caused to the lettuce was observed. In contrast, when the root-knot nematodes and *P. tracheiphilum* were combined there was a marked reduction of lettuce growth. The interaction was found to be additive.

Chapter 15

ORNAMENTAL, MEDICINAL, AROMATIC AND TUBER CROPS

15.1. ORNAMENTAL CROPS

India is known for growing of traditional flowers such as jasmine, marigold, chrysanthemum, tuberose, crossandra and aster. Commercial cultivation of cut flowers- rose, orchids, gladiolus, carnation, gerbera, anthurium and lilies-has also become popular.

With the liberalization of Indian economy, the domestic floricultural trade, which hitherto remained an unorganized sector, became an organized sector with the active participation of corporate giants. The establishment of state-of-art hi-tech floricultural units in early 90's for intensive cultivation of cut flowers contributes about Rs. 300 crores of valuable foreign exchange today. Various government agencies are actively involved in the promotion of this sector and are highly committed to resolve the teething problems. The world demand for floricultural products is approximately US $ 40 billions and cut flowers contribute nearly 60% (US$ 25 billions) to the global trade. India's share in the world flower market is about 0.7% which is contributed mainly from open cultivation.

Floriculture is estimated to cover an area of 161,000 ha with production of 870,000 MT of loose flowers during 2008. The area under cut flowers has increased significantly in recent years and the production has reached 4341 million numbers (Tables 15.1 and 15.2). Another 500 ha of climatic control greenhouse are available for growing quality flowers for export. Many corporate houses (more than 100)

Table 15.1: Area and production of ornamental crops in India (2007-08) (Bijay Kumar, 2009)

Ornamental Crops	*Area ('000 ha)*	*Production ('000 MT)*	*Productivity (MT/ha)*
Flowers Loose	161	870	5.4
Flowers Cut*	–	4341*	–

* Million numbers

Table 15.2: State-wise area and production of flowers (2007-08) Area and Production of Flowers State-Wise Area and Production of Flowers

State / UT's	*Area ('000 ha)*	*Production*			
		Loose ('000 MT)	*% of total production*	*Cut (million nos.)*	*% of total production*
Karnataka	22.34	169.12	19	555.000	13
Maharashtra	16.74	69.45	8	572.800	13
Uttar Pradesh	8.41	12.36	–	375.200	9
Uttaranchal	0.90	0.70	–	145.545	3
Gujarat	9.74	49.50	6	506.300	12
Delhi	5.50	5.70	–	103.800	2
West Bengal	27.42	48.45	6	1968.000	45
Tamil Nadu	26.74	214.38	25	–	–
Andhra Pradesh	23.52	126.27	14	6.782	–
Punjab	1.00	77.90	9	–	–
Haryana	6.11	61.76	7	1.053	–
Others	12.30	34.78	6	7.266	3
Total	**160.72**	**870.37**	–	**4341.746**	–

have set up 100% export oriented units with an investment of more than Rs. 1000 crores. India is bestowed with natural advantages like favourable weather conditions, soil, cheap labour which are likely to make a real mark in international trade. With this, India is emerging as one of the powerful forces in world flower trade. India exported floriculture products worth US $ 84.12 million during 2007.

Floriculture is an intensive type of agriculture wherein the income per unit area is much higher than any other agricultural product if it is done in a scientific way. The area used for production of ornamental plants is low, but the cash returns are high because of the intensive cropping. The flowers grown in large areas include jasmine, rose, marigold, chrysanthemum, crossandra, champak, tuberose, gladiolus and aster. New crops like lilies, tulips, anthurium, carnation, gerbera, etc., are also finding potential place in the foreign market. With the rapid development in urban planning and hotel industry, there has been an increasing demand for cut flowers and ornamentals. The cut flowers followed by live plants, cut foliage, seeds and bulbs have got an immense potential for export to Western countries where the cost of production is comparatively high. India exports roses, gladioli, carnations, chrysanthemums, lilies and jasmines to Gulf, Germany, France, England, Netherlands and Singapore. Our share in the export market has registered a sharp increase from Rs. 14.45 crores in 1991-92 to Rs. 299.41 crores in 2005-06.

15.1.1. Rose, *Rosa* spp.

Rose is the most ancient and popular flower grown the world over. In India, it is commercially grown for cut flowers and rose oil. Rose flowers without stems and loose flower petals are used in traditional markets for making garlands, for offering in temples, while the florist shops sell cut roses with stems mainly for bouquets and floral arrangements. In recent times, about 60 units have been established under joint ventures around Bangalore, Pune, Nasik, Mumbai, Hyderabad, Gurgaon (Haryana), Chandigarh and Saharanpur (Uttar Pradesh) for growing roses in greenhouses for export of cut flowers to Japan, The Netherlands, Germany and other European countries. Besides, the Damask rose (*Rosa damascena*) and Edward rose (*R. bourboniana*) are cultivated for rose *attar* and other products – *gulkand, gulabjal*, and *pankkurj*. It is cultivated in about 6,000 ha area in the states of West Bengal, Uttar Pradesh, Gujarat, Haryana, Jammu and Kashmir, Tamil Nadu, Karnataka, Rajasthan, Madhya Pradesh, Andhra Pradesh and Punjab.

15.1.1.1. *Lesion Nematode, Pratylenchus vulnus and Crown Gall, Agrobacterium rhizogenes Disease Complex*

Serious losses have been recorded in southern California apparently caused by a new strain of *A. rhizogenes* (cause hairy root of roses) in association with infestations of the lesion nematode, *P. vulnus* (Fig. 15.1) (Munnecke *et al.*, 1963).

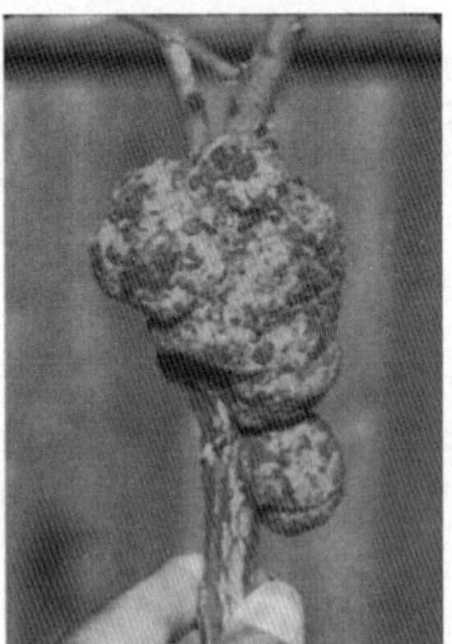

Fig. 15.1: Crown gall on rose

15.1.2. Carnation, *Dianthus caryophyllus*

Carnation flower is valued for its excellent keeping quality, wide array of colours and forms, and ability to rehydrate after continuous transportation. Moderate climatic control measures that are economical can deliver quality carnations at the internationally competitive prices year round. In India, Carnation and gerbera are commercially grown mostly under cover around big cites like Solan, Shimla, Mandi, Kullu, Chandigarh, Ludhiana, Guragaon, Kalimpong, Pune, Bangalore and Delhi. Carnation has tremendous potential for export market.

15.1.2.1. *Root-knot Nematode, Meloidogyne incognita and Wilt, Fusarium oxysporum f. sp. dianthi Disease Complex*

M. incognita in association with *F. oxysporum* f. sp. *dianthi* has been reported to cause wilt disease complex in carnation (Fig. 15.2).

Fig. 15.2. Root-knot nematode and *Fusarium* wilt disease Complex in carnation

In an experiment carried out to study the role of root-knot nematode, *M. incognita* in predisposing carnation to *Fusarium* wilt, it was observed that when both the pathogens were simultaneously inoculated the root galling index (RGI) was 3.1, while prior inoculation of with *F. oxysporum* showed a RGI of 2.8. When *M. incognita* was inoculated alone the RGI was 3.5, while prior inoculation of *M. incognita* recorded an RGI of 4.15 twelve weeks after inoculation.

The appearance of the wilt symptoms were accelerated when *M. incognita* was inoculated two weeks prior to *F. oxysporum* f. sp. *dianthi*. The rate of wilting was observed to be 2.1 and 4.6, respectively during 12th and 25th week of observation, while the root galling index (RGI) values were recorded to be 4.15 and 4.75, respectively (Table 15.3) (Shylaja, 2005).

Table 15.3: Effect of interaction of *M. incognita* and *F. oxysporum* f. sp. *dianthi* on wilt disease complex on carnation cv. Ivonne

Treatment	***Root-knot index (1-10 scale)***		***Wilt disease index (1-5 scale)***	
	12 weeks	***25 weeks***	***12 weeks***	***25 weeks***
M. incognita	7.0	8.2	0.0	0.0
F. oxysporum f. sp. *dianthi*	0.0	0.0	1.2	2.5
M. incognita + *F. oxysporum* f. sp. *dianthi* (simultaneous inoculation)	6.2	7.4	1.5	3.3

Table 15.3: (*Contd...*)

Table 15.3: *(Contd...)*

Treatment	*Root-knot index (1-10 scale)*		*Wilt disease index (1-5 scale)*	
	12 weeks	*25 weeks*	*12 weeks*	*25 weeks*
M. incognita 2 weeks prior to *F. oxysporum* f. sp. *dianthi*	8.3	9.5	2.1	4.6
F. oxysporum f. sp. *dianthi* 2 weeks prior to *M. incognita*	5.6	7.0	0.8	2.3
Uninoculated control	0.0	0.0	0.0	0.0
CD (P = 0.05)	**0.37**	**0.42**	**0.13**	**0.18**

It was observed that maximum mortality was recorded when *M. incognita* was inoculated 2 weeks prior to *F. oxysporum* f. sp. *dianthi* which was 33.4% and 79.2%, respectively, at 12th and 25th week of observation. When *F. oxysporum* was inoculated 2 weeks prior to *M. incognita* the mortality was found to be 8.6% and 40.2%, respectively at 12th and 25th weeks after inoculation. Simultaneous inoculation of both the pathogens showed 12% and 63% mortality at 12th and 25th weeks (Table 15.4) (Shylaja, 2005).

Table 15.4: Effect of interaction of *M. incognita* and *F. oxysporum* f. sp. *dianthi* on plant mortality of carnation cv. Ivonne

Treatment	*% plant mortality*	
	12 weeks	*25 weeks*
M. incognita	0.0	14
F. oxysporum f. sp. *dianthi*	6.0	45
M. incognita + *F. oxysporum* f. sp. *dianthi* simultaneously	12.0	63
M. incognita 2 weeks prior to *F. oxysporum* f. sp. *dianthi*	33.4	80
F. oxysporum f. sp. *dianthi* 2 weeks prior to *M. incognita*	8.6	40
Uninoculated control	0.0	0
CD (P = 0.05)	**4.3**	**7.56**

Ferraz and Lear (1976) studied the interaction of four plant parasitic nematodes (*Circonemoides curvatum, Pratylenchus dianthus, Rotylenchus robustus* and *Meliodogyne hapla*) and *F. oxysporum* f. sp. *dianthi* in carnations (Table 15.5). They reported that plants wilted earlier and more severe symptoms were present with the nematode-fungus combination than with the fungus alone. All nematodes except *R. robustus* showed synergistic interaction with *Fusarium.*

(i) *Integrated management*

(a) *Two bioagents*: The studies carried out to evaluate combination of bio-agents for the biological control of wilt (*F. o.* f. sp. *dianthi*) and root-knot nematode (*M.*

Table 15.5: Effect of interaction of *M. incognita* and *F. oxysporum* f. sp. *dianthi* on plant growth parameters of carnation cv. Ivonne

Treatment	***Plant height (cm)***	***Root length (cm)***	***Shoot weight (g)***	***Root weight (g)***
M. incognita	44.0	14.00	23.9	4.66
F. oxysporum f. sp. *dianthi*	40.0	13.60	23.5	4.30
M. incognita + *F. oxysporum* f. sp. *dianthi* (simultaneous inoculation)	22.4	13.72	23.38	2.50
M. incognita 2 weeks prior to *F. oxysporum* f. sp. *dianthi*	32.8	12.54	26.20	2.58
F. oxysporum f. sp. *dianthi* 2 weeks prior to *M. incognita*	17.4	14.00	20.22	1.28
Uninoculated control	71.4	34.6	55.26	7.32
CD (P = 0.05)	**7.56**	**3.45**	**3.20**	**0.53**

incognita) disease complex in carnation, revealed that a combination of *Pochonia chlamydosporia* + *Paecilomyces lilacinus* each at 20 g/m^2 gave significant increase in plant growth parameters (plant height, root length and root weight) and flower yield (stalk length, stalk weight and flower diameter) (Table 15.6) (Shylaja, 2005).

Table 15.6: Effect of biocontrol agents on the plant growth and flower yield of carnation infected with *M. incognita* and *F. oxysporum* f. sp. *dianthi*

Treatment	***Plant height (cm)***	***Stalk length (cm)***	***Root length (cm)***	***Root weight (g)***	***Stalk weight (g)***	***Flower diameter (cm)***
Formulations of *P. chlamydosporia* and *P. lilacinus* each at 20 g/m^2	54.0	48.8	21.3	20.5	17.24	5.85
Formulations of *P. chlamydosporia* and *T. harzianum* each at 20 g/m^2	52.2	43.4	15.7	16.44	15.90	5.15
Formulations of *T. harzianum* and *P. lilacinus* each at 20 g/m^2	47.8	39.2	18.0	18.06	14.66	4.99
Control	22.8	17.6	12.8	14.86	10.82	3.88
CD (P = 0.05)	**5.16**	**4.54**	**2.9**	**5.6**	**2.01**	**0.81**

The lowest root galling (1.64) and wilting index (2.0) and plant mortality (49.5%) was found in plants treated with *P. chlamydosporia* and *P. lilacinus*. The lowest number of nematodes in both soil and roots (44.4 J_2 per 100 cm^3 soil and 10 nematodes/5g roots) were also recorded in plants treated with *P. chlamydosporia* and *P. lilacinus* (Table 15.7) (Shylaja, 2005).

Table 15.7: Effect of biocontrol agents on root galling and wilt disease index in carnation infected with *M. incognita* and *F. oxysporum* f. sp. *dianthi*

Treatment	*Root galling index*	*Wilt disease index*	*% plant mortality*	*No. of nemas/100 cm³ soil*	*No. of nemas/ 5g roots*
Formulations of *P. chlamydosporia* and *P. lilacinus* each at 20 g/m²	1.64	2	49.5	44.4	10.0
Formulations of *P. chlamydosporia* and *T. harzianum* each at 20 g/m²	2.70	3	60.4	73.6	13.6
Formulations of *T. harzianum* and *P. lilacinus* each at 20 g/m²	2.17	3	63.9	62.4	21.0
Control	3.66	5	95.8	125.8	42.2
CD (P = 0.05)	**0.82**	**0.53**	**10.46**	**5.20**	**3.12**

15.1.2.2. *Root-knot Nematode, Meloidogyne spp. and Bacterial Wilt, Pseudomonas caryophylli Disease Complex*

Stewart and Schindler (1956) studied wilting of carnation cuttings infected with *Pseudomonas caryophylli* in association with *Meloidogyne* spp. or *Helicotylenchus nannus*. The results indicated that root-wounding by *Meloidogyne* spp. and *H. nannus* increased the rate of wilting in the presence of bacterial pathogens.

15.1.3. Gerbera, *Gerbera jamesonii*

Gerbera, commonly known as Transvaal daisy, Barberton daisy or African daisy, is an important flower grown throughout the world for beds, borders, pots and rock gardens. The flowers are of various colours and suit very well for floral arrangements. The cut blooms also have long vase life. In India, it is distributed in the temperate Himalayas from Kashmir to Nepal at altitudes of 1300 to 3200 metres above sea level. Gerbera belongs to the family Asteraceae and is native to South African and Asiatic regions. To meet the quality standards of export market, it has to be grown under naturally ventilated low cost polyhouses.

15.1.3.1. *Root-knot Nematode, Meloidogyne incognita and Foot Rot, Phytophthora parasitica Disease Complex*

Sustainable production of gerbera is seriously hampered by the disease complex caused by *M. incognita* and *P. parasitica*. These two pathogens reduce the productivity of gerbera significantly to the tune of 40-60%.

Management methods

Integrated management

Combined application of neem cake enriched with either *Trichoderma harzianum* and *Pseudomonas fluorescens* [mixing 50 g of *T. harzianum* (2×10^6 cfu/g) or *P.*

fluorescens (2×10^8 cfu/g) in 1 kg of neem cake] applied at 25 g/m^2 was found effective for the management of disease complex and increased the flower yield by 26% in gerbera cv. Debora (Manoj Kumar *et al.*, 2010).

15.1.4. Gladiolus, *Gladiolus* spp.

Gladiolus is very much liked for its majestic spikes containing attractive, elegant and delicate florets. These florets open in sequence over a longer duration and hence have a good keeping quality of cut spikes. It is cultivated for cut flowers, garden and interior decoration, making bouquets, bedding, rockeries and pots. West Bengal, Maharashtra, Uttar Pradesh, Punjab, Haryana and Andhra Pradesh are the major gladiolus-growing states.

15.1.4.1. *Root-knot Nematode, Meloidogyne incognita and Wilt, Fusarium oxysporum f. sp. gladioli Disease Complex*

The commercial production of gladiolus is limited by soil-borne pathogens like root-knot nematodes and Fusarium wilt. Their combined occurrence in cultivated soils aggravated the wilt problem causing high plant mortality in gladiolus fields (Fig. 15.3).

Fig. 15.3: Root-knot nematode and *Fusarium* wilt disease complex in gladiolus

(i) *Biomanagement*

(a) *Antagonistic bacteria*: Application of *P. fluorescens* effectively controlled the disease complex leading to significant improvement in plant growth and flowering (Mustafa and Khan, 2004).

(ii) *Integrated management*

(a) *Bioagents and botanicals*: Gladiolus plants treated with *P. lilacinus* + *T. harzianum* + neem cake and *P. lilacinus* + *T. viride* + neem cake combinations not only controlled *M. incognita* infection but also *Fusarium* wilt till the harvest of flower spikes. The corms and cormels obtained from plants treated with these combinations were free from *Fusarium* infection. Bioagent colonization of galled roots was maximum in *P. lilacinus* + *T. viride* combination (94%) followed by *P.*

lilacinus + *T. harzianum* combination (69%). Similarly, the parasitization of eggs was maximum in *P. lilacinus* + *T. viride* combination (44%) followed by *P. lilacinus* + *T. harzianum* combination (38%), while egg mass parasitization was same in both the combinations (58%) (Nagesh *et al.*, 1998) (Table 15.8).

Table 15.8: Effect of integration of antagonistic fungi with neem cake on root-knot and wilt disease complex on gladiolus.

Treatment/ Dose/ plant	*% Healthy plants*	*% Redn. in nema multipln.*	*% Infected corms & cormels*	*% Root coloni-zation*	*% Parasitization*	
					Egg masses	*Eggs*
P. lilacinus- 8×10^{10} spores	29	38	52	48	49	46
T. harzianum- 8.8×10^{10} spores	36	16	30	50	44	32
T. viride- 8.8×10^{10} spores	40	24	20	53	47	39
P. lilacinus + Neem cake 8×10^{10} spores + 20 g	34	64	32	62	61	58
P. lilacinus + *T. harzianum* + Neem cake (½ dose each)	78	26	4	69	58	38
P. lilacinus + *T. viride* + Neem cake (½ dose each)	88	47	0	74	58	44
CD (P = 0.05)	**4.36**	**6.22**	**8.54**	**3.88**	**4.11**	**3.56**

15.1.4.2. Root-knot Nematode, *Meloidogyne javanica* and Bacterial Wilt, *Pseudomonas marginata* Disease Complex

Root infection with *M. javanica* greatly increased the severity of gladiolus scab caused by *P. marginata* (El-Goorani *et al.*, 1974). This study shows the systemic nature of physiological changes brought about by root-knot nematodes in favouring infection of foliar bacterial plant pathogens.

15.1.5. Tuberose, *Polianthes tuberosa*

Tuberose is one of the most important bulbous ornamentals grown both for cut and loose flowers, and also for the extraction of its highly valued natural flower oil. It is also grown for garden decoration in pots, beds and borders. It is commercially grown in Karnataka, West Bengal, Maharashtra, Punjab, Andhra Pradesh and Tamil Nadu. At present, the total area under tuberose cultivation in the country is estimated at about 3,000 ha.

15.1.5.1. *Root-knot Nematode, Meloidogyne incognita and Wilt, Fusarium oxysporum f. sp. dianthi Disease Complex*

Rao *et al.* (2001) reported that root-knot nematodes accelerate and increase *Fusarium* wilt symptom development and ultimately increase the death rate of tuberose plants infected with both the pathogens. The fungal infection was observed to aggravate in the presence of *M. incognita*.

The root-knot nematode, *M. incognita* reduced flower yield considerably and the tuberose plants became highly susceptible to the attack by *F. oxysporum* f. sp. *dianthi* (Fig. 15.4) (Rao *et al.*, 2002).

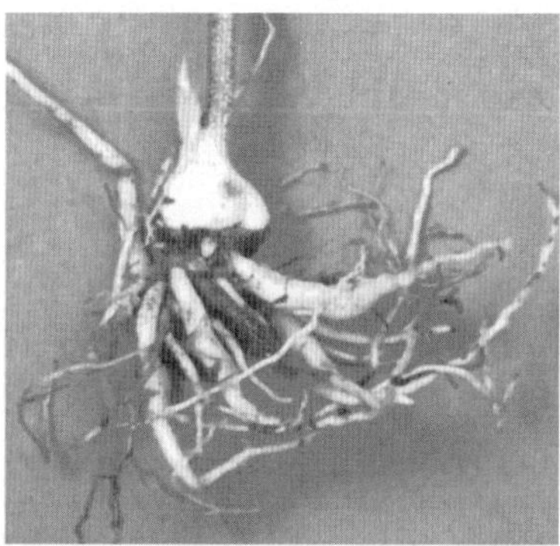

Fig. 15.4: Root-knot nematode and *Fusarium* wilt complex in tuberose

In an experiment undertaken to assess the impact of different densities of root-knot nematodes in inducing *Fusarium* wilt-root knot disease complex on the flower yield in tuberose, it was observed that the spike emergence was significantly delayed (65.07 and 68.33 days) due to the high population levels of nematodes at 100 and 150/100 cm^3 soil. The longest delay in spike emergence (81 days) was recorded when *Fusarium* was present along with the highest density of nematodes at 150/100 cm^3 soil. It was observed that the disease complex drastically reduced the yield of flowers in tuberose, thereby bringing down the production of flowers in this commercial crop. It was found that the damage was maximum in the presence of both pathogens *viz; M. incognita* and *F. oxysporum* f. sp. *dianthi*, when compared to the presence of either one of these pathogens (Table 15.9) (Shylaja, 2005).

(i) *Integrated management*

(a) *Two bioagents*: In course of the experiments carried out to evaluate combination of bio-agents for the control of wilt and root-knot nematode disease complex in tuberose, the best result was obtained in plants treated with *P. chlamydosporia* + *T. harzianum* (Rao *et al.*, 2003). The above treatment gave significant reduction in root galling, nematode population both in soil and roots, disease index, and increased egg parasitization and root colonization by bioagents and flower yield (Tables 15.10 and 15.11).

Table 15.9: Effect of interaction of *M. incognita* with *F. oxysporum* f. sp. *dianthi* on flower yield of tuberose

Treatment	*Days for spike emergence*	*No. of spikes/plot*	*Spike length (cm)*	*Days for Ist flowering*
F. oxysporum f. sp. *dianthi* at 10 ml (10^2 spores/ml)	63.00	17.33	74.33	74.00
M. incognita at 50 J_2/100 cm^3 soil	62.33	18.00	69.33	72.33
M. incognita at 100 J_2/100 cm^3 soil	65.07	17.00	63.67	75.33
M. incognita at 150 J_2/100 cm^3 soil	68.33	13.67	62.67	77.67
F. oxysporum f. sp. *dianthi* at 10 ml + *M. incognita* at 50 J_2/100 cm^3 soil	74.67	13.33	61.33	85.33
F. oxysporum f. sp. *dianthi* at 10 ml + *M. incognita* at 100 J_2/100 cm^3 soil	76.00	12.50	59.33	89.33
F. oxysporum f. sp. *dianthi* at 10 ml + *M. incognita* at 150 J_2/100 cm^3 soil	81.00	11.00	56.67	89.67
Control	60.67	24.00	80.67	72.00
CD (P = 0.05)	**4.00**	**2.31**	**5.85**	**4.33**

Table 15.10: Effect of integration of bioagents for the management of disease complex and flower yield in tuberose

Treatment/ Dose ($g/4m^2$)	*Root-knot (1-10 scale)*	*Nema popn./ 100 ml soil*	*Nema popn./ 5 g roots*	*Disease index (1-5 scale)*	*No. of spikes/ $4m^2$*	*Spike length (cm)*
P. chlamydosporia – 20	6.21	62	36	3.76	24	60.56
P. chlamydosporia – 40	4.26	54	31	3.11	26	65.32
T. harzianum – 20	6.56	69	39	2.82	20	63.37
T. harzianum – 40	5.21	58	35	2.33	23	66.87
P. chlamydosporia – 20 + *T. harzianum* – 20	4.79	46	17	2.18	29	71.79
P. chlamydosporia – 40 + *T. harzianum* – 40	4.20	41	10	1.59	24	75.73
Control	8.49	134	46	4.42	18	45.69
CD (P = 0.05)	**1.76**	**26.49**	**7.47**	**0.76**	**2.33**	**7.52**

Table 15.11: Effect of integration of bioagents on egg parasitization and root colonization by the bioagents in tuberose

Treatment/Dose (g/4m²)	***Root colonization (cfu/g)***		***% egg parasitization***	
	Pc	***Th***	***Pc***	***Th***
P. chlamydosporia – 20	30458	–	40.98	–
P. chlamydosporia – 40	38943	–	54.69	–
T. harzianum – 20	–	40369	–	50.68
T. harzianum – 40	–	45653	–	55.84
P. chlamydosporia – 20 + *T. harzianum* – 20	30879	38789	37.49	51.69
P. chlamydosporia – 40 + *T. harzianum* – 40	36278	44926	53.96	52.38
Control	–	–	–	–
CD (P = 0.05)	**4563.28**	**5219.75**	**6.47**	**7.42**

Shylaja (2005) found that integration of *P. chlamydosporia* with *T. harzianum* gave maximum increase in plant height, shoot weight and root length; while *P. chlamydosporia* + *P. lilacinus* gave maximum root weight (Table 15.12).

Table 15.12: Effect of bioagents on plant growth parameters of tuberose infected with *M. incognita* and *F. oxysporum* f. sp. *dianthi*

Treatment	***Plant height (cm)***	***Shoot weight (g)***	***Root length (cm)***	***Root weight (g)***
Formulations of *P. chlamydosporia* and *P. lilacinus* each at 20 g/m²	33.01	29.46	7.83	10.10
Formulations of *P. chlamydosporia* and *T. harzianum* each at 20 g/m²	33.39	34.80	8.63	7.20
Formulations of *T. harzianum* and *P. lilacinus* each at 20 g/m²	32.01	24.83	8.04	6.72
Control	20.60	22.52	4.80	4.15
CD (P = 0.05)	**4.94**	**3.51**	**3.03**	**1.82**

Integration of *P. chlamydosporia* with *T. harzianum* gave maximum flower yield characteristics (spike length, spike weight, number of flowers/spike, flower weight and number of spikes/plot) (Table 15.13) (Shylaja, 2005).

Table 15.13: Effect of bioagents on flower yield characteristics of tuberose infected with *M. incognita* and *F. oxysporum* f. sp. *dianthi*

Treatment	*Spike length (cm)*	*Spike weight (g)*	*No. of flowers/spike*	*Flower weight (g)*	*No. of spikes /plant*
Formulations of *P. chlamydosporia* and *P. lilacinus* each at 20 g/m^2	70.63	43.31	29.87	1.04	22.80
Formulations of *P. chlamydosporia* and *T. harzianum* each at 20 g/m^2	75.73	53.01	45.00	1.05	24.0
Formulations of *T. harzianum* and *P. lilacinus* each at 20 g/m^2	72.89	43.54	29.90	0.97	21.60
Control	45.69	25.51	23.65	0.50	18.40
CD (P = 0.05)	**5.09**	**4.08**	**6.22**	**0.15**	**2.64**

Integration of *P. chlamydosporia* with *P. lilacinus* gave least root galling and wilt disease index, while integration of *P. chlamydosporia* with *T. harzianum* gave least nematode population both in soil and roots (Table 15.14) (Shylaja, 2005).

Table 15.14: Effect of bioagents on *M. incognita* and *F. oxysporum* f. sp. *dianthi* disease complex of tuberose

Treatment	*Root galling index*	*Wilt disease index*	*No. of nemas/ 100 cm^3 of soil*	*No. of nemas/5g roots*
Formulations of *P. chlamydosporia* and *P. lilacinus* each at 20 g/m^2	2.18	1.50	36	9
Formulations of *P. chlamydosporia* and *T. harzianum* each at 20 g/m^2	2.40	2.28	322	5
Formulations of *T. harzianum* and *P. lilacinus* each at 20 g/m^2	3.36	3.10	41	10
Control	4.10	4.28	134	46
CD (P = 0.05)	**0.36**	**0.40**	**5.54**	**4.06**

15.1.6. Crossandra, *Crossandra undulaefolia*

Crossandra is mainly cultivated commercially for its loose flowers in Tamil Nadu, Karnataka and Andhra Pradesh. It is also grown in Sri Lanka, Tropical Africa and Madagascar. It is a perennial herb and belongs to the family Acanthaceae.

15.1.6.1. *Root-knot Nematode, Meloidogyne incognita and Foot Rot, Phytophthora nicotianae Disease Complex*

(i) *Integrated management*

(a) *Bioagents and chemicals*: Combined application of *Trichoderma harzianum* (20 g/plant) and Aliette (0.3%) did not show any mortality of plants and was also effective for the management of root rot disease and root-knot nematodes (Table 15.15) (Ramachandran and Sinha, 2000).

Table 15.15: Effect of *Trichoderma*, Aliette and neem cake on disease complex in crossandra

Treatment	***Mortality****	***Collar rot****	***Root rot***	***Root-knot index***
Nematode alone	Nil	Nil	1.60	2.80
Nematode + Neem cake + *P. nicotianae*	1	Nil	1.75	1.25
Nematode + *T. harzianum* + *P. nicotianae*	Nil	Nil	1.20	1.20
Nematode + Aliette + *P. nicotianae*	Nil	Nil	0.40	1.40
Nematode + *P. nicotianae*	2	2	3.50	1.70

* Denotes number of plants

Scale for root rot index: 0 = No root rot, 1 = 1-25%, 2 = 25-50%, 3 = 50-75%, 4 = 75-100%

Scale for root-knot index: 0 = No galls, 1 = 1-10%, 2 = 10-20%, 3 = 20-50%.

15.1.6.2. *Root-knot Nematode, Meloidogyne incognita and Wilt, Fusarium spp. Disease Complex*

M. incognita interact synergistically with wilt fungi *F. oxysporum* and *F. solani* and results in premature death of crossandra plants (Khan and Parvatha Reddy, 1992). *F. solani* and *F. oxysporum* are invariably associated with the lesions caused by *P. delattrei* in crossandra (Khan and Parvatha Reddy, 1992a; Srinivasan and Muthurksihanan, 1975). Srinivasan (1974) found that wilt symptoms appeared earlier when nematodes were added two weeks prior to *F. solani*.

15.1.7. Chrysanthemum, *Dandrathema grandiflora*

Garden chrysanthemum, also known as 'Queen of the East', is a highly versatile and accommodating ornamental with a wide range of type, size and colour in flowers. In India, large flowered varieties are grown for cut flower, making garland, wreaths and 'veni', religious offerings and for bedding and potting purposes. Its commercial cultivation is being done in Maharashtra, Rajasthan, Madhya Pradesh and Bihar. Small flowered varieties are commercially grown for loose flowers in Tamil Nadu, Karnataka and Maharashtra. In India, chrysanthemum is grown in about 4,000 ha.

15.1.7.1. *Lesion Nematode, Pratylenchus coffeae and Damping-off, Pythium aphanidermatum and Rhizoctonia solani Disease Complex*

Disease severity due to *P. aphanidermatum* or *R. solani* increased with increasing inoculum levels of *P. coffeae*. It increased further when nematode infested plants were inoculated with both the fungi together in chrysanthemum (Hasan, 1988).

15.1.7.2. *Sting Nematode, Belonolaimus longicaudatus and Damping-off, Pythium aphanidermatum Disease Complex*

Chrysanthemum roots inoculated with *B. longicaudatus* and *P. aphanidermatum* develop symptoms of root rot earlier and more extensively than those inoculated with the fungus alone.

15.1.7.3. *Root-knot Nematodes, Meloidogyne spp. and Fusarium Wilt Disease Complex*

Fusarium wilt symptoms appeared earlier and were more severe in Fusarium-susceptible chrysanthemum cultivar Yellow Delaware when infected with *Meloidogyne javanica* than with *M. hapla* and *M. incognita* (Johnson and Littrel, 1969).

15.1.8. Snapdragon, *Antirrhinum majus*

15.1.8.1. *Root-knot Nematode, M. incognita and Damping-off, P. aphanidermatum Disease Complex*

Methyl bromide reduced ($P<0.05$) root-knot nematode galling relative to the control. Root galling in treatments involving solarization was generally intermediate, and did not differ ($P>0.10$) from either the methyl bromide or control treatments. Methyl bromide reduced the number of dead plants compared to all other treatments in the first year. However, in the second year, when disease incidence was greater, the numbers of dead plants in control and methyl bromide plots were greater ($P<0.05$) than in plots that had been solarized (Table 15.16). However, plants that survived in the methyl-bromide-treated plots were larger than those in all other treatments, and as a result, marketable yield was greatest ($P<0.05$) in plots treated with methyl bromide. The addition of Kodiak or Biophos to solarization did not improve nematode or pathogen control over the treatment with solarization alone.

Table15.16: Impact of methyl bromide and alternative treatments on galling due to root-knot nematodes, plants dead from soil-borne disease, plant weight, and flower yield of snapdragons grown in infested site.

Treatment	*Galling*	*Dead plants*	*Plant wt.*	*Yield*
2006-07 season				
Control	–	–	–	–
Solarized	+	–	–	–
Solarized + Kodiak	±	–	–	–
Solarized + Biophos	±	–	–	–
Methyl bromide	+	+	+	+
2007-08 season				
Control	–	–	–	–
Solarized	±	+	–	–
Solarized + Kodiak	±	+	–	–
Solarized + Biophos	±	+	–	–
Methyl bromide	+	0	+	+

\+ Best treatment or statistically equal to best treatment
\- Worst treatment or statistically equal to worst treatment
+/- Intermediate; statistically equal to both best and worst treatments
0 Intermediate but statistically different from best and worst treatments

Plots that were solarized produced only 38.0-47.7% as much flower yield as the methyl bromide plots in the first year, and only 29.9-35.5% as much in the second year (Table 15.17). The disease complex between *M. incognita* and *P. aphanidermatum*, which increased in severity in the second year, was likely a major influence on crop yield. Although solarization treatments reduced plant mortality compared to methyl bromide, the plants that survived were larger in the plots treated with methyl bromide, and therefore produced more marketable yield.

Table 15.17: Effect of alternative treatments on flower yield, expressed relative to performance of methyl bromide treatment.

Treatment	*Flower yield*	
	2006-07	*2007-08*
Control	24.7%	4.0%
Solarized	40.5%	33.3%
Solarized + Kodiak	47.7%	29.9%
Solarized + Biophos	38.0%	35.5%
Methyl bromide	100.0%	100.0%

15.1.9. Marigold, *Tagetes* spp.

15.1.9.1. *Root-knot Nematode, Meloidogyne incognita and Root Rot, Rhizoctonia solani Disease Complex*

The marigold cvs. were found resistant to root-knot nematode and did not support gall formation or egg mass production. The marigold cv. Jafri was found highly susceptible (suppressed plant growth at 2g *R. solani*/pot), cv. Yellow Small expressed moderate susceptibility, whereas the cv. Hazara showed resistance against *R. solani*.

Presence of the nematode in concomitant inoculation treatment synergized the pathogenesis of *R. solani* resulting in greater root rot with corresponding decrease in plant growth and dry matter production. The presence of the nematode also broke down the resistance response of cv. Hazara when the two pathogens were present at higher inoculum levels (*M. incognita* at 5000 J_2/pot and *R. solani* at 4g/pot). The concentration of total phenolic compounds and salicylic acid was much greater in the cvs. expressing resistance against *R. solani* in the presence or absence of *M. incognita* compared to the cv. which was highly susceptible to the fungus and developed extensive rotting in the presence of root-knot nematode (Rizvi *et al.*, 2010).

15.1.10. Balsam

15.1.10.1. *Lesion Nematodes, Pratylenchus spp. and Wilt, Verticillium albo-atrum Disease Complex*

Müller (1977) studying interactions between five species of *Pratylenchus* and *V. albo-atrum* on balsam, reported that wilt was obtained when the fungus was combined with *P. penetrans* and *P. vulnus* but wilting did not occur when combined with *P. crenatus, P. thornei* and *P. fallax.*

15.2. MEDICINAL CROPS

In recent times, due to increasing realization of health hazards and toxicity caused by synthetic drugs (side effects), there has been a renewal of interest in the use of plant based drugs throughout the world. A campaign is going on worldwide to utilize more and more plant derived chemicals in the human health care/personal care system. As a result, the demand for plant-based raw materials has increased enormously in both national and international markets. Herbal plants are finding diverse uses as raw materials, not only for medicines but also as biopesticides.

In medicinal plant sector, the WHO has estimated that about 80% of the population in developing countries rely on traditional medicines mostly plant drugs for primary health-care needs. Even modern medicines contain about 25% drugs derived from plants. A large number of people are earning their livelihood from different activities of medicinal plants such as collection, cultivation, marketing, processing, etc. The area under cultivation of the above crops is low, but the cash returns are high because of intensive cropping.

Medicinal crops form one of the important groups due to their demand in various pharmaceutical industries and also to earn foreign exchange by way of export. The potential for earning foreign exchange by India from the exports of medicinal and aromatic plants is estimated over US$ 3,000 million per annum. The export of these plants and their products has a tremendous potential in advanced countries like Europe, USA and Japan. The international market for medicinal plants related trade is estimated at US$ 60 billion/year having a growth rate of 7% per annum. Destination of foreign tourists is now aimed at India for health treatment. Kerala and Gujarat have become their favourite destinations since Ayurvedic treatment is very popular in these two states.

Unless we start cultivating the important medicinal plants, it will not be possible to meet the increased demands. However, cultivation of medicinal plants has to be taken up in polyculture models rather than in monoculture models. There is a huge task to be carried out at every level to make the Indian system of medicine a viable, sustainable and modern. For this task, urgency is felt which can only be established through regular and systematic scientific research on collection, conservation, evaluation and development of good agricultural practices to compete in the world market.

15.2.1. Dioscorea, *Dioscorea floribunda*

Various species of Dioscorea are the source of diosgenin, a steroidal sapogenin, and is the most commonly used as precursor for synthesis of certain drugs which include cortisones, sex hormones and oral contraceptives.

15.2.1.1. *Root-knot Nematode, M. incognita and Wilt, F. oxysporum f. sp. dioscorca Disease Complex*

A positive interaction between *M. incognita* and *F. oxysporum* f. sp. *dioscorea* occurs in *Dioscorea rotundata* cv. Habanero.

15.2.2. *Coleus, Coleus forskohlii*

Coleus belongs to the Tulsi family Lamiaceae and is indigenous to India. It is used in Ayurvedic medicines and as a condiment. It shot into prominence in modern medicine with the isolation of the diterpinoid, forskolin, from its tubers. The therapeutical properties of forskolin in the treatment of glaucoma, congestive cardiomyopathy, asthma and certain cancers combined with its use in cosmetics has enhanced importance of coleus in modern medicine. Coleus proved to be the exclusive source of forskolin and therefore, the sudden spurt in the demand for its tubers led to exploitative collection from natural stand. The drug is claimed to improve appetite, facilitate digestion, increase vitality and is useful in anemia and inflammation. Recently it is also being used in the treatment of alcoholic addiction. Forskolin has a multiple biological activities like positive iontropic, antihypertensive, branchospasomolytic, antithrombotic, platelet aggregation inhibiting and adenylate cyclase stimulation. The species is now considered endangered and a prime candidate for cultivation, more especially captive cultivation.

15.2.2.1. *The Root-knot Nematode, Meloidogyne incognita and Wilt, Fusarium chlamydosporum, Ralstonia solanacearum Disease Complex*

(i) *Biomanagement*

(a) *Antagonistic bacteria*: Greenhouse studies using the plant growth promoting rhizobacterial (PGPR) fluorescent pseudomonads (bioformulations containing RB 50 and RB 31 strains) showed significant increase in seedling biomass besides reduction in coleus wilt complex due to combination of pathogens. RB 50 and RB 31 strains decreased root-knot index and incidence of disease complex while increasing biomass and tuber yield. Biochemical analyses in the above treatments showed elevated expression of defense enzymes (peroxidase, polyphenol oxidase and phenylalanine ammonialyase) and higher accumulation of phenolic compounds (activity being highest in respect of RB 50 and RB 31 treated plants) compared to respective inoculated checks. The suppression of disease complex in coleus is largely due to aforementioned biocontrol mechanism as well as induction of systemic resistance by efficient PGPR strains (Lingaraju and Mallesh, 2010).

15.2.2.2. *Root-knot Nematode, Meloidogyne incognita and Root Rot, Macrophomina phaseolina Disease Complex*

Simultaneous inoculation of *M. incognita* and *M. phaseolina* as well as nematode inoculation followed by fungus 15 days later caused significant reduction in tuber yield and 100% root rot disease in medicinal coleus. The nematode multiplication was adversely affected when fungus was inoculated prior to nematode (Table 15.18) (Senthamarai *et al.*, 2008).

15.2.3. Ashwagandha, *Withania somnifera*

Ashwagandha is a plant of immense medicinal importance growing all over North-Western and Central India. Roots of this plant are the major source of alkaloids and steroidal lactones (withanoids), which are intensively used in various pharmaceutical industries. It has adaptogenic, immuno-modulator, aphrodisiac, anti-stress and mildly sedative properties. Its roots yield valuable drugs which are employed in rheumatic pain, inflammation of joints, nervous disorders, cough, cold, epilepsy, and cardio and nerve tonic. At present, 4,000-5,000 ha area is under its cultivation mainly in Madhya Pradesh (Manasa Neemuch and Jawad tehsil of Mandsaur district) and neighbouring districts of Rajasthan (Kota and Charu).

Table 15.18: Effect of *Meloidogyne incognita* and *Macrophomina phaseolina* on root galling and yield of coleus

Treatment	*Tuber yield/ plant (g)*	*No. of galls/ plant*	*% disease incidence*
M. incognita (1 J2/g soil)	44.00	768	0
M. phaseolina (5 g/kg soil)	61.50	0	50
M. incognita (prior) + *M. phaseolina* (15 days later)	34.50	373	100
M. phaseolina (prior) + *M. incognita* (15 days later)	52.50	110	50
M. incognita + *M. phaseolina* (simultaneously)	13.00	316	100
Uninoculated control	84.38	0	0
CD (P = 0.05)	**10.07**	**2.08**	**43.99**

15.2.3.1. *The Root-knot Nematode, Meloidogyne incognita and Wilt, Fusarium chlamydosporum Disease Complex*

(i) *Biomanagement*

(a) *Antagonistic bacteria*: Greenhouse studies using the plant growth promoting rhizobacterial (PGPR) fluorescent pseudomonads (bioformulations containing RB 50 and RB 31 strains) showed significant increase in seedling biomass besides reduction in ashwagandha wilt complex due to combination of pathogens. RB 50 and RB 31 strains decreased root-knot index and incidence of disease complex while increasing biomass and tuber yield. Biochemical analyses in the above treatments showed elevated expression of defense enzymes (peroxidase, polyphenol oxidase and phenylalanine ammonialyase) and higher accumulation of phenolic compounds (activity being highest in respect of RB 50 and RB 31 treated plants) compared to respective inoculated checks. The suppression of disease complex in ashwagandha is largely due to aforementioned biocontrol mechanism as well as induction of systemic resistance by efficient PGPR strains (Lingaraju and Mallesh, 2010).

15.3. AROMATIC CROPS

In recent years, there has been an increased interest in the cultivation of aromatic plants to meet the requirements of cosmetic, flavouring and perfumery industries. Aromatic crops are estimated to cover an area of 386,100 ha with production of 325,000 MT during 2008 (Bijay Kumar, 2009) (Table 15.19). The aromatic plants provide raw material for the production of flavours, herbal cosmetics, perfumery etc. Aromatic plants and their products, particularly the essential oils, are now becoming one of the more important export items from many developing countries of Asia. Chemically essential oils are terpenes, which act as carriers of the aromatic substances. Most essential oils also contain camphors, and the more odiferous compounds present in them consist of oxygen derivatives of terpenes, alcohols, esters, aldehydes and ketones. India has enjoyed a prominent position in the manufacture of superior perfumes and aromatics by using essential oils. The important aromatic plants are lemon grass, vetiver, patchouli, palmarosa, citronella, mints, geranium, lavender, basil, jasmine etc.

Table 15.19: State-wise area and production of aromatic crops (2007-08) (Bijay Kumar, 2009)

State	*Area ('000 ha)*	*Production ('000 MT)*	*Productivity (MT/ha)*
Madhya Pradesh	19.6	117.6	6.0
Rajasthan	198.2	94.1	2.1
Chhattisgarh	11.5	65.6	5.6
Andhra Pradesh	9.6	13.8	1.4
Uttar Pradesh	133.7	13.4	0.1
Tamil Nadu	7.6	11.2	1.4
Karnataka	2.0	7.2	3.6
Punjab	2.4	1.1	0.5
Haryana	1.4	1.0	0.7
Jharkhand	0.1	0.0	–
Total	**386.1**	**325.0**	**0.8**

In the world market, India stands in second position. Most of the essential oils produced though marketed within the country, but a sizeable amount has also been exported. Among the essential oils exported from India are Japanese mint oil, ambrette seed essence, sandal wood oil, citronella oil, lemon grass oil, palmarosa oil, ginger oil, clove oil, eucalyptus oil, vetiver oil, pepper oil, etc. The annual exports of the derivatives of aromatic plants are to the tune of Rs. 600-700 million.

15.3.1. Mint, *Mentha* spp.

Mints are a group of aromatic herbs belonging to the family Lamiaceae, which are considered to be the most important cash crops in Indo-Gangetic plains. Cultivation of mints has been done in a large scale in many tropical and sub-tropical countries of the world including India, China, Brazil, Japan and USA. Indian farmer grow it as a bonus crop on an area of more than 1,50,000 ha in Central and Northern parts of Indo-Gangetic plains. The crop fit well in the traditional food based cropping system with other crops like paddy, wheat, potato, mustard, maize, okra, carrot, onion, spinach, pigeon pea, cowpea, etc. Similarly it can be grown as an intercrop with sugarcane and some legumes. In India, the highest area under commercial cultivation of mint has been in Uttar Pradesh (Terai region), but during last few years it has spread through out the country, especially in North and North-Western states including Madhya Pradesh. The mint products export for the year 2007-08 was 21,000 MT which was worth Rs. 12805 million.

Among different oil yielding species of mint, Japanese mint (*M. arvensis*) occupies the highest area under mint cultivation followed by Peppermint (*Mentha spicata*), Spear mint (*M. piperata*), Bergamot mint (*M. citrata*) and Scotch spearmint (*M. cardiaca*).

Japanese mint essential oil is used as a source of menthol, menthyl acetate, menthone and terpenes which finds wide use in medicine, perfumes, food and cosmetics. The oil of peppermint, which contains less menthol but has a sweeter aroma and taste, is used in tooth paste, chewing gums, candies, high grade liquors, medicines and other pharmaceutical preparations. Spearmint oil having carvone and limonene, is mainly used for flavouring; whereas, Bergamot mint oil is mainly used in perfumery and flavouring due to higher content of linayl acetate and linalool.

15.3.1.1. *The Lesion Nematode, Pratylenchus* spp. *and Wilt, Verticillium albo-atrum Disease Complex*

Bergeson (1963) reported that when *P. penetrans* infected *Mentha piperita* plants were inoculated with *V. albo-atrum,* diagnostic symptoms of *Verticillium* wilt appeared approximately two weeks earlier than the plants inoculated with the *Verticillium* fungus in the absence of nematode. Physiological changes evidently occur in fungus infected plants, making them more susceptible or attractive to the nematode pathogen (Fig. 15.5) (Mountain and McKeen, 1962).

Fig. 15.5: Lesion nematode and *Verticillium* wilt complex in mint.

Presence of *P. miniyus* increased both incidence and severity of *Verticillium* wilt (*V. dahliae* f. sp. *menthae*) disease, reduced the period for the development of wilt by 2 to 3 weeks, increased the reproduction of nematode and reduced the dry weight of peppermint plants up to 68% when both the pathogens were present (Faulkner and Skotland, 1965). The optimum soil temperature for disease development when the fungus is present alone is 24°C, as compared to 27°C in plants exposed to both pathogens (Faulkner and Bolander, 1969). Wilt increased at all test temperatures when the nematode was present, and the optimum temperature for nematode reproduction also changed (Faulkner and Skotland, 1965). When the root systems of peppermint plants were split and each separately inoculated with *Verticillium* and nematodes, an effect on disease severity and changes in incubation periods occurred. These results indicate that the nematode causes a physiological change in the host that makes it more susceptible to the fungus.

15.3.2. Patchouli, *Pogostemon patchouli*

Patchouli is a highly aromatic bushy under-shrub. It is cultivated for its highly fragrant leaves which contain a very sweet smelling oil of lasting sticky odour. The crop is grown in small pockets in Karnataka, Kerala and Tamil Nadu. It has a very characteristic aroma and blends well with other essential oils. The oil is used in very low concentration (2 ppm) in scenting soaps, cosmetics, after-shave lotions, detergents and many fancy products. The oil is also used to flavour foods, beverages, candy and baked products. In combination with sandal wood oil, it is used in blending tobacco and making incense sticks.

15.3.2.1. *Root-knot Nematode and Root Rot Disease Complex*

(i) *Integrated management*

(a) *Bioagents and botanicals*: Combined mortality due to root rot and root-knot nematode could be minimized by the application of *Trichoderma harzianum* + karanj cake at 5 MT/ha.

15.3.3. Basil, *Ocimum basilicum*

Among several species of basils, sweet basil (*Ocimum basilicum*) is considered to be the most important crop for its high quality of essential oil. Major components of its oil are linalool (43-50%), menthyl chavicol (18-33%), eugenol and isoeugenol (5-6%) and the minor constituents are alpha and beta pinene, camphor, geraniol, etc. The oil of sweet basil owes its importance to its extensive use in condimentary products, cosmetics, toiletry, perfumery and confectionary industries, particularly in European countries.

15.3.3.1. *The Root-knot Nematode, Meloidogyne incognita and Wilt, Fusarium oxysporum Disease Complex*

Both the pathogens caused significant reduction in all growth parameters when inoculated alone as well as in various inoculation sequences. Highest percent reduction in shoot/root length, shoot/root fresh and dry weight (74.2, 76.0, 67.0, 69.0, 67.5 and 73.9%, respectively) was observed in plants inoculated simultaneously with nematode and fungus, followed by nematodes 10 days prior to fungus, fungus 10 days prior nematode, nematode alone and fungus alone, respectively. Root colonization by the fungus was significantly increased in the presence of nematodes. Similarly, reproduction factor of nematode was significantly greater when nematode was present alone (10.5) followed by nematode 10 days prior to fungus (6.4), fungus 10 days prior to nematode (5.8), and nematode and fungus simultaneously (5.0). Root infection by the fungus was highest in simultaneously inoculated plants with nematode and fungus (82.3%), followed by nematode 10 days prior to fungus (74.1%), fungus 10 days prior to nematode (68.3%), and fungus alone (41.2%), respectively (Haseeb *et al.*, 2010).

15.4. TUBER CROPS

Tuber crops form an important part of the diet in many areas of the tropics and subtropics. They are most important after cereals and grain legumes. They are mostly grown for the carbohydrates stored in underground storage organs *viz.*, roots or rhizomes. In addition, the leaves or leaf stalks of some of the tuber crops are rich in protein, vitamins and minerals and are used for either human consumption or cattle feed. Tuber crops are a major source of food and calories in many tropical countries. The annual tropical root crop production (sweet potato and cassava) is in the range of 10.200 million MT in 0.396 million ha with productivity of 20.5 MT/ha (Table 15.20). Root and tuber crops are predominantly grown in the upland or rain fed areas, mainly by resource poor small and marginal farmers. The major tuber crops include cassava (tapioca), sweet potato, colocasia, amorphophyllus, yams and coleus. Cassava and sweet potato are grown to a larger extent in southern and eastern India. Their cultivation has spread to Maharashtra, Gujarat and North-Eastern states as well. The Salem belt in Tamil Nadu and the Samalkot belt in Andhra Pradesh are known for cassava as an industrial crop. Yams and Colocasia are popular in Andhra Pradesh, Tamil Nadu, West Bengal, Uttar Pradesh and Orissa states.

Table 15.20: Area, production and productivity of tuber crops in India (2007-08) (Bijay Kumar, 2009)

Tuber crop	*Area ('000 ha)*	*Production ('000 MT)*	*Productivity (MT/ha)*
Sweet potato	126	1146	9.1
Cassava (Tapioca)	270	9054	33.5
Total	**396**	**10200**	**20.5**

Tuber crops, especially cassava can be successfully substituted as alternative source of feed for livestock in both tropical and subtropical countries. Cassava has also been used as a substrate for the production of ethyl alcohol. The tropical root crops, in general, have a great potential in meeting basic food and energy needs of the developing world. Tuber crops are very efficient in solar energy transfer, having a clear superiority over cereals in biological efficiency and are comparatively free from severe pests and diseases. Cassava and sweet potatoes account for about 30% of the total production of root crops from developing countries. The potential of root crops as nutritionally rich sources of β-carotene, anti-oxidants, dietary fibre and minerals like calcium have begun to be recognized as a result of the multifarious research programmes worldwide.

15.4.1. Colocasia, *Colocasia esculenta*

Colocasia, also known as cocoyam, taro, dasheen and eddoe, is a member of the family Araceae and occurs wild in S.E. Asia. It is cultivated throughout the humid tropics and is of great importance in the Pacific Islands. It is mostly a staple food or subsistence crop grown commercially in some countries.

15.4.1.1. *Rice Root Nematode, Hirschmanniella miticausa and Root Rot, Corticium solani Disease Complex*

A possible interaction occurred with *Hirschmanniella miticausa* and *Corticium solani* causing 'mitimiti' disease of taro, reported by Bridge *et al.* (1983).

Chapter 16

PLANTATION AND SPICE CROPS

16.1. PLANTATION CROPS

Plantation crops serve a variety of human needs such as food, oil, industrial raw materials, beverages and confectionary items. They generate huge employment opportunities directly or indirectly to several million people in their production, processing, marketing and international trade sectors. Plantation crops are a large group of crops. These are grown over an area of 3.226 million ha with production of 12.045 million MT in India (Bijay Kumar, 2009) (Tables 16.1 and 16.2). The plantation crops contributed 5.82% of total horticulture production, in 2007-2008. India is the largest producer and consumer of areca nut, cashew nut, tea, third largest producer of coconut, fourth largest producer and consumer of rubber and sixth largest producer of coffee in the world. While the major plantation crops include coconut, areca nut, oil palm, cashew, coffee, tea and rubber, minor plantation crops are cocoa and betel vine. They play an important role in export as well as domestic requirement, employment generation and poverty alleviation particularly in rural sector. The annual income generated is more than Rs.5,52,590 million and export earning is approximately Rs. 93,350 millions.

Table 16.1: Crop-wise area and production of plantation crops in India (2007-08) (Bijay Kumar, 2009)

Plantation Crops	*Area ('000 ha)*	*Production ('000 MT)*	*Productivity (MT/ha)*	*% Share of India production*
Coconut	1939.9	19893.9	5.6	90.4
Cashew nut	868.0	665.0	0.8	5.5
Areca nut	386.6	476.0	1.2	4.0
Cocoa	31.8	10.6	0.3	0.1
Total	**3226.4**	**12045.5**	**3.7**	

There is a good scope for switching over the production of plantation crops *viz.*, coffee, tea, coconut and areca nut to organic farming, as it is possible to realize higher returns from the unit quantity exported when grown organically. The plantation crops are generally grown in ecologically fragile hilly tracts,

Table 16.2: State-wise area, production and productivity of plantation crops in India (2007-08) (Bijay Kumar, 2009)

State	*Area in '000 ha*	*Production in '000 MT*	*Productivity in MT/ha*	*% Share of India production*
Kerala	1073.7	43.59.9	4.0	36.2
Tamil Nadu	503.9	3810.6	7.5	31.6
Karnataka	679.2	1401.0	2.0	11.6
Andhra Pradesh	289.0	1021.2	3.5	8.5
Maharashtra	190.2	334.1	1.7	2.8
Orissa	182.0	279.8	1.5	2.3
West Bengal	44.1	279.1	6.3	2.3
Assam	104.3	186.1	1.7	1.5
Goa	82.2	120.8	1.4	1.0
Gujarat	20.4	99.2	4.8	0.8
Andaman & Nicobar	25.5	66.9	2.6	0.6
Others	71.8	86.7	1.2	0.7
Total	**3226.3**	**12045.5**	–	

adopting organic farming methods would protect environment and also prevent contamination thereby producing products of high nutritional quality. Further, the plantation crops are amenable for organic farming as they produce huge amount of waste biomass for recycling, which can meet major portion of nutrients required.

Export of Plantation Produce

India is the largest producer in the world of cashew nuts, coconuts and tea. India is largest producer, importer, processor, exporter and second largest consumer of cashew. Indian exports during 2006-2007 were US $ 866 million in tea and coffee (Table 16.3).

Table 16.3: Export of plantation crops (in US $)

Commodity	*Apr-Mar 2006*	*Apr-Mar 2007*	*% Growth*	*% Share*
Tea	390.92	432.00	10.51	0.34
Coffee	358.83	434.77	21.16	0.34
Cashew	585.76	553.41	-5.52	0.44

India is the largest producer and consumer of tea in the world. During 2006-07, India produced 0.95 million MT of tea of which 0.218 million MT was exported. Tea accounts for Rs. 80 billion in turnover in India.

India produced 0.29 million MT of coffee accounting for 4% of the global production. Over 80% of the coffee produced in India is exported.

India is the fourth largest producer of natural rubber with a share of 8.8% in world production in 2006. The yield/ha in India at 1879 kg/ha, was amongst the highest in the world. India produced 0.852 million MT of natural rubber during 2006-07 and exported 56545 MT.

16.1.1. Coffee, *Coffea arabica, C. canephora*

Coffee is an important commercial plantation crop mostly grown on the hilly slopes of the traditional coffee tracts in South India (Karnataka, Kerala, Tamil Nadu and Andhra Pradesh) and non-traditional Orissa, Maharashtra, Madhya Pradesh and North-Eastern Council States. The coffee industry is well established and flourishing enterprise. The major coffee plantations in India are in the Southern States with Karnataka being the major coffee producing state. Recently some coffee plantations have come up in the North Eastern States of India as well. Besides being one of the major foreign exchange earner to a tune of around Rs. 2000 million annually, coffee industry supports more than four lakh people to earn their livelihood. The two main species of coffee in cultivation are *Coffea arabica* (Arabica) and *Coffea canephora* (Robusta). *C. arabica* produces finer seeds which are more valued and grows best at higher elevations (1000 and 1500 m MSL). *C. canephora* is sturdier and better adapted to the warm climate prevailing in the lower elevations (500 and 1000 m MSL). Robusta, in general, is more tolerant to many diseases and pests including nematodes, as compared to Arabica. Coffee in India is grown under a canopy of shade trees of various species along with other plantation crops like black pepper, cardamom, banana, cocoa, etc. The present area under coffee in India is above 342,313 hectares with production of 262,000 MT (2007-08), nearly three-fourth of which is under the fine quality Arabica species with most of the remaining area under Robusta. The productivity of coffee is 0.765 MT/ha. During 2007-08, 218,996 MT of coffee was exported earning Rs. 205 crores of foreign exchange.

India produced 290,000 tonnes of coffee accounting for approximately 4% of the global production. Over 80% of the coffee produced in India is exported. Although over 50 varieties are produced, the popular ones are the Indian Robusta which is used for blending and Arabica. There is a strong demand in the world market for production of high value, high quality, especially organic coffee. Today, the area under certified organic coffee is 2500 ha with an estimated production of 1500 MT of which 1000 MT is being exported currently.

16.1.1.1. *Corky-Root, Meloidogyne arabicida and Fusarium oxysporum Disease Complex*

Coffee corky-root disease, also called corchosis, was first detected in 1974 in a small area of Costa Rica where the root-knot nematode *M. arabicida* is the dominant species. An epidemiological study revealed a constant association between *Meloidogyne* spp. and *Fusarium* sp. in cases of corky root. No corky root appears to have been reported in association with *M. exigua*, which is the prevalent root-knot nematode on coffee in Costa Rica. *Fusarium* spp. are often cited as components of disease complexes in association with nematodes. Combined inoculations using *M. arabicida* or *M. exigua* with *F. oxysporum* under controlled conditions showed that only the combination with *M. arabicida* produced corky-root symptoms on *Coffea arabica* cvs Caturra or Catuai. *F. oxysporum* alone was nonpathogenic. *M. exigua* or *M. arabicida* alone caused galls and reduction in shoot height, but no corky-root symptoms. When cultivars susceptible and resistant to *M. arabicida* were studied under field conditions for 5 years, all the susceptible cultivars exhibited corky-root symptoms on 40-80% of their root systems. Cultivars that were resistant to *M. arabicida* but not to *M. exigua* showed no corky root. These observations lead to the conclusion that corky-root disease has a complex etiology, and emphasize the dominant role of *M. arabicida* as a predisposing agent to subsequent invasion by *F. oxysporum*.

Management methods

Cultivars that were resistant to *M. arabicida* but not to *M. exigua* showed no corky root. Consequently, genetic resistance to *M. arabicida* appears to provide an effective strategy against the disease.

16.1.1.2. *Root-knot Nematode, Meloidogyne incognita and Fusarium oxysporum f. sp. coffeae Disease Complex*

Chlorosis, root necrosis, wilting and stunting of coffee plants were greater in plants inoculated with the fungus, *F. oxysporum* f. sp. *coffeae* four weeks after the root-knot nematode, *M incognita*. Similar but less severe symptoms were observed in plants to which the fungus was added two weeks after the nematode inoculation. Significant differences among treatments were found in height and dry weight of roots and shoots.

Histological studies of root sections revealed giant cell development and hyphal penetration of giant cells, xylem vessels and the female nematode (Negron and Acosta, 1989). They found that *F. oxysporum* f. sp. *coffeae* caused increased root necrosis and chlorosis on the foliage of coffee plants (cv. Bourbón) if the plants had been inoculated with *M. incognita* 2 or 4 weeks previously. Sections taken from the roots of plants preinoculated with *M. incognita* were found to be colonized by *F. oxysporum* f. sp. *coffeae* in a uniquely different way from those where the fungus and nematode were either inoculated simultaneously, or where the fungus was inoculated alone. In the former case, the hyphae of *F. oxysporum* f. sp. *coffeae* were found to be abundant in the xylem vessels, giant cells and female nematodes. Giant cells, colonized by the fungus were in varying states of disrepair, with depleted or partially depleted contents. In comparison, plants that had been simultaneously inoculated with *M. incognita* and *F. oxysporum* f. sp. *coffeae* had fewer giant cells colonized by the fungus and no hyphae within the xylem.

16.1.1.3. *Coffee Decline*

Coffee decline, also known as leaf yellowing of coffee, is a disease caused by infection from a number of plant pathogens. It has caused major economic losses since 1996. Symptoms develop at the beginning of the dry season, when the available soil moisture falls and the damaged root system can no longer support plant growth.

A survey conducted in 1999 found that the organisms associated with this disease include *Pythium vexans* and other *Pythium* species, *Fusarium oxysporum* and the plant parasitic nematodes *Criconomella magnifica, Rotylenchulus reniformis, Pratylenchus coffeae, Helicotylenchus dihystera, Xiphinema difussum* and *Meloidogyne* sp.

Symptoms

- The older leaves of the infected plants turn yellow and eventually drop, with yellowing later progressing to the younger leaves (Fig. 16.1).
- The yield and the quality of the coffee beans is greatly reduced on the branches showing symptoms of decline.
- The feeder roots develop root knots, symptomatic of *Meloidogyne* infection. These infected feeder roots become necrotic, and later the necrosis extends into the tap root. There may be pink discolouration of the collar region of the trunk.
- Adventitious roots develop from the base of the tree along the soil surface, and the plant will die at onset of the dry season.
- Young trees are susceptible to infection and can die even in the wet season.

Fig. 16.1: Decline affected coffee plant

***Geographical distribution*:** Coffee decline has been observed in all of the highland coffee growing regions, but is most serious in Nghe an, Dac lac, Lam dong and Quang tri provinces in Vietnam.

***Host range*:** The Arabica cultivars of coffee are more susceptible than the Robusta group.

***Epidemiology*:** When the nematode population in the soil is very high, there is increased disease pressure on the trees as the nematodes provide an entry point for fungal infection.

Infection occurs through the roots during the rainy season. Young coffee trees planted in the place of old trees are more prone to the disease. This often occurs when 20-year old trees are removed and the second crop cycle begins.

The root system can be weakened during the rainy season from strong winds. The soil surrounding the base of the tree loosens and washes away. The cavity fills with water to provide conditions conducive for disease development.

Management

- When replanting the coffee field, select resistant varieties.
- Rotation with sugarcane, corn or peanuts for 3 years between coffee crops.
- Avoid damaging the root system when applying fertilizer to the base of the coffee trees.
- In the nursery, chemical control of the nematodes or fungi may result in significant reductions in diseased plants, with the combination of nematicide and fungicide being most effective.

16.1.2. Coconut, *Cocos nucifera*

The coconut palm, originated in South East Asia, is now cultivated widely throughout the tropics of the world. The largest producers of coconut are the Philippines, Indonesia, India and Sri Lanka. In all the coconut producing countries, it is an important source of export earning despite being a regular constituent of the food. Coconut is an important small holder's plantation crop. The total area under coconut cultivation in India is 1.939 million ha with an annual production of 10.894 million tonnes and the average productivity of 5.6 tonnes/ha is the highest in the world (Bijay Kumar, 2009) (Table 16.4). It provides proteins, fats and some vitamins. Coconut contributes 700 billion rupees to the GDP of the country. Coir and coir products exported from India earned Rs. 2260 million during the above period. The contribution of the crop to the total edible oil pool in India is around 6 per cent. It is supporting the livelihood of 10 million people across the country. The four Southern States – Kerala, Tamil Nadu, Karnataka and Andhra Pradesh together accounts for 90% of the total area and production. Coconut is predominantly grown under rainfed conditions in Kerala and parts of coastal Karnataka, Maharashtra and Tamil Nadu. In rest of the country, it is mainly grown under irrigated conditions.

16.1.2.1. *Red Ring Nematode, Bursaphelenchus cocophilus*

Introduction

Bursaphelenchus cocophilus (syn. *Rhadinaphelenchus cocophilus*) causes the lethal red ring disease of coconut and other palms. The common name, the red ring

Table 16.4: State-wise area, production and productivity of coconut in India (2007-08)

State	*Area in '000 ha*	*Production in '000 MT*	*Productivity in MT/ha*
Kerala	870.9	4165.9	4.783
Tamil Nadu	374.6	3736.4	9.974
Karnataka	401.0	1118.2	2.787
Andhra Pradesh	105.0	912.4	8.689
West Bengal	25.1	247.1	9.844
Orissa	51.0	189.8	3.721
Maharashtra	21.0	120.5	5.738
Assam	19.0	105.3	5.542
Gujarat	16.4	95.2	5.195
Goa	25.5	87.2	3.419
Andaman & Nicobar	21.4	61.2	2.859
Lakshadweep	2.7	36.5	13.592
Pondicherry	2.1	13.2	6.285
Tripura	3.3	4.8	1.454
Nagaland	0.9	0.1	0.111
Total	1939.9	10893.9	5.615

nematode, is derived from its distinguishing symptom. Symptoms of red ring disease were first described on Trinidad coconut palms in 1905. Palms are important landscape plants in subtropical areas of the United States and the introduction of this nematode would cause great concern for the landscape and tourism industries.

Red ring disease can appear in several species of tropical palms, including date, Canary Island date and Cuban royal, but is most common in oil and coconut palms. The red ring nematode parasitizes the palm weevil *Rhynchophorus palmarum*, which is attracted to fresh trunk wounds and acts as a vector for *B. cocophilus* to uninfected trees.

Distribution

This nematode is distributed in central and south America, and some of the islands in the Caribbean (CAB, 1999), Belize, Brazil, Colombia, Costa Rica, Ecuador, El Salvador, French Guyana, Grenada, Guatemala, Guyana, Honduras, Mexico, Nicaragua, Panama, Peru, Saint Vincent and the Grenadines, Surinam, Trinidad and Tobago, and Venezuela. In some areas, mainly from Mexico to South America and in the lower Antilles, *B. cocophilus* is co-distributed with its primary vector, *R. palmarum*. Although the nematode has been reported to be present in Barbados, Dominica, and Jamaica, these reports have not been confirmed, and EPPO considers it to be absent from these countries. The nematode has not been reported in the continental US, Virgin Islands, or Hawaii. CAB International

(2002) reports *B. cocophilus* in Puerto Rico, however, recent surveys for the red ring nematode in Puerto Rico negated this report. Although some sources report that the nematode is present in the Bahamas, Dominican Republic, and Haiti, this is questionable and requires confirmation. *R. palmarum* has been found in Central and South America and east from some of the West Indies to Cuba.

Host plants

Hosts of *B. cocophilus* are confined to the family Palmae where the nematode is known to infect over 17 species. Most palm species appear to be susceptible to inoculation by red ring nematode but disease severity and symptoms are variable. The most economically important species with red ring disease susceptibility are coconut palm, *Cocos nucifera*, the African oil palm, *Elaeis guineensis*, and the date palm, *Phoenix dactylifera*. Under greenhouse conditions, West Indian royal palm (*Roystonea oleracea*), gru-gru palm (*Acrocomia aculeata*), Moriche palms (*Mauritia flexuosa*), and cucurite palm (*Maximiliana maripa*) were infected artificially.

Economic importance

The nematode can cause losses up to 80%, however, the losses typically range from 10-15% on coconut palms (Esser and Meredith, 1987). This nematode causes serious damage to coconut palms, which are stunted and eventually killed by the nematode infection. Red ring disease is one of the most important wilt diseases of coconut palm in the Neotropics causing up to 10-15% annual losses.

In Trinidad, red ring disease kills 35 percent of young coconut trees. In nearby Tobago, one plantation lost 80 percent of its coconut trees. In Grenada, 22.3 percent of coconut palms were found to be infected. Of those infected, 92 percent had been invaded by palm weevils. It is estimated that 72 percent of those weevils were carrying *B. cocophilus* (Esser and Meredith, 1987).

Considering that more than eight million acres of coconut palms are grown, red ring nematode is one of the most important pests in the tropics. Although *B. cocophilus* and *R. palmarum* are not found in Florida, some other potential beetle vectors of the red ring nematode - *Metamasius hemipterus* and *Rhynchophorus cruentatus* - are common in Florida. If the nematode were introduced to Florida, an epidemic could potentially occur. Therefore, this nematode is of great regulatory concern.

Etiology

Kingdom	:	Animalia
Phylum	:	Nematoda
Class	:	Secernentea
Subclass	:	Tylenchina
Order	:	Aphelenchida

Superfamily	:	Aphelenchoidoidea
Family	:	Parasitaphelenchidae
Subfamily	:	Bursaphelenchinae
Genus	:	*Bursaphelenchus*
Species	:	*Cocophilus*
Binomial name	:	*Bursaphelenchus cocophilus*

Biology and life cycle

The red ring nematode follows a typical plant parasitic life cycle, having 4 molts before becoming an adult. The whole life cycle lasts approximately ten days. The survival stage is the J3 (Blair and Darling, 1968). The dissemination of this nematode depends on its relationship with its vector.

B. cocophilus is associated with the palm weevil, *R. palmarum* which transmits it to the coconut palm. The red ring nematode is co-distributed with the palm weevil in the lower Antilles, and Mexico southward into South America. Reports of detection of *R. palmarum* in Texas and California have not been substantiated by recent survey results. A schematic drawing of the association of *B. cocophilus* with its weevil and coconut hosts is presented in Fig. 16.2. Adult females, which are internally infested by *B. cocophilus*, disperse the dauer juvenile nematodes to

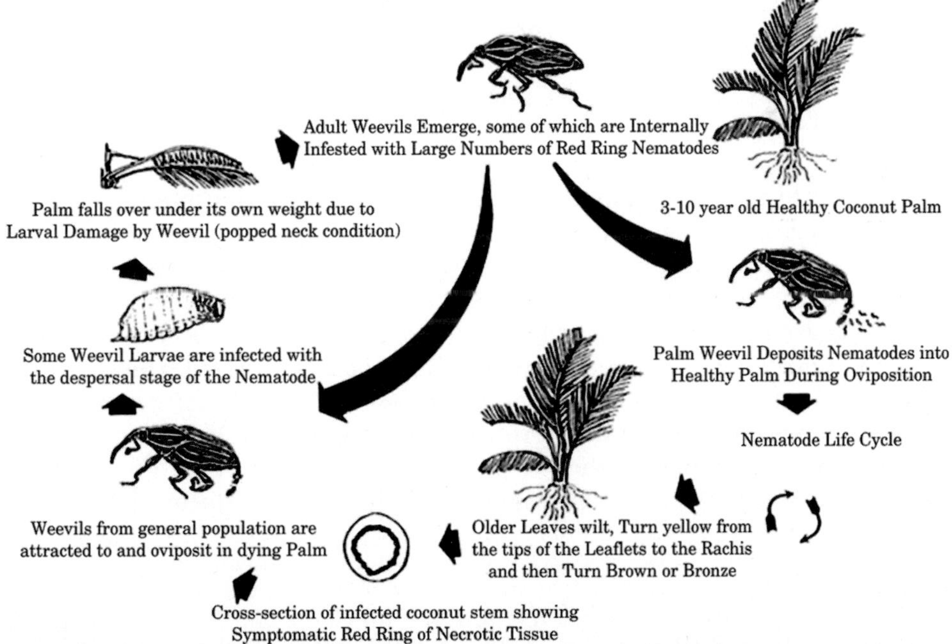

Fig. 16.2: Association of the red ring nematode, *Bursaphelenchus cocophilus* with its vector weevil, *Rhynchophorus palmarum* and coconut host.

healthy coconut palms and deposit the juvenile stage of the nematode during oviposition, usually at the leaf axils or internodes. Nematodes enter oviposition wounds, feed, and reproduce in the palm tissues, causing the death of the infected trees. The weevil larvae are associated with as many as 10,000 juveniles of *B. cocophilus*, which persist in the insect through metamorphosis, apparently without molting, and appear to aggregate around the genital capsule of the adult weevil. The adult weevils emerge from their cocoons in the rotted palm and disperse to apparently healthy or stressed and dying palms, completing the life cycle. Other beetles, *Dynamis borassi* and *Metamasius hemipterus* are also reported to vector the red ring nematode.

The rate of disease development is affected by the host palm species, age, and other factors. The first clear internal symptom in *C. nucifera* is a red ring in the stem that occupies about 70 cm around the inoculation point and occurs between 14 and 21 days post-inoculation. The first external symptoms occur at about 28 days post-inoculation and include leaf yellowing in two to three of the older leaves and premature death of the oldest leaf. The first external symptoms occur in about 30 days regardless of whether the inoculation is made in a wound at the stem base or just below the crown. At 42 days, the entire stem is infected with *B. cocophilus* at close to peak population levels and some of the petioles and roots are infested with nematodes. *B. cocophilus* occurs intercellularly in the ground parenchymal cells adjacent to and within the red ring in the stem, discoloured tissue in the petioles, and in the cortex of the roots. Red ring nematodes do not occur in the xylem or phloem tissues, but xylem vessels become occluded with tyloses where they pass through the red ring. Physical damage caused by nematode feeding compromises the storage capacity and energy dynamics of the palm host and vascular occlusion prevents normal water relations. Experimental inoculations into nuts on inflorescences or onto unwounded stems of coconut do not lead to red ring disease. Inoculations onto wounded or unwounded roots of 3-7 year-old coconut palms do lead to red ring disease. Infested root to healthy root transmission and contamination of wounds in healthy coconut by red ring nematodes carried on the body or feces of *R. palmarum* are occasional modes for the spread of red ring disease. However, *B. cocophilus* inoculation by internally infested *R. palmarum* females during oviposition into healthy, pruned, or wounded palms is probably the most common route for transmission. In a study in Trinidad, over 47% of newly-emerged adult *R. palmarum* from red ring-diseased coconut palms were internally infested with more than 1,000 dauer juveniles of *B. cocophilus*.

Symptoms

Symptoms of red ring disease vary widely with palm species and age, cultivar, and environmental conditions. Palms younger than 2.5 years old cannot be experimentally infected with *B. cocophilus* and red ring disease has not been recorded from palms of this age in the field. Host specific differences in symptoms were consistent in both coconut and African oil palms in cross inoculations, regardless of the host of inoculum origin.

The symptoms produced by this nematode are chlorosis beginning in the oldest leaves and a distinct red/brownish ring in the trunk of the tree. In *C. nucifera*, classical red ring symptoms include premature nut fall (except for mature nuts), withering of inflorescences, and yellowing, bronzing, and death of progressively younger leaves. Yellowing of leaves usually starts at the tips of the pinnae and moves inward to the rachis and then to the base of the petiole. Several of the dying or dead leaves will often break close to the petiole and remain hanging from the stem (Fig. 16.3). A stem transverse section will reveal a discrete brick to brownish-red ring that is 2-6 cm wide and occurs 2-6 cm within the stem periphery (Fig. 16.3). The leaf petioles and cortex of roots can also be discoloured yellow to brownish-red. In longitudinal section, discolouration is usually continuous throughout the length of the stem, appearing as two bands that unite at the base and form discontinuous lesions near the crown.

Fig. 16.3: Left – Red ring disease affected coconut palm. Right – Cross section of coconut stem showing red ring symptoms

Coconut palms 3-10 years-old usually die within 2-4 months of infection. Severe damage to the crown of red ring-diseased coconut palms is caused by larval feeding of the large weevil vector, *R. palmarum*. Older coconut palms (> 20 years old) have been reported with red ring disease displaying less definitive symptoms with a more prolonged death. Dauer juveniles can be harvested from the discoloured tissue of the ring from coconut (up to 11,000 nematodes/g of tissue), from leaf petioles, or roots to confirm disease diagnosis from symptoms. Nematode recovery from African oil palms is highly variable. Chronic little leaf symptoms caused by *B. cocophilus* have also been reported for coconut palms, especially in older trees. Coconut palms begin abnormal production of very short leaves which give crowns the unusual appearance of a feather duster. As the disease progresses, there can be a decrease in leaf size and surface area to the point where the leaf is reduced to a leafless rachis with suberized lesions over most of its surface. New leaves and inflorescences are aborted and palms become unproductive. Red ring nematodes can be recovered from necrotic lesions in the middle and distal parts of unpresented leaves.

Diagnosis or Identification

The distinguishing characteristics of this nematode are a well developed metacarpus from J2 through adult, a short stylet 11-15 µm in adults, adults typically 1 mm in length. Females have the vulva located two-thirds body length and have a vulval flap. Females have a long post uterine sac and a rounded tail. Males have seven papillae in the tail region, distinct spicules, and bursa shaped as a spade (Goodey 1960).

Females and males of *B. cocophilus* are 60-139 and 65-179 times longer than wide, respectively, with the greatest body width being less than 15.5 µm and total length ranging from 812-1369 µm from coconut palms with typical red ring symptoms. The metacarpus and stylet in the second-stage juveniles and adults are well developed (Fig. 16.4). Stylet length is between 11-15 µm in adults. Females have a vulval flap which appears bowed posteriorly when viewed ventrally, a long post-uterine sac (extending about 75% of the vulva-anal distance), and an elongate tail (62-117 µm) with a rounded terminus. Males have seven caudal papillae; one ventral preanal papilla, one pair of subventral preanal or adanal papillae, and two pairs of subventral postanal papillae. The distal ends of the spicules in the males are heavily sclerotized and the caudal alae form a spade-shaped flap (= bursal flap). Third-stage dauer juveniles from coconut palm usually range from 700-920 µm and have a pointed tail with or without a mucron. The metacarpus is usually not well developed in dauer juveniles from the palm or the weevil vector and the stylet is not visible. Survival of dauer juveniles of *B. cocophilus* is very poor under unsterile conditions in soil or water at room temperature or in the refrigerator (100% mortality in < 7 days). Survival can be prolonged for 70-80 days at room temperature when nematodes are surface-sterilized and stored in autoclaved red ring nematode-diseased stem tissue extract (R) or R plus D-glucose or lactose. Red ring nematodes can be inoculated into and cultured in coconut palms older than 2.5 years old, in husks of nearly mature coconut fruits, or in excised leaf stalks (leaf 6 to 13 on a nut-bearing coconut palm) which has been trimmed of the pinnae and the cut ends have been paraffin coated. Cultures in immature fruits or leaf stalks must be subcultured about once every 4 weeks.

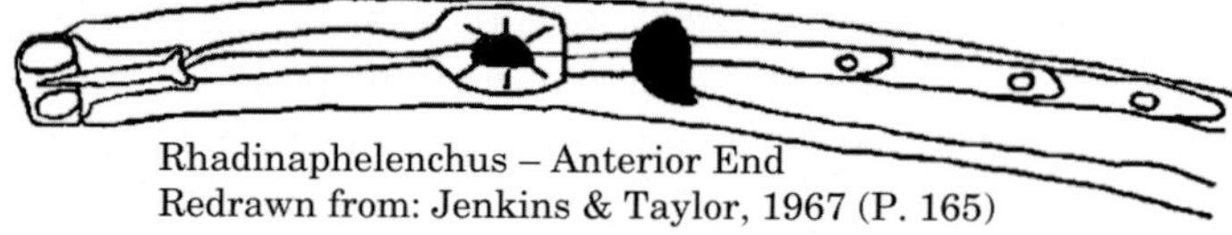

Fig. 16.4: Anterior end of *Bursaphelenchus cocophilus*

Insect Vector relationship

The vector, *R. palmarum* (the palm weevil), carries the J3 stage to healthy palms. Female weevils are internally infested around the oviducts, when they lay their eggs in the palm they also disseminate the nematode (Chinchilla, 1991).

Dissemination and Infection

The main vector of red ring disease is the palm weevil, *R. palmarum*, although vectors such as ants, spiders and other types of weevils also have been reported. The palm weevils are attracted to wounds or cuts in the trunks of the palms. Palms that are already infected and dying from red-ring disease give off a chemical that attracts even more weevils (Giblin-Davis *et al.*, 1996). At an infected palm, a weevil ingests the red ring nematodes or picks them up on the surface of its body. Those nematodes are then left behind at the next palm, usually transmitted as the weevil lays its eggs. The nematodes also can be transmitted by tools that have been used to cut down infected trees. On the body of a weevil or in the soil, red ring nematodes survive less than a week, but they can survive 16 weeks in nut husks and 90 weeks in seedling tissue. The nematodes may also live for long periods within the weevil. On their own, red ring nematodes can move 5.6 mm an hour in soil and almost 0.25 mm an hour in roots (Esser and Meredith, 1987).

Red ring nematodes invade both palm tissue and roots. In leaves, stem and roots, they block water pathways, reducing the palm's water absorption. The heaviest concentration of nematodes can be found within a foot of the highest part of the internal red ring that is a classic symptom of red ring disease; as many as 50,000 have been found in 10 grams of infected stem tissue (Esser, 1969). In the surrounding soil, nematode concentration is generally low. They have been found as deep as 80 cm, but most of those in the soil are 30 cm to 40 cm deep (Chinchilla, 1991).

Management

To manage this disease scouting is the most important aspect; early detection of infected trees may save plantations. If an infected tree is found it must be removed, treated with herbicide and cut down. Leaving the stump behind can lead to vector reproduction and spread of the nematode. Trapping the vector is another strategy; it reduced the disease incidence from 10% to 1% (Oehlschlager, 2002).

Controlling the vector *R. palmarum* can help reduce red ring nematode infestation. Incidence of the disease fell from 10 percent to 1 percent in Mexico with palm weevil control (University of California, Davis, Department of Nematology). Insecticides can reduce palm weevil infestations.

The most useful and most important method for management for red ring nematode is the early removal and destruction of red-ring infested palms. This aggressive phytosanitation is the best chance to halt the spread of red ring disease to other nearby trees. Infested palms should be sprayed with an insecticide and then destroyed as soon as possible once the presence of red ring nematodes has been confirmed. Sometimes weevil larvae will remain in the tissue of palms that are killed with herbicide. These trees should be cut into sections and treated with insecticide or burned (Giblin-Davis, 2001).

Eradication/Suppression Management Strategies

Phytosanitation is still the best method of red ring disease management. This strategy is directed at reducing the vector population as well as the number of sources for nematode inoculum. As soon as palms with red ring disease or *B. cocophilus*-induced little leaf symptoms have been detected they should be destroyed. In coconut palm, the disease can be confirmed by examining stem tissue extracted with a coring device for evidence of discoloured tissue and red ring nematodes. Trees should be sprayed with an insecticide (*i.e.* methomyl) and killed with 100-150 ml (48.3% *a.i.*) of the herbicide monosodium acid methanearsonate (MSMA) or other herbicide that is injected or placed into the trunk. Occasionally trees injected with MSMA will harbour weevil larvae. Therefore, the tree should be cut and sectioned to make sure that weevils are not present. Palms that are heavily infested with weevils should be cut, sectioned, and treated with an insecticide such as methomyl, trichlorfon, monocrotophos, carbofuran, cabaryl, or lindane. Injections of systemic nematicides, such as fenamiphos, oxamyl, and carbofuran into little leaf symptomatic palms can help with palm recovery. However, because of the damage to the very young leaves in little leaf palms the recovery can take between 6-8 months.

The concerted efforts towards aggressive phytosanitation and mass-trapping with traps baited with sugarcane and synthetic aggregation pheromone (Rhyncholure; racemic 6-methyl-2-hepten-4-ol; ChemTica International) reduce the numbers of *R. palmarum* and change their distribution patterns (from highly aggregated to random) while reducing red ring disease incidence. At mass trapping onset, most *R. palmarum* were captured in "border" traps of the test site suggesting removal of potential immigrants into the study area. A combination of perimeter and "internal" traps is most effective for mass trapping. More than 62,500 weevils (94 weevils per ha per month) were captured during one study, with red ring disease incidence decreasing by a factor of > 2.

Research Needs

It is important to determine if *R. cruentatus*, which is endemic to southeastern USA is able to vector the red ring nematode on economically important palms. There is a need for systematic periodic surveys for *B. cocophilus* in the Caribbean and South American countries where the nematode is not known to be present, especially in Jamaica, which is a source of certified seed nuts. Palm weevil pheromones should allow for determining weevil vector presence and red ring nematode contamination at ports of entry in countries without *R. palmarum* and/or *B. cocophilus* and could be useful in settling disputed claims of weevil and/or red ring nematode presence in countries such as the Bahamas, Jamaica, Puerto Rico, Dominican Republic, and Haiti.

16.1.3. Arecanut, *Areca catechu*

Areca nut, otherwise known as the betel nut, has its origin in the humid regions of Asia and Malay Islands. India continues to dominate the world in area, production and productivity of arecanut and has achieved self sufficiency in arecanut production. It is an important cash crop of India. India is the largest producer of areca nut in the world covering an area of 3.866 lakh ha with annual production of 4.760 lakh tonnes (Bijay Kumar, 2009) (Table 16.5). Most of the production is domestically consumed. The major areca nut producing states include Karnataka, Kerala, Assam, West Bengal, Meghalaya and Tamil Nadu. It is masticatory and the ripe fruits are sometimes used as an antihelminthic and astringent in Europe. It has a toning effect and stimulates the nerves. Areca nut industry forms the economic backbone of nearly 6 million people in India and for many of them it is the sole means of livelihood. The extract after boiling the nuts contains tannins which are used for tanning leather.

Table 16.5: State-wise area, production and productivity of arecanut in India (2007-08)

State	*Area in '000 ha*	*Production in '000 MT*	*Productivity in MT/ha*
West Bengal	9.0	22.0	2.444
Meghalaya	12.1	16.5	1.363
Tamil Nadu	4.9	9.0	1.836
Tripura	3.4	6.9	2.029
Andaman & Nicobar	4.1	5.7	1.390
Mizoram	2.0	5.3	2.750
Maharashtra	2.2	3.6	1.636
Goa	1.7	2.6	1.529
Nagaland	0.2	1.3	6.500
Andhra Pradesh	0.3	0.2	0.666
Pondicherry	0.1	0.1	1.000
Total	**386.6**	**476.0**	**1.231**

16.1.3.1. *Burrowing Nematode, Radopholus similis and Root Rot, Cylindrocarpon obtusisporum Disease Interactions*

A direct correlation between levels of *R. similis* present in soil and growth and intensity of root lesions in areca nut seedlings was recorded. The nematode in combination with the fungus *C. obtusisporum* caused more damage in inoculated plants where the nematode was introduced three weeks prior to the fungus (Sundararaju and Koshy, 1987).

16.1.4. Oil Palm, *Elaeis guineensis*

16.1.4.1. *Red Ring Disease*

The most important disease of oil palm in Central America is caused by the nematode *Bursaphelenchus cocophilus*: losses of 5-15% in plantations have been rather common in several countries of Tropical America. Oil palms showing typical symptoms of the red ring disease were found for the first time in Quepos, Costa Rica, in 1984.

The symptomatology of the Little Leaf disease in oil palm was not recognized as associated with *B. cocophilus* in Central America until 1986 (Chinchilla and Richardson, 1987). Nevertheless, these symptoms were observed since the early eighties, and it is believed that the disease has been present since the beginning of the seventies. About the same time (1980), it was also observed that some oil palm plants showed signs of little leaf in plantations in the Pacific coast, but no mention was made as to the age of the plants, nor was the symptomatology detailed.

Symptoms

One of the characteristics of this nematode is the great variability of symptoms it can cause in the oil palm plants affected. Generally, in oil palms, the disease is more common in plants over 5-6 years of age. Several attempts to infect nursery plants of oil palm have failed (Malagutti, 1953; Dao and Oostembrink, 1967).

The nematode *B. cocophilus* has been associated with at least three types of symptoms in oil palms in Costa Rica, Honduras and other Latin american countries. Nevertheless, any combination of symptoms may occur in a particular infected palm.

(i) *Classic or typical symptomatology*

In oil palm, the symptomatology that is considered as classic is produced when older and sometimes intermediate leaves, turn yellowish and progressively become dry (Fig. 16.4). These symptoms then advance, affecting younger leaves. The oldest leaves usually break at the petiole, a short distance from the trunk, and the part of the leaf farthest away from the trunk remains hanging down for a long period. When the trunk of these palms is cut transversely, a brownish, pink, rose or beige ring, a few centimeters in width and located near the periphery of the trunk, may be seen. In some cases, the ring is not continuous throughout the length of the trunk (Fig. 16.5), and it may appear at the top

Fig. 16.5: Left - Oil palm plant infected with red ring nematode. Right-cross section of oil palm stem showing red ring caused by *Bursaphelenchus cocophilus*

third of the trunk, but it is apparently nonexistent at its intermediate region, although it may reappear at the bottom part. The ring may also be evident in a small section of the top or bottom parts of the trunk. Generally, the new leaves are of a yellowish pale green color and shorter than usual. Nevertheless, when the palm has been attacked severely by the palm weevil (*R. palmarum*), the apical region is partially destroyed. Once the plant is infected, the palm may die in 2-3 months.

(ii) *Little leaf symptoms*

Another symptom observed is the condition known as "Little Leaf", where most leaves preserve their green colour and no type of necrosis is present in the stems of the affected palms. Initially, the plant starts producing very short leaves and the center of the crown takes a compact appearance. Eventually, as new short leaves are produced, which may be simple stumps, the central part of the crown takes the appearance of a funnel. As the disease progresses, all new leaves issued are short and deformed, showing different degrees of necrosis of the leaflets, and abnormal degrees of suberization of the rachis (Fig. 16.6). The production of small leaves as another symptom caused by *R. cocophilus* has also been described in Surinam, Brazil and Venezuela (Malagutti, 1953; Van Hoof and Seinhorst, 1962; Schuiling and Dinther, 1981).

Fig. 16.6: Oil palm plants showing little leaf symptoms

The largest population of nematodes is usually located in the young leaves, during their stage of rapid elongation (negative leaves -11 to -2), and they frequently appear in greater numbers in the leaflets of the central and apical parts of the leaf. These leaves, when issued, present a necrosis (drying) of the leaflets much more generalized around the central and/or apical parts, which corresponds to the zone where the greatest population of nematodes are located. These leaflets, besides being partially necrosed, do not open completely and remain partially folded along the rachis. Active nematodes were located in leaves as young as those in position -25 to -28. Many nematodes in these negative leaves appear to have a semi-ectoparasitic behaviour.

While the disease is progressing, all new leaves in the plant are small and deformed, which makes the plant appear as a giant duster. In cuts from the petioles and rachis of the infected negative leaves, many yellowish-orange spots may be seen. At the zone immediately below the main meristem, dispersed yellowish stains may also be seen.

The disease presents itself chronically, and the palm may remain in this condition for several years. A very low percentage of the plants recover, producing some normal-sized leaves. Nevertheless, most of these palms become diseased again, and they initiate a new cycle of small-sized leaf production. The disease causes a pronounced retardation of the longitudinal growth of the stem, wherefore, the palms which have been diseased for three or more years are notoriously shorter than their healthy neighbours.

The development of inflorescences in the leaf axis abort. When the plant has bunches already formed at the time of the first symptoms, these continue their development, but, as the disease progresses, bunch failure invariably makes its appearance. In some of these palms, when their trunk is cut near the base, necrotic dark brown, almost black spots, and (sometimes) a similarly-coloured ring more or less defined, may be seen. Generally, this ring only takes up a very limited portion of a longitudinal section of the stem. In oil palm, it is evident that age is not involved in the development of a particular type of symptomatology.

The presence of small leaves must not be taken as the sole indicator *for R. cocophilus* infection, as there are other causes that may bear forth this symptom. Among these causes are: recovery after an attack of common spear rot, Fusarium wilt, attack at the whorl region by some insects, boron deficiency, etc.

Maas (1970) in Surinam performed several controlled inoculations in oil palm, and observed that the disease was lethal only in very vigorous plants. Less vigorous plants developed the little leaf symptomatology. This type of relation is not apparent in Central America, where the Little Leaf symptomatology is developed, regardless of the apparent vigour of the attacked plant.

Plants with little leaf, five or more years old, are frequently observed as a consequence of an attack of common spear rot and, in some of these cases, the presence of small leaves is a sign of recovery of the diseased plant. Nevertheless, many of these plants will never produce normal leaves again, but will enter a chronic phase of the little leaf disease. It has sometimes been observed that there are cases where a previous common spear rot attack attracts the weevil, which in turn inoculates the plant with the nematode. When the infection by the nematode occurs in a plant with a severe attack of common spear rot, the plant may or may not show any external symptom that indicates the presence of the nematode. In some rare cases, the shoot of these plants may be totally destroyed, and the plant may even die following the sequence common to this type of disease (common spear rot), which may occur in 2 or 3 months. When the trunk is cut, the development of a ring or stains, from where the nematode can be obtained, may be seen.

(iii) *Red ring - little leaf symptomatology*

A combination of the symptoms described earlier, can be seen when the younger leaves are of a pale green colour, shorter and more erect than usual, and appear forming a compact mass. The primordial inflorescences are necrosed and the bunches in formation start to rot, or else are small and of irregular ripening. Some of the youngest leaves are extremely small, or are reduced to mere stumps, and the leaflets present different degrees of a necrosis that develops starting from the tips, especially in the leaflets located in the central part of the leaf. Often, some small leaves will not present apparent necrosis, but an arching of the rachis near the apical end. The oldest leaves may remain green for a long time, but a yellowing of the intermediate leaves will eventually occur, and the symptoms will become generalized, thus causing the death of the plant.

In a lengthwise cut, the rachis of the youngest leaves, especially the negative ones, presents a yellowish-orange tinge that increases its intensity after the cutting is performed. Some of the intermediate and lower leaves-petioles show necrotic dark brown stains in their internal parts. Palms that present this symptomatology often develop an intense orange colouration in the external part of the petioles of the older leaves, but similar types of colourations are also observed in plants with nutritional problems.

When the trunk of these palms is transversally cut, different types of internal necrosis can be seen. Depending on the section of the trunk in which the cut is performed, one or more discontinuous concentric rings, or small necrotic spots without a definite distribution pattern, can be seen. Sometimes, the central part of the trunk is occupied by a dark brown or yellowish necrosis several centimeters in radius with a darker brown border. This symptomatology probably results from the death of the tissues within the ring. In severe cases, this central necrosis of the trunk can be found in different degrees of decomposition, degenerating itself eventually into an aqueous and pestilent rotting. The disintegration of the tissues within the ring is frequently present in older palms that have been affected for several years. The disintegration of the tissue within the ring in the stem or trunk was also observed by Maas (1970) in Surinam.

By performing a lengthwise cut on the stem, it is possible to observe why, depending on where the cut was made, the trunk may appear: a) apparently healthy, b) with a defined ring, c) one or more discontinuous rings, d) a necrosed central area, or else e) dark spots within the central region. The development of the symptoms in the trunk may apparently progress upwards or downwards, which makes it necessary in some cases to make transversal cuts in different sections of the trunk in order to be able to observe the symptoms. This lack of relation between the type and extension of the necrosis in the stem and the intensity of the external symptoms in oil palm, was also observed by Schuiling and Dinther (1981) in Brazil, who also noted that the presence of a few internal stains within the trunk did not necessarily imply the posterior development of a ring in this particular tissue. In some occasions, the palms thus affected enter a chronic stage of the disease in which the leaves do not become yellow nor desiccated, but the tree continues producing little leaves for three years or more.

Distribution and incidence

The disease associated with *B. cocophilus* is considered potentially one of the most serious threats to oil palm in Central America, and it is generalized enough in adult plantations. In general, within areas of young palms (belonging to the age group 10 years and under), it is found that the disease is not that common, (0.1% diseased palms/hectare or less) and in older areas (20 years) it normally does not progress much. Nevertheless, the incidence of diseased plants may be elevated in areas 11-16 years of age. Due to the fact that in most cases the genetic origin of the older plantations in Central America is not known, it is not possible to characterize genetically the more susceptible progenies. The biggest increase in the incidence of the disease occurred in Honduras, specifically regarding some material brought in directly from Africa and planted between 1973-74. Some of these lots have 10-20% diseased palms in some specific areas.

The vector: *Rhynchophorus palmarum*

The adults of the palm weevil *R. palmarum* (Coleoptera: Curculionidae) present an ample variation in size, ranging from 20 mm to 41 mm in length (body length determined from the last abdominal segment to the anterior part of the head, excluding the rostrum). The mean body length may depend on the area where the insect is collected: 34 mm in Costa Rica and 31 mm in Honduras (Chinchilla *et al*, 1991; Morales and Chinchilla, 1991).

The sexes may generally be differentiated, the male bearing a pubescent tuft over its rostrum. This characteristic, nevertheless, can not be used with the smaller specimens (usually less than 29 mm) as it is absent in the male. The separation of sexes must be made, in these cases, checking the genitalia.

Depending on the type of trap used to capture the adults in the field, there may be a prevalence of males or females. When pieces of oil palm stems were used more males were captured, but in field trials with the aggregation pheromone of the male, more females were obtained. Nevertheless, in all studies, mean body length in males was larger than in females. The life cycle, from egg to adult, occurs within 80-160 days and the adult may live for three months. Both the mating period and the egg-laying period occur within 14 days (Griffith, 1968). According to Hagley (1963), approximately 73% of the eggs were fertile. The oviposition period takes 9-11 days; some females lay up to 60 eggs, during the first three days. Copulation may occur between insects recently emerged from the pupa, and takes about 3 minutes. Egg hatching occurs in 3 days, and then follow 9 larval stages (60 days), a pre-pupal stage, and finally the pupal stage. Adults are more active early in the morning and late in the afternoon.

The importance of *R. palmarum* as a primary pest of oil palms is sometimes questionable, and it is generally found attacking palms that have been physically injured (tools, rat attacks, etc.), or have been affected by some type of disease that causes fermentation of the tissues, as is the case of Wet Basal Rot and Common Spear Rot. The development of a high number of larvae within these plants may aggravate the disease symptoms and accelerate the death of the palm. In the case of the spear rot in adult palms, where the population of *R. palmarum* is elevated, the insect invades the shoot of the affected plant and, depending on the number of larvae, may cause the death of the palm without allowing the problem to be detected in time.

Natural enemies

Very few studies have been done on the natural regulating factors of the different stages of the life cycle of *R. palmarum*. It is known that elevated populations of larvae found in a trunk provoke cannibalism. Also, a high mortality rate in larvae is produced after the attacked tissue begins to rot.

Some lizards, and perhaps toads, feed off the adults. In Trinidad, Griffith (1969b) found that a bacteria *Micrococcus* sp. attacked larvae, causing their death in a few days. It is also known that the nematode *Neoaplectana* sp. destroys the adults, pupae and even larvae of the insect (Blair, 1970).

The nematode *Rhabditis* sp. has been found to multiply saprophytically in decomposing diseased coconut palm tree tissue, and it is then acquired *by R. palmarum*, adversely affecting its longevity (Griffith, 1968). Another nematode, *Praecocilenchus* sp, is considered a true parasite of *R. cocophilus* (Nickele, 1974; Morales and Chinchilla, 1991).

Causal agent

The causal agent of the disease is the nematode *R. cocophilus*. This is a nematode approximately 1 mm in length, very slender and transparent. Its life cycle comprises an egg stage and four larval stages. The complete life cycle from egg to adult is one of the shortest in the animal kingdom, occurring in only 9-10 days (Blair and Darling, 1968).

It has been observed that the nematodes associated with the little leaf symptoms in oil palm, differ in several morphological characteristics from those associated with the typical symptomatology in coconut palm. The greatest difference between populations is the length of the nematode, which is greater in specimens from coconut palms (Salazar and Chinchilla, 1988). These morphometric differences may well represent only infraspecific variability (Giblin-Davis *et al.*, 1989).

Location of the nematode within the palm, within the vector and in the soil

In oil palms with red ring symptoms in the stem, the nematode is located in the discoloured tissue and within the internal adjacent tissue, which is still apparently healthy. Nevertheless, it is common to find the nematode absent from the trunk, especially where necrotic discrete spots are present, which could indicate the failure of the nematode to establish itself completely within those tissues. Inside the ring area, Schuiling and Dinther (1981) found that the symptoms. In general, in the necrotic tissue of the petioles, the concentration of nematodes was greater (20-8400 per gram of tissue) and the number of dead individuals was smaller.

The number of nematodes found in the roots and in the soil around diseased trees is generally very low or none at all (Kastelein, 1987). A quantity of nematodes between 0 and 20/100 g of soil was found in the area around infected coconut palms. The nematode was located as deep as 80 centimeters, but most of them were 30-40 centimeters deep in the soil. During the rainy season, the nematodes were located nearer the soil surface.

Several other studies performed in oil palms, have not detected the presence of the nematode in roots and in the soil near diseased palms (Schuilin and Dinther, 1981). In Quepos (Costa Rica) and San Alejo (Honduras) repeated sampling of the roots and soil adjacent to the palms suffering Little Leaf was carried out, and not one specimen of *B. cocophilus* was found.

Survival of the nematode

Healthy coconut and wild palm trees recently felled may be easily colonized by *R. cocophilus* if visited by *R. palmarum* contaminated with the nematode (Maas, 1970). This situation explains why the occurrence of coconut or oil palms with red ring is not essential to the survival neither of the nematode nor for the contamination of the vector. Wild palm trees are frequently felled and the trunk abandoned as a normal practice in Central America (to extract the heart of the palm or to use the trunk or foliage for other purpose).

The most persistent form of the nematode is the third larval stage, characterized by a tapered terminal end of the body. This larval form (0.84 mm) survives in tissues in decomposition for up to three months, during which it may be acquired by adults or larvae of *R. palmarum*. The third larval stage is also present within the vector and it is the infective stage. The nematode does not suffer changes nor does it multiply inside *R. palmarum*, but it can survive the insect metamorphosis (Griffith, 1968).

The nematode may be located inside the intestines, the body cavity and in the excrement of the vector. Externally it can be transported in pieces of infected tissue caught in the hairs of the insect (Hagley, 1963, 1965; Griffith, 1968; Blair, 1970). The nematodes present inside the body of the insect are generally alive and are all the same size; but those that appear on the external part of the vector represent different development stages (Griffith, 1968; Schuiling and Dinther, 1981).

Griffith (1968) found in Trinidad that 80% of the adults *of R. palmarum* transported the nematode within the body cavity and all belonged to the third larval stage. Approximately 67% of the nematodes inside the insect larvae were located in the trachea and the majority were expelled during a molt. Nearly 50% of the nematodes inside the body of the vector survived the molt to adulthood (Griffith, 1969b).

R. palmarum as a vector and beginning of the infection

There is no doubt about the role of *R. palmarum* as an active vector *of B. cocophilus*, but the presence of insects contaminated with the nematode do not necessarily imply the presence and development of the red ring disease in oil palms. Some authors have informed about the disease without the apparent presence of *R. palmarum* within the plantation (Fig. 16.7) (Malagutti, 1953; Dao and Oostenbrink, 1967). Kraaijenga and Ouden (1966) found that in Surinam, the insects were frequently infected with *R. cocophilus*, but the disease was rare in oil palms, which indicated that the nematode was probably endemic in wild palms. The low incidence of the disease was explained by the presence of a prolonged dry period, which is repeated twice a year in Surinam, and which could be detrimental for nematodes that are transported externally by the vector.

Fig. 16.7: The palm weevil, *Rhynchophorus palmarum*

The percentage of adult insects that carry the nematode internally and/or externally varies amply from one place to another and appears to be strongly influenced by environmental conditions and by the quantity of cultivated or wild palm trees that act as reservoirs for the nematode (Table 16.6).

Table 16.6: The percentage of adult insects that carry the nematode internally and/or externally

Percent	***Type of contamination***	***Location***	***Observations***	***Reference(s)***
38.5	External	Trinidad	Up to 71 nematodes /insect	Hagley, 1963
50.0	External + internal	Trinidad		Cobb cited by Hagley, 1963
72.0	External	Trinidad		Fenwik, 1962
76.0			13% with 50 or more nematodes	Fenwik, 1962
72.0		Grenada		Singh, 1972
3.9		Brazil		Luchini, s.f.
16.3		Brazil	6.1% with 2 or more nematodes	Fenwick cited by Hagley, 1963
40	Internal + external		Blair, 1970b	
10.9	5 times more internal contamination than external	Brazil	0.5-3% insects infected by trap (5 day collect) 40-6500 nematodes in adult insects and 50-100 in larvae. 16.5-28.1 number of adult insects per trap in five days	Schuiling & Dinther, 1981 Kastelein, 1987
1.76	Internal + external	Costa Rica	Population peak in the dry season. Higher % of contaminated insects, last months of rainy season	Morales & Chinchilla, 1991
30-68	Internal + external	Honduras	Population behavior like in Costa Rica. Most insects trapped in palms of intermediate age.	Chinchilla, *et al*, 1991.

Fenwick (1967) found that reducing the population of weevils by applying insecticides to the axillae of the coconut palm leaves reduced not only the population of the insect, but the appearance of new cases of the disease as well. Nevertheless, the reduction in the vector population was not always related to a proportional fall in the incidence of the disease. A clear relationship was not found between the fluctuations in the population of the insect and the appearance of new cases of red ring. Other observations indicated that the total amount of insects present in a plantation and the percentage of the insects that were contaminated, were not necessarily correlated with the incidence of red ring disease in oil palms. In Surinam, Maas (1970) found 27% of insects to be contaminated, but the incidence of the disease was extremely low during the study period in 1969. In 1962, Van Hoof and

Seinhorst (1962) had found only 7% of insects infected and also a very low incidence of the disease. Considering the incubation period (3.5 months), there might be a close relationship between the fluctuation of the population of the infected vectors, and disease incidence (Morales and Chinchilla, 1991).

There is still a fair amount of controversy on the form of contamination of a healthy oil palm by the nematode. The fact that the disease is more common in low and poorly drained areas, and that the nematode may be found in the root area and in wet soil, induced several authors in the past to suggest a possible root transmission of the nematode. This possibility was supported by the fact that the inoculation of one sole root caused the development of the typical symptomatology in an originally healthy tree (Fenwick, 1968).

Malagutti (1953) in Venezuela, performed crossed inoculations with populations of nematodes obtained from coconut and oil palm, thus obtaining the development of the typical symptomatology of the disease in 80-90 days in both plants. Inoculations performed in one year old palms did not reproduce the disease.

Other palms that may be attacked by *B. cocophilus* are: *Attalea* sp., *Mauritia flexuosa*, *Maximiliana maripa*, *Roystonea oleracea*, *Acrocomia aculeata*, and *Oenocarpus distichus* (Schuiling and Dinther, 1981).

Disease epidemiology

Schuiling and Dinther (1981) measured the changes in populations of *R. palmarum* in Paricatuba, Brazil in an oil palm plantation, by means of traps, and they found a negative correlation between the abundance of insects contaminated by the nematode and rainfall: 9.7% contamination in the dry season and 3.9% in the rainy season. Peaks of maximum incidence of the red ring disease were observed approximately five months after the appearance of an increase in the number of insects contaminated with *R. cocophilus*. These five months were considered by the authors as the incubation period of the disease.

In Central America, the *R. palmarum* population associated with oil palm reaches its peak during the dry season and the percentage of insects that carry the nematode is greater during the last months of the rainy season (Chinchilla *et al*, 1991; Morales and Chinchilla, 1991). A lower proportion of contaminated, field collected adults, during the last months of the dry season, may be the consequence of this being a population of old insects, that have lost most of their nematodes in the environment.

Schuiling and Dinther (1981) did not find evidence of the development of the disease within infection foci in the field and only in 5% of the cases were the affected palms adjacent to another palm bearing the symptoms. The incidence of the disease in the study area was between 0.4 and 6 diseased palms per hectare. In Central America, the most common situation is a uniform distribution of new cases in the plantation. Normally, a group of three or four, non contiguous plants (separated by one or two healthy trees) form a new foci. Studies performed in Honduras and Costa Rica have shown that the population density of *R. palmarum* in palms 5-6 years old was considerably lower than the population found in areas 10 years and older. Higher populations of the vector in adult palms, are responsible in part of a higher disease incidence in these areas.

Management methods

Disease control must be integrated and directed as much towards decreasing the vector population, as towards the reduction of the sources of nematodes within the plantation and its surroundings.

In the case of plants with severe symptoms, it is recommended to poison the plant with a systemic herbicide injected in the trunk, and felling the plant after it has dried. It has been observed that older plants can be easily poisoned with 100 ml of MSMA (Bueno 6), but higher doses are required (125-150 ml) to kill some of the intermediate-aged plants with a vigorous growth rate. A mixture of Picloran and 2,4-D has also been recommended for plant elimination. Even though the insecticide does not kill the nematode directly, it does help interrupt the transmission cycle making the treated palm less attractive to *R. palmarum*. In spite of this unattractiveness, it has been occasionally observed that some palms injected with MSMA were colonized by weevil larvae, wherefore the treated palm should be felled once they are dry, and be checked for the presence of larvae for their elimination. As the tissues rot, the nematode will eventually die.

When a palm is severely attacked by the weevil, it must be felled and cut in sections, which are then cut open lengthwise; an insecticide is then applied, such as Furadan, Sevin (carbaryl), Dipterex (Triclorfon), Lannate (Methomyl), Vydate (Oxamyl), Nemacur (phenamiphos), Azodrin (monocrotophos), etc. In the case of palms which present symptoms of little leaf without extensive necrosis of the trunk, there is, apparently, the possibility of recovery by means of using trunk-injected systemic nematicides, which may be also applied directly in the whorl of the plant. In several preliminary tests, it was observed that products like Carbofuran (Furadan), Oxamyl (Vydate) and Fenamifos (Nemacur) permitted the apparent recovery from the symptoms of little leaf in a percentage of the treated palms. The visible response took around 6-8 months to appear, due to the fact that the young leaves that were affected during their active growth period always came out deformed. The lack of response to the treatment may be due to several factors, but it is obvious that, if there is extensive damage done to the trunk of the palm by the disease, recovery will be literally impossible.

In several tests performed on oil palms in Honduras, treatments with soil-applied Temik 10 G (80-200 g/plant), trunk-applied (10 g/plant), and axilla-applied (30 g/plant) (for plants with symptoms of little leaf) were entirely ineffective.

However, when Temik was applied directly to the whorl, many plants recovered from the little leaf symptoms.

The attraction of *R. palmarum* adults by certain types of recently exposed internal tissues, especially pieces of stem from diverse palm trees, has been used against weevil populations very successfully. Adults that reach these traps may be collected by hand, or else the pieces of tissue may be impregnated with an insecticide such as Lannate, Sevin, Furadan, Vydate, Nemacur, Dipterex, etc. Pieces of oil palm trunks are generally more efficient as weevil traps than coconut palm trunk pieces (Kraaijenga and Ouden, 1966) and, according to Fenwick (1966) it is better to use pieces of non-infected trees. Traps made with pieces of plantain or banana pseudo-stems, crushed sugar cane stalks, or ripe fruits, are not so attractive to *R. palmarum*. Some strong smelling insecticides (*i.e.* Cytrolane) have a deterrent effect on adults of *R. palmarum*. These insecticides should not be used in traps, but can be very useful to protect the cuts of recently fell diseased palms.

Whereas the weevil is attracted by any type of wounds, these must be avoided at any cost, especially during harvest. Spear rot must also be given special attention, especially in palms that have entered the stage of susceptibility to the nematode (5-6 years); in these cases, it is advisable to treat the diseased part with an insecticide to avoid risking visits from the vector insect.

Natural recovery of diseased plants and resistance

It has been observed that some plants with the little leaf symptoms begin a cycle of normal-sized leaf production and that, after a certain number of months, they renew all their foliage. In some cases, this "recovery" is temporary and the plant again initiates a little-leaf production cycle. Van Hoof and Seinhorst (1962) also observed the phenomenon of the recovery of the symptomatology of little leaf in oil palms in Surinam. Oil palms with the typical symptomatology of red ring does not recover and the plant invariably dies after a few months.

In Honduras, it has been observed that the nematode may be found in oil palm even in leaves as young as the -25 leaf, which indicates that a natural or induced visible process of recovery could even take a year, in a palm with a rate of foliar emission of 2 leaves per month. Van Hoof and Seinhorst (1962) found the first signs of nematode injury in leaves 2.2 to 4 centimetres long, all of which presented small yellowish patches over the petiole and the base. The nematode was found even in leaves 1.75 metres long with the leaflets still folded and protected from the sun. The nematode lives in these tissues ectoparasitically. This type of behaviour could explain why these nematodes are more susceptible to prolonged dry periods.

There are no concrete studies that show the resistance or tolerance to the disease of oil palm (*E. guineensis* and *E. oleifera*).

Schuiling and Dinther (1981) observed in Brazil that in some plants, the development of the symptoms was detained immediately after the youngest leaves came out shorter and with a more compact appearance. The oldest leaves remained green and the inflorescences and bunches continued their development. These plants maintained this condition for one year or more, and the authors considered it a tolerance to the nematode. It was also found that the frequency of this type of response was augmented with age.

16.1.5. Betel Vine, *Piper betel*

Betel vine is a perennial, dioecious, evergreen creeper grown in India. It is a chewing stimulant of brain, lungs and heart. The leaves contain B and C vitamins. It is an important cash crop in Andhra Pradesh, Assam, Bihar, Karnataka, Kerala, Madhya Pradesh, Maharashtra, Orissa, Tamil Nadu, Tripura, Uttar Pradesh and West Bengal with an annual turnover of about Rs. 700 crores. Its leaves are exported to Pakistan, Bangladesh, Indonesia, Malaysia, Burma and Thailand.

16.1.5.1. *Root-knot Nematode, Meloidogyne incognita and Foot Rot, Phytophthora nicotianae var. parasitica Disease Complex*

The population of root-knot nematode had been found to be positively correlated with *Phytophthora* wilt disease incidence in diseased betel vine gardens (Fig. 16.8) (Sitaramaiah and Parvathi Devi, 1994).

Fig. 16.8: Root-knot nematode and foot rot disease complex in betel vine

Management methods

(i) *Chemical methods*

Venkata Rao *et al.* (1973) used nematicides like D-D and DBCP for the control of betel vine wilt caused by an association of *M. incognita acrita* and *P. nicotianae* var. *parasitica*.

(ii) *Integrated methods*

(a) *Bioagents and botanicals*: Spot application of neem cake enriched with *P. lilacinus* at 3 MT/ha in 3 split doses–1st dose at 45 DAP and remaining 2 splits at 45 days interval during North-East monsoon season (Oct-Dec) gave effective control of the disease complex.

(b) *Two bioagents*: Combined application of *Pseudomonas* sp. (Pfbv 22) and *Bacillus* sp. (Bbv 57) gave significant reduction in nematode infestation (gall index, no. of egg laying females and soil population), wilt disease incidence and increase in leaf yield. The treatment also enhanced the biochemical markers responsible for induced systemic resistance such as peroxidase, polyphenol oxidase and phenylalanine ammonia lyase (Jonathan *et al.*, 2006) (Table 16.7).

Table 16.7: Effect of rhizobacterial formulations on leaf yield, nematode and wilt incidence in betel vine under glasshouse conditions

Treatment/Dose (2.5 x 10^8 cfu/g)	*Plant height (cm)*	*No. of leaves/ vine*	*Nema popn. in 250 ml soil*	*Gall index (0-5 scale)*	*Wilt index (0-5 scale)*
Pseudomonas spp. – Pfbv 22	154	195	168	3.0	2.4
Bacillus spp. – Bbv 57	148	201	160	3.0	2.6
P. fluorescens – Pf 1	135	190	158	3.5	2.4
Pfbv 22 + Bbv 57	165	225	140	2.5	1.7
Pfbv 22 + Pf 1	148	197	160	3.0	2.2
Bbv 57 + Pf 1	160	190	174	3.5	2.4
Pfbv 22 + Bbv 57 + Pf 1	157	191	173	3.0	2.4
Metalaxyl (0.2%) + Carbofuran (2 g/vine)	160	211	149	2.4	1.9
Control	115	163	515	5.0	3.3
CD (P = 0.05)	**8.2**	**6.5**	**6.9**	**0.5**	**0.4**

Integrated methods

- Summer ploughing and exposing the field to sunlight during May prior to sowing of *Sesbania* sp. (live standard) minimizes the initial load of inoculum of both the nematode and the fungus.
- Selection of healthy seed vines from nematode-free and disease-free mother plants for planting.
- Dipping seed vines in 0.25% Bordeaux mixture solution for 5 min before planting.
- Application of FYM at 30 MT/ha to promote multiplication of antagonistic microbes which in turn kills nematodes.

- Spot application of neem cake enriched with *P. lilacinus* at 3 MT/ha in 3 split doses –1st dose at 45 DAP and remaining 2 splits at 45 days interval during North-East monsoon season (Oct-Dec).
- Rotation of betel vine crop with rice.

16.1.5.2. *Nematodes, Meloidogyne incognita, Rotylenchulus reniformis and Foot Rot, Phytophthora capsici Disease Complex*

Jonathan *et al.* (1997) studied the interactions of *M. incognita* and *R. reniformis* with *P. capsici* in betel vine in Tamil Nadu and observed a synergistic interaction between the nematode and fungal pathogens as well as increase in vine mortality.

16.1.5.3. *Root-knot Nematode, Meloidogyne incognita and Foot Rot, Sclerotium rolfsii/Xanthomonas beticola Disease Complex*

Maximum reduction in height and weight of shoot and root of betel vine cv. Godibangla was observed in simultaneous inoculations of *M. incognita* with and *X. beticola/S. rolfsii*. Root rot was observed in all combined treatments (Acharya *et al.*, 1987).

16.2. SPICE CROPS

From time immemorial, India is regarded by the rest of the world as the "magic land of spices". Spices are natural plant products used to improve flavour, aroma, taste and colour of food products, beverages, liquors, pharmaceutical, cosmetic and perfumery products. No other country in the world has such a diverse variety of spice crops as in India. Indian spices are also noted for their excellent aroma, flavour and pungency not so easily matched by any other country. Even in minute quantities, spices are a real delight to all our sensory organs, making the food dishes more palatable. Usage of spices in food processing industries is increasing quite fast in all the countries. Hence, the world demand for spices in recent years is continuously on the rise.

India is the largest producer, consumer and exporter of spices in the world. India is the largest producer in the world of ginger, turmeric and black pepper. The annual production of spices is 4.103 million MT valued at US$ 1,500 million. India's share of the world spices trade is estimated at 45 to 50% by volume and 25% by value. The present annual production of spices in the country is 4.103 million MT from over 2.603 million hectares (Bijay Kumar, 2009) (Tables 16.8 and 16.9). The spices contributed 1.98% of total horticulture production in 2007-2008. About 8.5% of India's export earnings from agricultural and allied products come from spices. Indian spices flavour foods in over 134 countries. The important spice crops include black pepper, cardamom, ginger, garlic, turmeric and chilli.

Table 16.8: Crop-wise area and production of spice crops in India (2007-08) (Bijay Kumar, 2009)

Spice crop	*Area in '000 ha*	*Production in '000 MT*	*Productivity in MT/ha*	*% Share of India production*
Chilli	765	1244	1.6	30
Turmeric	180	830	4.6	20
Garlic	168	810	4.8	20
Ginger	106	376	3.5	9
Coriander	398	243	0.6	6
Tamarind	60	192	3.2	5
Cumin	429	172	0.4	4
Black pepper	242	68	0.3	2
Fennel	49	62	1.3	2
Fenugreek	55	55	1.0	1
Others	152	51	0.3	1
Total	**2603**	**4103**	**1.6**	

Table 16.9: State-wise area, production and productivity of major spices in India (2007-08) (Bijay Kumar, 2009)

State	*Area in '000 ha*	*Production in '000 MT*	*Productivity in MT/ha*	*% Share of India production*
Andhra Pradesh	317.8	1235.2	3.9	30.1
Rajasthan	556.4	520.6	0.9	12.7
Gujarat	299.8	356.8	1.2	8.7
Karnataka	235.2	344.9	1.5	8.4
Tamil Nadu	126.8	279.2	2.2	6.8
Madhya Pradesh	194.6	249.8	1.3	6.1
Orissa	147.0	199.2	1.4	4.9
Uttar Pradesh	56.8	166.9	2.9	4.1
Kerala	312.3	159.3	0.5	3.9
West Bengal	79.0	114.4	1.4	2.8
Maharashtra	114.3	100.2	0.9	2.4
Meghalaya	18.4	80.9	4.4	2.0
Others	144.6	295.3	2.0	7.2
Total	**2603.0**	**4102.7**	**1.6**	

Since spices form a part of traditional/ethnic medicines, the demand for organically produced spices is growing. Export of organic spices from India has started in right earnest. The country at present exports around 50 MT of different varieties of organic spices. Export will get a significant boost in the coming years as more and more farmers switch to organic methods.

Export of spice produce

India is the largest producer, consumer and exporter of spices and spice products in the world and produces more than 50 spices. Spice export figures for the year 1996-97 and 2006-07 shows the phenomenal growth of the Indian spice industry. The Indian spice export for the year 1996-97 was 225,000 MT which was worth $350 million; in the year 2007-08 the spice exports touched the figure of 444,250 MT which was worth US$ 1102 million. Thus, these figures show the exponential growth of the Indian spice industry in a very short time. Spice exports from India have touched US $ one billion mark during 2007-08. The major destination of Indian spice exports is USA followed by European Union, Sri Lanka, Japan and Middle East Countries.

16.2.1. Black Pepper, *Piper nigrum*

Black pepper, known as the "King of spices", has its origin in the South-Western hills of India. It is a branching and climbing perennial shrub belonging to the family Piperaceae and is grown in the hot and humid parts of the world. Pepper is loved for its spicy factor, piperine. It is used in various culinary seasonings, as a preservative for meat and fish, in medicine and perfumery (pepper oil). In India, it is a common practice to use live trees as standards for climbing pepper. Pepper is the most important spice produced in the country and accounts for 50-60% of the total export earnings. It is being cultivated in 0.236 million hectares producing 0.05 million tonnes of black pepper with an average yield of 4.72 tonnes per hectare (Table 16.10). The black pepper export for the year 2007-08 was 35,000 MT which was worth Rs.5195 million.

Table 16.10: State wise area and production of black pepper (2006-07)

States	*Area (ha)*	*Production (MT)*	*Productivity (MT/ha)*
Karnataka	15150	12000	0.792
Kerala	216710	33950	0.156
Tamil Nadu	3700	4000	1.081
Nagaland	7	7	1.000
Andamans	610	40	0.655
Total	**236177**	**49997**	**0.211**

16.2.1.1. *Nematodes, Meloidogyne spp., Radopholus similis and Foot Rot, Phytophthora capsici Disease Complex*

Increased susceptibility of *M. incognita* and *M. javanica* infested cultivars of black pepper to *Phytophthora* infestation has been reported (Fig. 16.9).

(i) *Integrated management*

(a) *Bioagents, botanicals and physical methods*: Integrated management of foot

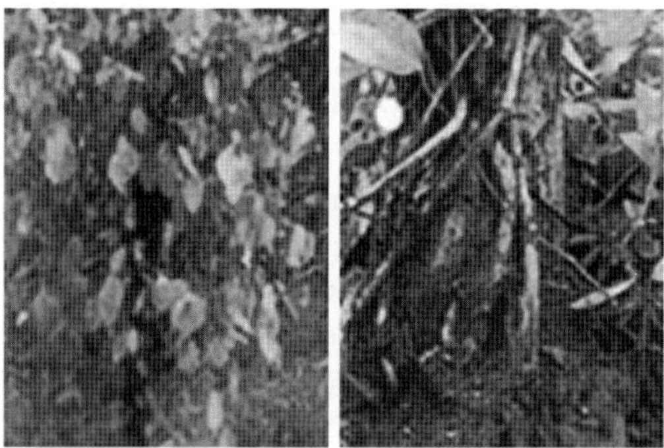

Fig. 16.9: Root-knot nematode and foot rot disease complex in black pepper

rot (*P. capsici*) and nematodes (*M. incognita* and *R. similis*) on black pepper was achieved by:

- Mixing AMF and *Trichoderma harzianum* in solarized nursery mixture to raise healthy and robust seedlings.
- Application of *T. harzianum* and FYM in planting pit.
- Field application of neem cake at 1 kg/vine mixed with 50 g of *T. harzianum* during August.

16.2.1.2. *Pepper Yellows, Meloidogyne incognita and Wilt, Fusarium oxysporum Disease Complex*

Sheela and Venkitesan (1990) found that simultaneous inoculation of *M. incognita* and *Fusarium* spp. led to the suppression of black pepper vine growth.

Management methods

The disease complex was managed by an integration of application of fertilizers of NPK (15: 15: 15) at 250 g/plant/year in combination with aldicarb at 50 g/plant and/or mancozeb at 12 g/plant. Additionally berry yield was also increased (Mustika *et al.*, 1984).

16.2.2. Cardamom, *Elettaria cardamomum*

Cardamom, also known as the "Queen of Spices" enjoys a unique position in the world market has its origin in the evergreen rain forests of Western Ghats of South India. India enjoyed near monopoly in the field till recently. Guatemala has emerged as a competitor to India. It is a perennial plant with an underground stem (rhizome) and aerial shoots and valued for their fruits (capsules). The fruits are small, trilocular capsules containing 15-20 seeds. India and Guatemala are the main producers and exporters of cardamom. Other cardamom growing countries are Tanzania, Sri Lanka, El Salvador, Vietnam, Laos, Kampuchea and Papua New Guinea. The area under cultivation in India during 2006-07 was 73,228 ha, with an annual production of 11,235 metric tonnes (Table 16.11). The export earning during 2007-08 was Rs. 247.5 million by exporting 500 million tonnes of small cardamom. Kerala, Karnataka, and Tamil Nadu produce cardamom in India. While Kerala accounts for 60 per cent of the production, Karnataka contributes 30 per cent, and Tamil Nadu 10 per cent. In Karnataka, cardamom is cultivated mostly in Kodagu, Hassan, and Chikmagalur districts. The plants last for 15 years. They grow in moist, elevated, and shady forest loams. They require a fairly distributed annual rainfall ranging from 2,500 mm. to 5,000 mm. and are found within an altitude ranging between 600 m and 1,200 m. above the mean sea level. It is used for flavouring various food preparations, confectionery, beverages, liquors and medicines.

Table 16.11: State-wise area and production of small cardamom (2006-07)

State	*Area (ha)*	*Production (MT)*	*Productivity (MT/ha)*
Kerala	41362	8545	0.206
Karnataka	26611	1725	0.064
Tamil Nadu	5255	965	0.183
Total	**73228**	**11235**	0.153

16.2.2.1. *Root-knot Nematode, Meloidogyne sp. and Rhizome Rot, Rhizoctonia solani Disease Complex*

M. incognita was found to predispose cardamom seedlings to *R. solani* infection, which causes damping off and rhizome rot in the primary nursery (Ali and Venugopal, 1992, 1993).

Management methods

(i) *Chemical methods*: When both fungicide (metalaxyl) and nematicide (carbofuran) were applied for the control of rhizome rot disease and nematodes, the mortality of seedlings was least.

(ii) *Integrated management*

(a) *Two bioagents*: *P. lilacinus* in combination with *Trichoderma* spp. suppressed *Meloidogyne* spp. and rhizome rot disease (*R. solani*) complex when incorporated in solarized cardamom nursery beds (Eapen and Venugopal, 1995).

(b) *Bioagents, chemicals and physical methods*: Soil solarization alone enhanced the germination by 25.5% and suppressed weed growth by 82.0%. Solarization also enhanced the growth and vigour of cardamom seedlings. The disease complex was suppressed by incorporation of *P. lilacinus*/*T. harzianum* and phorate into the solarized nursery beds. This approach is being adopted on a large scale for the production of nematode-free cardamom seedlings.

16.2.3. Ginger, *Zingiber officinale*

Ginger, an herbaceous perennial, belonging to the family Zingiberaceae, is valued for its rhizome. The country of its origin is not definitely known but is presumed to be either from India or China. India is the largest producer (721539 metric tonnes) and exporter of dry ginger in the world (exported 6700 MT of dry ginger earning of Rs. 280 million). The other ginger producing countries are Jamaica, Sierra Leone, Nigeria, Southern China, Japan, Taiwan and Australia. In India, It is being cultivated in 129014 hectares producing 0.721 million tonnes of ginger with an average yield of 5.5 tonnes per hectare (Table 16.12).

Table 16.12: State-wise area and production of ginger (2006-07)

State	*Area (ha)*	*Production (MT)*	*Productivity (MT/ha)*
Karnataka	20489	198180	9.672
Assam	18180	123990	6.820
Meghalaya	9640	57280	5.941
Kerala	11080	42496	3.835
Arunachal Pradesh	5500	40000	7.272
Gujarat	2969	38140	12.812
Sikkim	6680	35630	5.338
Orissa	16070	31400	1.953
Uttaranchal	2250	27340	12.151
Himachal Pradesh	2396	20590	8.593
West Bengal	9820	20060	2.034
Mizoram	2950	17220	5.837
Tamil Nadu	669	12470	18.639
Nagaland	1225	12250	10.000
Andhra Pradesh	2430	10300	4.238
Bihar	491	7590	15.458
Madhya Pradesh	6020	7410	1.230
Manipur	4270	7050	1.646
Tripura	1410	4170	2.957
Uttar Pradesh	890	2566	2.883
Chhattisgarh	1888	2089	1.106
Andamans & Nicobar	205	1850	9.024
Maharashtra	1280	1230	0.960
Rajasthan	150	170	1.133
Jammu & Kashmir	4	68	17.000
Haryana	58	0	–
Total	**129014**	**721539**	**5.592**

16.2.3.1. *Root-knot Nematode, Meloidogyne incognita and Wilt, Fusarium solani Disease Complex*

Rotting of ginger roots by *F. solani* became more severe in the presence of *M. incognita* (Fig. 16.10) (Doshi and Mathur, 1987).

Fig. 16.10: Rhizome rot of ginger

16.2.3.2. *Lesion Nematode, Pratylenchus coffeae and Fusarium Wilt, Fusarium oxysporum f. sp. zingiberi Disease Complex*

(i) *Biomanagement*

(a) *Antagonistic fungi*: *Trichoderma* spp. and *T. virens* were found to be highly effective in suppressing *Fusarium* yellows and the lesion nematode, *P. coffeae*.

(ii) *Integrated management*

(a) *Bioagents and physical methods*: The efficacy of biocontrol agents like *Aspergillus niger*, *Trichoderma harzianum*, *T. hamatum* and *Gliocladium virens* either alone or in combination with soil solarization has been reported for the management of yellows disease (*Fusarium* + *Pratylenchus coffeae*) in Himachal Pradesh (Dohroo, 1995). Similar results have been reported by Eapen and Ramana (1996) from Kerala.

16.2.3.3. *Root-knot Nematode, Meloidogyne incognita and Damping-off, Pythium aphanidermatum Disease Complex*

Among several nematodes reported on ginger, the root-knot nematode, *M. incognita* is most predominant in all ginger growing areas. The incidence of rhizome rot caused by *P. aphanidermatum* was found to be more severe when rhizomes were infected with *M. incognita* in Himachal Pradesh (Dohoro *et al.*, 1987).

M. incognita had synergistic effect in rhizome rot (*P. aphanidermatum)* of ginger, particularly when nematodes were inoculated to the plants prior to fungal inoculation. The damage to the crop was more and the onset of the disease was early when *M. incognita* was inoculated 40 days prior to fungal inoculation (Table 16.13) (Ramana *et al.*, 1999).

Table 16.13: Effect of *Meloidogyne incognita* and *Pythium aphanidermatum* on yield, nematode multiplication and disease incidence in ginger

Treatment	*Rhizome weight (g)*	*Nema popn./g root*	*Nema popn./100 ml soil*	*% disease incidence*	*Days reqd for max dis devp*
Control	270	–	–	–	–
M. incognita (Mi)	182	1806	124	–	–
P. aphaniderm-atum (Pa)	181	–	–	80	105
Mi + Pa (simultaneously)	119	2218	14	60	105
Mi + Pa (20 days later)	195	1607	41	60	75
Mi + Pa (40 days later)	167	1038	39	80	44
Pa + Mi (40 days later)	127	1416	10	68	105

16.2.3.4. *Root-knot Nematode, Meloidogyne incognita and Bacterial Wilt, Ralstonia solanacearum Disease Complex*

Bacterial wilt of ginger, caused by *R. solanacearum* is influenced by *M. incognita* (Samuel and Mathew, 1983).

REFERENCES

Abawi, G.S. and Barker, K.R. 1984. Effects of cultivar, soil temperature and population levels of *Meloidogyne incognita* on root necrosis and *Fusarium* wilt of tomatoes. *Phytopathology* **74**: 433-438.

Abd-El-Ghafar, N.Y. and Abd-El-Kader, K.M. 1997. Interaction between root-knot nematode (*Meloidogyne incognita*) and *Pseudomonas* (=*Burkholderia*) *solanacearum* on tomato. *Ann. Agri. Sci. Cairo* **42**: 607-617.

Abu-Elamayem, M.M., Shehata, M.R.A., Tantaway, G.A., Ibrahim, I.K. and Schuman, M.A. 1978. Effect of CGA 12223 and benomyl against *Meloidogyne javanica* and *Rhizoctonia solani*. *Phytopathol. Zeist* **92**: 289-293.

Abuzar, S. and Haseeb, A. 2006. Efficacy of carbofuran, bavistin, neem, *Trichoderma harzianum* and *Aspergillus niger* against *Meloidogyne incognita* and *Fusarium oxysporum* f. sp. *vasinfectum* disease complex on okra. *Indian J. Nematol.* **36**: 299-301.

Acharya, A., Dash, H.C. and Padhi, N.N. 1987. Pathogenic association of *Meloidogyne incognita* with *Sclerotium rolfsii* and *Xanthomonas bataticola* on betel vine. *Indian J. Nematol.* **17**: 196-198.

Acosta, N. and Ayala, A. 1976. Effects of *Pratylenchus coffeae* and *Scutellonema bradys* alone and in combination on Guinea yam (*Dioscorea rotunda*). *J. Nematol.* **8**: 315-317.

Akiew, E., Trevorrow, P.R. and Waite, C.A. 1991. Tobacco bacterial wilt investigations in North Queensland. *ACIAR Bacterial Wilt Newsletter* **8**:11-12.

Alam, M.M., Samad, A. and Anver, S. 1990. Interaction between tomato mosaic virus and *Meloidogyne incognita* in tomato. *Nematol. Medit.* **18**: 131-133.

Al Banna, L. and Gardner, S.L. 1993. Three new species of nematodes associated with endemic grape (*Vitis*) in California. *J. Helminth. Soc. Wash.* **60**: 243-249.

Alfaro, A. and Goheen, A.C. 1974. Transmission of strains of grapevine fanleaf virus by *Xiphinema index*. *Plant Dis. Reptr.* **58**: 549-552.

Al-Hazmi, A.S. 1985. Interaction of *Meloidogyne incognita* and *Macrophomina phaseolina* in root disease of French bean. *Phytopathol. Zeist* **113**: 311-316.

Ali, M.A., Trabulsi, I.Y. and Ebd-Elsamea, M.E. 1981. Antagonistic interaction between *Meloidogyne incognita* and *Rhizobium leguminosarum* on cowpea. *Plant Dis.* **65**: 432-435.

Ali, S.S. 1988. Influence of 'katte' mosaic virus of cardamom on the population of *Meloidogyne incognita*. *Nematol. Medit.* **17**: 121-122.

Ali, S.S. and Venugopal, M.N. 1992. *Nematol. Medit.* **20**: 65-66.

Ali, S.S. and Venugopal, M.N. 1993. *Curr. Nematol.* **4**: 19-24.

Allen, W.R., Van Schagen, J.G. and Ebsary, B.A. 1984. Comparative transmission of the peach rosette mosaic virus by Ontario population of *Longidorus diadecturus* and *Xiphinema americanum* (Nematoda: Longidoridae). *Can. J. Plant Pathol.* **6**: 29-32.

Alphey, T.J.W. 1973. Annual Report (1972). *Scot. Hort. Res. Inst.*, p. 20

Anandraj, M., Venugopal, M.N., Veena, S.S., Kumar, A. and Sarma, Y.R. 2001. Ecofriendly management of diseases of spices. *Indian Spices* **38(3)**: 28-31.

Andret-Link, P., Schmitt-Keichinger, C., Demangeat, G., Komar, V., Bergdoll, M., Vigne, E., and Fuchs, M. 2003. RNA2-encoded proteins for the exclusive transmission of Grapevine fanleaf virus by its nematode vector *Xiphinema index*. *In*: *Proceedings of 8th International Congress of Plant Pathology*, Christchurch, pp. 237.

Andret-Link, P., Schmitt-Keichinger, C., Demangeat, G., Komar, V. and Fuchs, M. 2004. The specific transmission of Grapevine fanleaf virus by its nematode vector *Xiphinema index* is solely determined by the viral coat protein. *Virology* **320**: 12-22.

Andrivon, D., Lucas, J.M., Guerin, C. and Jouan, B. 1998. Colonization of roots, stolons, tubers and stems of various potato (*Solanum tuberosum*) cultivars by the black dot fungus *Colletotrichum coccodes*. *Plant Pathol.* **47**: 440-445.

Argelis, A. 1987. Present situation of grapevine virus diseases with reference to the problems which they cause in Greek vineyards. *In*: *Integrated Pest Control in Viticulture* (R. Cavalloro, *ed*), pp. 309-312. A.A. Balkema, Rotterdam, Netherlands, 395 pp.

Armstrong, G.M. and Armstrong, J.K. 1981. Formae speciales and races of *Fusarium oxysporum* causing wilt diseases. *In*: *Disease, Biology and Taxonomy* (P.E. Nelson, Toussoun and R.J. Cook, *eds.*), pp. 391-399. Pennsylvania State Univ. Press, University Park.

Arulpragasam, P.V. and Addaickan, S. 1983. Soft root rot: A new root disease of tea? *Tea Quarterly* **52**: 52-55.

Arya, R. and Saxena, S.K. 1999. Influence of certain rhizosphere fungi together with *Rhizoctonia solani* and *Meloidogyne incognita* on germination of 'Pusa Ruby' tomato seeds. *Indian Phytopath.* **52**: 121-126.

Astier, S., Albouy, J., Maury Y. and Lecoq, H. 2001. *Principes de Virologie Végétale*. INRA, Paris, France.

Ateka, E.M., Mwang'ombe, A.W. and Kimenju, J.W. 2001. Studies on the interaction between *Ralstonia solanacearum* (smith) and *Meloidogyne* spp. in potato. *African Crop Sci. J.* **9**: 527-535.

Atilano, R.A., Menge, J.A. and Van Gundy, S.D. 1981. Interaction between *Meloidogyne arenaria* and *Glomus fasciculatum* in grape. *J. Nematol.* **13**: 52-57.

Ayala, A. and Allen, M.W. 1966. Transmission of the California tobacco rattle virus by three species of the nematode genus *Trichodorus*. *Nematologica* **12**: 87.

Ayala, A. and Allen, M.W. 1968. Transmission of the California tobacco rattle virus (CTRV) by three species of the nematode genus *Trichodorus*. *J. Agric. Univ. Puerto Rico.* **52**: 101-125.

Ayala, A., Allen, M.W. and Noffsinger, E.M. 1970. Host range, biology, and factors affecting survival and reproduction of the stubby root nematode. *J. Agr. Univ. Puerto Rico* **64**: 341-369.

Back, M.A., Haydock, P.P.J. and Jenkinson, P. 2002. Disease complexes involving parasitic nematodes and soilborne pathogens. *Plant Pathol.* **51**: 683-697.

Back, M.A., Jenkinson, P. and Haydock, P.P.J. 2000. The interaction between potato cyst nematodes and *Rhizoctonia solani* diseases in potatoes. *In*: *Proceedings of the Brighton*

Crop Protection Conference, Pests and Diseases. Farnham, UK: British Crop Protection Council, pp. 503-506.

Barker, K.R. and Hussey, R.S. 1976. Histopathology of nodule tissues of legumes with certain nematodes. *Phytopathology* **66**: 851-855.

Barua, L. and Bora, B.C. 2008. Comparative efficacy of *Trichoderma harzianum* and *Pseudomonas fluorescens* against *Meloidogyne incognita* and *Ralstonia solanacearum* complex in brinjal. *Indian J. Nematol.* **38**: 86-89.

Barua, L. and Bora, B.C. 2009. Compatibity of *Trichoderma harzianum* and *Pseudomonas fluorescens* against *Meloidogyne incognita* and *Ralstonia solanacearum* complex on brinjal. *Indian J. Nematol.* **39**: 29-34.

Baujard, P. 1989. Remarques sur les genres des sous-familles Bursaphelenchidae Paramanov, 1964 et Radinaphelenchinae Pramanov, 1964 (Nematoda: Aphelenchoididae). *Rev. de Nématol.* **12**: 323-324.

Belin, C., Schmitt, C., Demangeat, G., Komar, V., Pinck, L., and Fuchs, M. 2001. Involvement of RNA2-encoded proteins in the specific transmission of Grapevine fanleaf virus by its nematode vector *Xiphinema index*. *Virology* **291**: 161-171.

Ben-Yephet, Y. and Szmulewich, Y. 1985. Inoculum levels of *Verticillium dahliae* in the soils of the hot semi-arid Negev region of Israel. *Phytoparasitica* **13**: 193-200.

Bergeson, G.B. 1963. Influence of *Pratylenchus penetrans* alone and in combination with *Verticillium albo-atrum* on growth of peppermint. *Phytopathology* **53**: 1164-1166.

Bergeson, G.B. 1972. Concepts of nematode-fungus associations in plant disease complexes: A review. *Exptl. Parasitol.* **32**: 301-314.

Bergeson, G.B. 1975. The effect of *Meloidogyne incognita* on the resistance of four musk melon varieties to *Fusarium* wilt. *Plant Dis. Reptr.* **59**: 410-413.

Bergeson, G.B., Van Gundy, S.D. and Thomason, I.J., 1970. Effect of *Meloidogyne javanica* on rhizosphere microflora and *Fusarium* wilt of tomato. *Phytopathology* **60**: 1245-1249.

Bernard, E.C. and Laughlin, C.W. 1976. Relative susceptibility of selected cultivars of potato to *Pratylenchus penetrans*. *J. Nematol.* **8**: 239-242.

Bertrand, B., Nunez, C. and Sarah, J.L. 2000. Disease complex in coffee involving *Meloidogyne arabicida* and *Fusarium oxysporum*. *Plant Pathol.* **49**: 383-388.

Bhagawati, B., Choudhury, B.N. and Sinha, A.K. 2009. Management of *Meloidogyne incognita-Rhizoctonia solani* complex on okra through bioagents. *Indian J. Nematol.* **39**: 156-161.

Bhagawati, B., Das, B.C. and Sinha, A.K. 2007. Interaction of *Meloidogyne incognita* and *Rhizoctonia solani* on okra. *Ann. Plant Prot. Sci.* **15**: 533-535.

Bhagawati, B. and Goswami, B.K. 2000. Interaction of *M. incognita* and *Fusarium oxysporum* f. sp. *lycopersici* on tomato. *Indian J. Nematol.* **30**: 93-94.

Bhagawati, B., Goswami, B.K. and Singh, C.S. 2000. Management of disease complex of tomato caused by *Meloidogyne incognita* and *Fusarium oxysporum* f. sp. *lycopersici* through bioagents. *Indian J. Nematol.* **30**: 16-22.

Bhattarai, S., Haydock P.P.J., Back, M.A., Hare, M.C. and Lankford, W.T. 2009. Interactions between the potato cyst nematodes, *Globodera pallida*, *G. rostochiensis*, and soil-borne fungus, *Rhizoctonia solani* (AG3), diseases of potatoes in the glasshouse and the field. *Nematology* **11**: 631-640.

Bird, A.F. 1972. Quantitative studies on the growth of syncytia induced in plants by root-knot nematodes. *Int. J. Parasitol.* **2**: 157-170.

Bird, A.F. 1974. Plant response to root-knot nematode. *Ann. Rev. Phytopath.* **12**: 69-85.

Bird, A.F. and Bird, J. 1986. Observations on the use of insect parasitic nematodes as a means of biological control of root-knot nematodes. *Int. J. Parasitol.* **16**: 511-516.

Bird, G.W. 1981. Management of plant parasitic nematodes in potato production. *In*: *Advances in Potato Pest Management* (J.H. Lashomb and R. Casagrande, *eds.*), pp. 223-243. Hutchinson Ross Publishing Company, Strondsburg, PA., 288 pp.

Birpee, L.L. and Bloom, J.R. 1978. The influence of *Pratylenchus penetrans* on the incidence and severity of *Verticillium* wilt of potato. *J. Nematol.* **10**: 95-99.

Blair, G.P. 1969a. The problem of control of red ring disease. *In*: *Nematodes of Tropical Crops* (J.E. Peachy, *ed.*), pp. 99-108. Tech. Communication No. 40, Commonwealth Bureau of Helminthology.

Blair, G.P. 1969b. Studies on red ring disease of coconut palm. *Proc. of the Symp. on Trop. Nematol.*, Univ. of Puerto Rico, pp. 89-106.

Blair, G.P. and Darling, D. 1968. Red ring disease on the coconut palm, inoculation studies and histopathology. *Nematologica* **14**: 395-403.

Blake, C.D. 1966. *Nematologica* **12**: 129-132.

Bodine, E.W., Blodgett, E.D. and Lott, T.B. 1951. Rasp leaf. U.S. *Dep. Agric. Handb.* 10, pp. 71-80.

Booth, C. and Stover, R.H. 1974. *Cylindrocarpon musae* sp. nov., commonly associated with burrowing nematode (*Radopholus similis*) lesions on bananas. *Trans. British Mycol. Soc.* **63**: 503-507.

Botseas, D.D. and Rowe, R.C. 1994. Development of potato early dying in response to infection by two pathotypes of *Verticillium dahliae* and coinfection by *Pratylenchus penetrans*. *Phytopathology* **84**: 275-282.

Bowers, J.H., Nameth, S.T., Riedel, R.M. and Rowe, R.C. 1996. Infection and colonisation of potato roots by *Verticillium dahliae* as affected by *Pratylenchus penetrans* and *P. crenatus*. *Phytopathology* **86**: 614-621.

Bowman, P. and Bloom, J.R. 1966. Breaking the resistance of tomato varieties to *Fusarium* wilt by *Meloidogyne incognita. Phytopathology* **56**: 871.

Boydston, R.A., Mojtahedi, H., Crosslin, J.M., Thomas, P.E., Anderson, T. and Riga, E. 2004. Evidence for the influence of weeds on corky ringspot persistence in alfalfa and Scotch spearmint rotations. *Am. J. Potato Res.* **81**. 215-225.

Brathwaite, C.W.D. and Siddiqi, M.R. 1975. *Rhadinaphelenchus cocophilus*. C.I.H. Description of Plant Parasitic Nematodes, Set 5, No. 72.

Braun, A.L., Mojtahedi, H. and Lownsbery, B.F. 1975. Separate and combined effects of *Paratylenchus neoamblycephalus* and *Criconemoides xenoplax* on Myrobalan plum. *Phytopathology* **6**: 328-330.

Breece, J.R. and Hart, W.H. A possible association of nematodes with the spread of peach yellow bud mosaic virus. *Plant Dis. Reptr.* **43**: 989-990.

Bridge, J., Mortimer, J.J. and Jackson, G.V.H. 1983. *Hirschmanniella miticausa* n. sp. (Nematoda: Pratylenchidae) and its pathogenicity on taro (*Colocasia esculenta*). *Rev. de Nematol.* **6**: 285-290.

Brittain, J.A. and Miller, R.W. (*eds.*). 1978. Managing peach tree short life in the Southeast. South Carolina Extn. Circ. 585.

Brodie, J. and Cooper, W.E. 1964. Relation of parasitic nematodes to post emergence damping-off of cotton. *Phytopathology* **54**: 1023-1027.

Brown, C.R., Mojtahedi, H., Santo, G.S., Hamm, P., Pavek, J.J., Corsini, D., Love, S., Crosslin, J.M. and Thomas, P.E. 2000. Potato germplasm resistant to corky ringspot disease. *Am. J. Potato Res.* **77**: 23-27.

Brown, D.J.F. and Boag, B. 1975. *Longidorus macrosoma. Commonwealth Institute of Helminthology, Descriptions of Plant-parasitic Nematodes. No. 67.*

Brown, D.J.F., Halbrendt, J.M., Robbins, R.T. and Vrain, T.C. 1993. Transmission of nepoviruses by *Xiphinema americanum*-group nematodes. *J. Nematol.* **25**: 349-354.

Brown, D.J.F., MacFarlane, S.A., Hernandez, C. and Bol, J.F. 1995. Investigations on the genetic determinants of nematode transmissibility of tobraviruses. *Russian J. Nematol.* **4**: 81.

Burpee, L.L. and Bloom, J.R. 1978. The influence of *Pratylenchus penetrans* on the incidence and severity of *Verticillium* wilt of potato. *J. Nematol.* **10**: 95-99.

CABI/EPPO. 1999. *Rhadinaphelenchus cocophilus*. Distribution Maps of Plant Diseases No. 786, CAB International, Wallingford, UK.

Candia, J.D. and Simmonds, F.J. 1965. A tachnid parasite of the palm weevil, *Rhyncophorus palmarum. Commonwealth Inst. of Biol. Control Tech. Bull. No. 5,* pp. 127-128.

Caperton, C.M., Martyn, R.D. and Starr, J.L. 1986. Effects of *Fusarium* inoculum density and root-knot nematodes on wilt resistance on summer squash. *Plant Dis. Reptr.* **70**: 207-209.

Cappaert, M.R., Powelson, M.L., Christensen, N.W. and Crowe, F.J. 1992. Influence of irrigation on severity of potato early dying and tuber yield. *Phytopathology* **82**: 1448-1453.

Cappaert, M.R., Powelson, M.L., Christensen, N.W., Stevenson, W.R. and Rouse, D.I. 1994. Assessment of irrigation on severity of potato early dying. *Phytopathology* **84**: 792-800.

Carling, D.E. and Brown, M.F. 1980. Relative effect of vesicular *Arbuscular mycorrhizal* fungi on growth and yield of soybeans. *Soil Sci. Soc. America J.* **44**: 528-532.

Carter, G.E., Jr. 1976. Effect of soil fumigation and pruning date on the indoleacetic acid content of peach trees in a short life site. *Hortsci.* **11**: 594-595.

Carter, G.E., Jr. 1978. Effect of soil fumigation and pruning date on the resumption of growth of the vascular cambium of peach trees in a short life site. *Hortsci.* **13**: 156-158.

Caubel, G. and Samson, R. 1984. Effect of the stem nematode, *Ditylenchus dipsaci* on the development of 'Café au lait' bacteriosis in garlic (*Allium sativum*) caused by *Pseudomonas fluorescens*. *Agronomie* **4**: 311-313.

Caveness, F.E., Gilmer, R.M. and Williams, R.J. 1975. Transmission of cowpea mosaic by *Xiphinema basiri* in Western Nigeria. *In*: *Nematode Vectors of Plant Viruses* (F. Lamberti, C.E. Taylor and J.W. Seinhorst, *eds.*), pp. 289-290. Plenum, New York.

Chahal, P.P.K. and Chhabra, H.K. 1984. Interaction of *Meloidogyne incognita* with *Rhizoctonia solani* on tomato. *Indian J. Nematol.* **14**: 56-57.

Chahal, V.P.S. and Chahal, P.P.K. 1998. Interactions between nematodes and fungi. *In*: *Plant Nematode Management: A Biocontrol Approach* (P.C. Trivedi, *ed.*), pp. 200-207. Scientific Publishers, Jodhpur.

Chaitali, L., Singh, S. and Goswami, B.K. 2003. Effect of cakes with *Trichoderma viride* for the management of disease complex caused by *Rhizoctonia bataticola* and *Meloidogyne incognita* on okra. *Ann. Plant Prot. Sci.* **11**: 178-180.

Chandler, W.A., Owen, J.H. and Livingston, R.L 1962. Sudden decline of peach trees in Georgia. *Plant Dis. Reptr.* **46**: 831-834.

Chandrasekar, V. and Johnson, J.E. 1997. The structure of Tobacco ringspot virus: a link in the evolution of icosahedral capsids in the picornavirus superfamily. *Structure* **6**: 157-171.

Chang, H.Y. and Raski, D.J. 1972. *Hemicriconemoides chitwoodi* on grapevines. *Plant Dis. Reptr.* **56**: 1028-1030.

Chapman, R.A. 1965. Infection of single root systems by larvae of two coincident species of root-knot nematodes. *Nematologica* **12**: 89.

Charlton, B.A. 2006. Effects of oxamyl on suppression of the Tobacco rattle virus vector *Paratrichodorus allius* and corky ringspot disease of potato in the Klamath Basin of south-central Oregon. MS Agronomy Thesis, Oregon State University.

Chaya, M.K., Rao, M.S. and Ramachandran, N. 2010. Biomanagement of disease complex in okra (*Abelmoschus esculentus* L. Moench) using microbial antagonists. *National Conf. on Innovations in Nematological Research for Agricultural Sustainability–Challenges and A Roadmap Ahead*, Tamil Nadu Agri. Univ., Coimbatore, pp. 96-97.

Chhabra, H. K. 1972. Citrus nematode and its control. *Pesticides* **6**(2): 88-89.

Chhabra, H.K. and Sharma, J.K. 1981. Combined effect of *Meloidogyne incognita* and *Rhizoctonia bataticola* on pre-emergence damping-off of okra and brinjal. *Sci. Cult.* **47**: 256-257.

Chhabra, H.K., Sidhu, A.S. and Singh, I. 1977. *Meloidogyne incognita* and *Rhizoctonia solani* interaction on okra. *Indian J. Nematol.* **7**: 54-57.

Chinchilla, C. 1988. El Sindrome del anillo rojo-hoja pequeña en palma aceitera y cocotero. *Boletin Technico* **2**: 113-136.

Chinchilla, C., Menjivar, R. and Arias, E. 1991. Variación estacional de la población de *Rhynchophorus palmarum* y su relación con la enfermedad del anillo rojo/ hoja pequeña en una plantación comercial de *Elaeis guineensis* en Honduras. *Turrialba.* **40(4)**:

Chinchilla, C. and Richardson, D.L. 1987. Four potentially destructive diseases of the oil palm in Central America. *Int. Oil Palm\Palm Oil Conferences. Progress and Prospects.* (Unedited). Kuala Lumpur, Mimeo, pp. 7.

Chinchilla, C.M. 1991. The red ring little leaf syndrome in oil palm and coconut palm. ASD Oil Palm Papers No. 1, 1-17.

Chitamber, J.J. and Raski, D.J. 1984. Reactions of grape rootstocks to *Pratylenchus vulnus* and *Meloidogyne* spp. *J. Nematol.* **16**: 166 170.

Choi, D.R., Ishibashi, N. and Tanaka, K. 1988. Possible integrated control of soil insect pests, soil-borne diseases, and plant nematodes by mixed application of fungivorous and entomogenous nematodes. *Bull. Faculty of Agri.* **65**: 27-35.

Chovatiya, K.P., Patel, B.N. and Patel, H.R. 2005. Studies on interaction between brinjal mosaic virus and root-knot nematode on brinjal. *Indian J. Nematol.* **35**: 148-150.

Clayton, C.E. 1977. Peach tree survival. *Fruit South* **1**: 53-58.

Cohn, E. and Mordechai, M. 1969. Investigations on the life cycles and host preference of some species of *Xiphinema* and *Longidorus* under controlled conditions. *Nematologica* **15**: 295-302.

Cohn, E., Tanne, E. and Nitzany, F.E. 1970. *Xiphinema italiae* a new vector of grapevine fanleaf virus. *Phytopathology* **60**: 181-182.

Conroy, J.J. and Green, R.J., Jr. 1974. Interactions of the root-knot nematode *Meloidogyne incognita* and the stubby root nematode *Trichodorus christiei* with *Verticillium albo-atrum* on tomato at controlled inoculum densities. *Phytopathology* **64**: 1118-1121.

Conroy, J.J., Green, R.J., Jr. and Ferris, J.M. 1972. Interaction of *Verticillium albo-atrum* and the root lesion nematode, *Pratylenchus penetrans*, in tomato roots at controlled inoculum densities. *Phytopathology* **62**: 362-366.

Cook, R.J. and Baker, K.F. 1983. *The Nature and Practice of Biological Control of Plant Pathogens.* Ame. Phytopath. Soc., St. Paul.

Cooper, J.I. and Thomas, P.R. 1970. *Trichodorus nanus* a vector of tobacco rattle virus in Scotland. *Plant Pathol.* **19**: 197.

Cooper, J.I. and Thomas, P.R. 1971. Chemical treatment of soil to prevent transmission of tobacco rattle virus to potatoes by *Trichodorus. Ann. Appl. Biol.* **69**: 23-24.

Cooper, K.M and Grandison, G.S. 1986. Interaction of vesicular arbuscular mycorrhizal fungi and root-knot nematode on cultivars of tomato and white clover susceptible to *Meloidogyne hapla. Ann. Appl. Biol.* **108**: 555-565.

Corbett, D.C.M. and Hide, G.A. 1971. Interactions between *Heterodera rostochiensis* Woll. and *Verticillium dahliae* Kleb. on potatoes and the effect of CCC on both. *Ann. Appl. Biol.* **68**: 71-80.

Cremer, M.C. and Kooistra, G. 1964. Investigations on notched leaf ("kartleblad") of *Gladiolus* in its relation to tobacco rattle virus. *Nematologica* **10**: 69-70.

Cropley, R. 1964. Further studies on European rasp leaf and leaf roll diseases of cherry trees. *Ann. Appl. Biol.* **53**: 333-371.

Crosse, J.E. and Pitcher, R.S. 1952. Studies in the relationship of eelworms and bacteria to certain plant diseases. 1. The etiology of strawberry cauliflower disease. *Ann. Appl. Biol.* **39**: 475-484.

Crosslin, J.M., Thomas, P.E. and Brown, C.R. 1999. Distribution of tobacco rattle virus in tubers of resistant and susceptible potatoes and systemic movement of virus into daughter plants. *Ame. J. Potato Res.* **76**:191-197.

Crow, W.T. 2005. Diagnosis of *Trichodorus obtusus* and *Paratrichodorus minor* on turf grasses in the Southeastern United States. Online. *Plant Health Progress* doi: 10.1094/PHP-2005-0121-01-DG.

Cullen, D.W., Lees, A.K., Toth, I.K. and Duncan, J.M. 2002. Detection of *Colletotrichum coccodes* from soil and potato tubers by conventional and quantitative real-time PCR. *Plant Pathol.* **51**: 281-292.

Dale, M.F.B, Robinson, D.J. and Todd, D. 2004. Effects of systemic infections with Tobacco rattle virus on agronomic and quality traits of a range of potato cultivars. *Plant Pathol.* **53**: 788-793.

Dallwitz, M.J. 1980. A general system for coding taxonomic descriptions. *Taxon* **29**: 41-46.

Dallwitz, M. J., Paine, T.A. and Zurcher, E.J. 1993. *User's Guide to the DELTA System: A General System for Processing Taxonomic Descriptions*. 4th edition. (CSIRO Division of Entomology: Canberra), pp. 136.

Daniell, J.W. 1973. Effects of time of pruning on growth and longevity of peach trees. *J. Amer. Soc. Hort. Sci.* **98**: 383-386.

Dao, F. and Oostenbrink, M. 1967. An inoculation experiment in oil palm with *R. cocophilus* from coconut and oil palm. *In*: *Symposium over Fytofarmacie in Fytiatre*, 19, Gent.

Davide, R.G. 1972. Influence of root-knot nematodes on the severity of bacterial wilt and Fusarium wilt of tomato. *Philippine Phytopathol.* **8**: 78-81.

Davis, J.R. 1981. *Verticillium* wilt of potato in southeastern Idaho. *Univ. Idaho Curr. Inf. Ser.* 564.

Davis, J.R. and Everson, D.O. 1986. Relation of *Verticillium dahliae* in soil and potato tissue, irrigation method and N-fertility to Verticillium wilt of potato. *Phytopathology* **76**: 430-436.

Davis, J.R. and Huisman, O.C. 2001. Verticllium wilt. *In*: *Compendium of Potato Diseases*. 2nd edn. (WR Stevenson *et al., eds.*), pp. 45-46. American Phytopathological Society Press, St Paul, MN.

Davis, J.R., Huisman, O.C., Everson, D.O. and Schneider, A.T. 2001. Verticillium wilt of potato: a model of key factors related to disease severity and tuber yield in southeastern Idaho. *Ame. J. Potato Res.* **78**: 291-300.

Davis, J.R., Huisman, O.C., Westerman, D.T., Hafez, S.L., Everson, L.H., Sorensen, L.H. and Schneider, A.T. 1996. Effects of green manures on Verticillium wilt of potato. *Phytopathology* **86**: 444-453.

Davis, J.R. and McDole, R.E. 1979. Influence of cropping sequences on soil-borne populations of *Verticillium dahliae* and *Rhizoctonia solani*. *In*: *Soil-Borne Plant Pathogens* (B. Schipper and W. Gams, *eds.*), pp. 399-405. Academic Press, London, pp. 685.

Davis, J.R., Sorensen, L.H., Stark, J.C. and Westerman, D.T. 1990. Fertility and management practices to control Verticillium wilt of Russet Burbank potato. *Ame. Potato J.* **67**: 55-65.

Davis, R.A. and Jenkins, W.R. 1963. Effects of *Meloidogyne* spp. and *Tylenchorhynchus claytoni* on pea wilt incited by *Fusarium oxysporum* f. sp. *pisi* race 1. *Phytopathology* **53**: 745.

Dean, C.G. 1979. Red ring disease of *Cocos nucifera* L. caused by *Rhadinaphelenchus cocophilus* (Cobb, 1919) Goodey, 1960. An annotated bibliography and review. Technical Communication No. 47, Commonwealth Inst. Helminthol., pp. 70.

de Bokx, J.A. (*ed.*). 1972. *Viruses of Potatoes and Seed-potato Production*. Centre for Agricultural Publishing and Documentation, Wageningen.

Decraemer, W. 1995. *The Family Trichodoridae: Stubby Root and Virus Vector Nematodes*. Kluwer Academic Publ., Boston, MA, pp. 360.

Dehne, H.W. 1982. Interaction between vesicular *Arbuscular mycorrhizal* fungi and plant pathogens. *Phytopathology* **72**: 1115-1119.

DeMarree, A. and Riekenberg, R. 1998. Budgeting. *In*: *Strawberry Production Guide for the Northeast, Midwest, and Eastern Canada* (M. Pritts and D. Handley, *eds.*), pp. 118-131. Northeast Regional Agricultural Engineering Service, Ithaca, N.Y.

De Moura, R.M., Echandi, E. and Powell, N.T. 1975. Interaction of *Corynebacterium michiganense* and *Meloidogyne incognita* on tomato. *Phytopathology* **65**: 1332-1335.

Denner, F.D.N., Millard, C.P. and Wehner, F.C. 2000. Effect of soil solarisation and mould board ploughing on black dot of potato, caused by *Colletotrichum coccodes. Potato Res.* **43**: 195-201.

DeVay, J.E., Lownsberry, B.F., English, W.H. and Lembright, H. 1967. Activity of soil fumigants in relation to increased growth and control of decline and bacterial canker in trees of *Prunus persica* (Abstr.). *Phytopathology* **57**: 809.

Devi, T.P. and Goswami, B.K. 1992. Effect of VA mycorrhiza on the disease incidence due to *Macrophomina phaseolina* and *Meloidogyne incognita* on cowpea. *Ann. Agric. Res.* **13**: 253-265.

Dhawan, S.C. and Sethi, C.L. 1977. Inter-relationship between root-knot nematode, *Meloidogyne incognita* and little leaf of brinjal. *Indian Phytopath.* **30**: 55-63.

Dhawan, S.C., Singh, S. and Kamra, A. 2008. Biomanagement of root-knot nematode, *Meloidogyne incognita* by *Pochonia chlamydosporia*. *Indian J. Nematol.* **38**: 119-121.

Dobinson, K.F., Harrington, M.A., Omer, M. and Rowe, R.C. 2000. Molecular characterisation of vegetative compatibility group 4A and 4B isolates of *Verticillium dahliae* associated with potato early dying. *Plant Dis.* **84**: 1241-1245.

Dohroo, N.P., Shyam, K.R. and Bhardwaj, S.S. 1987. Distribution, diagnosis and incidence of rhizome rot complex of ginger in Himachal Pradesh. *Indian J. Plant Pathol.* **5**: 24-25.

Doshi, A. and Mathur, S. 1987. Symptomatology, interaction and management of rhizome rot of ginger by xenobiotics. *Korean J. Plant Pathol.* **26**: 261-265.

Dowler, W.M. and Petersen, D.H. 1966. Induction of bacterial canker of peach in the field. *Phytopathology* **56**: 989-990.

DuCharme, E.P. 1968. *In*: *Tropical Nematology* (G.C. Smart and V.G. Perry, *eds.*), pp. 20-37. Univ. of Florida Press, Gainesville.

Duncan, L. and Ferris, H. 1983. Validation of a model for prediction of host damage by two nematode species. *J. Nematol.* **15**: 227-234.

Dunn, E. 1968. *8th Int. Symp. Nematol.,* Antibes, pp. 115.

Dunn, E. and Hughes, W.A. 1964. Interrelationships of the potato root eelworm, *Heterodera rostochiensis* Woll., *Rhizoctonia solani* Kuhn and *Colletotrichum atramentarium* (B. & Br.) Toub., on the growth of the tomato plant. *Nature, London* **201**: 413-414.

Dunn, E. and Hughes, W.A. 1967. Interactions of *Oospora pustulans, Rhizoctonia solani* and *Heterodera rostochiensis* on the potato. *European Potato J.* **10**: 327-328.

Eapen, S.J. and Venugopal, M.N. 1995. Field evaluation of *Paecilomyces lilacinus* and *Trichoderma* spp. in cardamom nurseries for the control of root-knot nematodes and rhizome rot disease. *National Symp. on Nematode Problems of India–An Appraisal of the Nematode Mangmt. with Eco-friendly Approaches and Biocomponents.* Indian Agri. Res. Inst., New Delhi.

Edmunds, J.E. and Mai, W.F. 1966. *Phytopathology* **56**: 1320-1321.

Edwards, M. 1991. Control of plant parasitic nematodes in sultana grapevines (*Vitis vinifera*) using systemic nematicides. *Australian J. Expt. Agri.* **31**: 579-584.

Egunjobi, O.A., Akonde, P.T. and Caveness, F.E. 1986. Interaction between *Pratylenchus sefaensis, Meloidogyne javanica* and *Rotylenchulus reniformis* in sole and mixed crops of maize and cowpea. *Rev. de Nematol.* **9**: 61-70.

Eisenback, J.D. 1985. Interaction among concomitant populations of nematodes. *In*: *An Advanced Treatise on* Meloidogyne, *Vol. I: Biology and Control* (J.N. Sasser and C.C. Carter, *eds.*), pp. 193-213. North Carolina State Univ. Graphics, Raleigh.

Eisenback, J.D. and Griffin, G.D. 1987. Interactions with other nematodes. *In*: *Vistas on Nematology* (J.A. Veech and D.W. Dickson, *eds.*), pp. 313-320. Soc. of Nematologists, Hyattsville.

El-Goorni, M.A., Abo-Eldahar, M.E. and McNair, F.F. 1974. Interaction between root-knot and *Pseudomonas marginata* on gladiolus corms. *Phytopathology* **64**: 271-272.

Elmer, W.H. and LaMondia, J.A. 1995. Influence of mineral nutrition on strawberry black root rot. *Adv. Strawberry Res.* **14**: 42-48.

Elmer, W.H. and LaMondia, J.A. 1999. Influence of ammonium sulfate and rotation crops on strawberry black root rot. *Plant Dis.* **83**:119-123.

El-Sherif, A.G. and Elwakii, M.A. 1991. Interaction between *Meloidogyne incognita* and *Agrobacterium tumefaciens* or *Fusarium oxysporum* f. sp. *lycopersici* on tomato. *J. Nematol.* **23**: 239-242.

Endo, B.Y. 1971. Nematode induced syncytia (giant cells). Host-parasite relationship of Heteroderidae. *In*: *Plant Parasitic Nematodes,* Vol. II (B.M. Zuckerman, W.F. Mai and R.A. Rohde, *eds.*), pp. 91-117. Academic Press, New York.

Epps, J.M. and Chambers, A.Y. 1962. Effect of seed inoculation, soil fumigation, and cropping sequences on soybean nodulation in soybean-cyst nematode-infested soil. *Plant Dis. Reptr.* **46**: 48-51.

Esmenjaud, D. 1986. Les nématodes de la vigne. *Phytoma* **374**: 24-27.

Esmenjaud, D., Walter, B., Minot, J.C., Voisin, R. and Cornuet, P. 1993. Biotin-avidin ELISA detection of grapevine fanleaf virus in the vector *Xiphinema index*. *J. Nematol.* **25**: 401-405.

Esmenjaud, D., Walter, B., Valentin, G., Guo, Z.T. and Cluzeau, D. 1992. Vertical distribution and infectious potential of *Xiphinema index* (Thorne and Allen, 1950) (Nematoda: Longidoridae) in fields affected by grapevine fanleaf virus in vineyards in the Champagne region of France. *Agronomie* **12**: 395-399.

Esser, R.P. 1969. *Rhadinaphelenchus cocophilus* a potential threat to Florida palms. *Nematology Circular No. 9,* Division of Plant Industry, Florida Department of Agriculture, Gainesville, Florida, USA.

Estores, R.A and Chen, T.A. 1970. Interaction of *Pratylenchus penetrans* and *Meloidogyne incognita acrita* as cohabitants on tomatoes. *Phytopathology* **60**: 1291.

Estores, R.A and Chen, T.A. 1972. Interaction of *Pratylenchus penetrans* and *Meloidogyne incognita acrita* as cohabitants on tomatoes. *J. Nematol.* **1**: 219-222.

Evans, K. 1987. The interactions of potato cyst nematodes and *Verticillium dahliae* on early and main crop potato cultivars. *Ann. Appl. Biol.* **110**: 329-339.

Fassuliotis, G. and Rau, G.J. 1969. The relationship of *Meloidogyne incognita acrita* to the incidence of cabbage yellows. *J. Nematol.* **1**: 219-222.

Faulkner, L.R. and Bolander, W.J. 1969. Interaction of *Verticillium dahliae* and *Pratylenchus miniyus* in *Verticillium* wilt of peppermint: Effects of soil temperature. *Phytopathology* **59**: 868-870.

Faulkner, L.R., Bolander, W.J. and Skotland, C.B. 1970. Interaction of *Verticillium dahliae* and *Pratylenchus miniyus* in *Verticillium* wilt of peppermint: Influence of the nematode as determined by a double root technique. *Phytopathology* **60**: 100-103.

Faulkner, L.R. and Skotland, C.B. 1965. Interaction of *Verticillium dahliae* and *Pratylenchus miniyus* in *Verticillium* wilt of peppermint. *Phytopathology* **55**: 583-586.

Fawcett, H.S. 1931. The importance of investigations on effects of known mixtures of organisms. *Phytopathology* **21**: 545-550.

Feder, W.A. and Feldmesser, J. 1961. The spreading decline complex: The separate and combined effects of *Fusarium* spp. and *Radopholus similis* on the growth of Duncan grapefruit seedlings in the greenhouse. *Phytopathology* **51**: 724-726.

Feldmesser, J. and Goth, R.W. 1970. Association of a root-knot with bacterial wilt of potato. *Phytopathology* **60**: 1014.

Ferraz, S. and Lear, B. 1976. Interaction of four plant parasitic nematodes and *Fusarium oxysporum* f. sp. *dianthi* on carnation. *Experientiae* **22**: 272-277.

Ferree, M.E. and McGlohon, N.E. 1975. You can prevent peach tree short life. *Fruit Grower*, Aug. 1975, **20**: 27.

Ferris, H., Lau, S. and Venette, R.C. 1994. Population energetics of bacterial-feeding nematodes: Respiration and metabolic rates based on carbon dioxide production. *Soil Biol. Biochem.* (In press).

Ferris, H. and McKenry, M.V. 1974. Seasonal fluctuations in the spatial distribution of nematode populations in a California vineyard. *J. Nematol.* **6**: 203-210.

Ferris, H. and McKenry, M. V. 1975. Relationship of grapevine yield and growth to nematode densities. *J. Nematol.* **7**: 295-304.

Ferris, H., McKenry, M.V. and McKinney, H.E. 1976. Spatial distribution of nematodes in peach orchards. *Plant Dis. Reptr.* **60**: 18-22.

Finlay, R.D. 1985. Interaction between soil micro-arthropods and endomycorrhizal associations of higher plants. *In*: *Ecological Interactions in Soil: Plants, Microbes and Animals* (A.H. Fitter, D. Atkinson, D.J. Read and M.B. Usher, *eds.*), pp. 319-331. Blackwell Scientific Publications, Oxford.

Flegg, J.J.M. 1969. Tests with potential nematode vectors of cherry leaf roll virus. Rep. E. Malling Res. Sta. 1968, pp. 155-157.

Florini, D.A. and Loria, R. 1990. Reproduction of *Pratylenchus penetrans* on potato and crops grown in rotation with potato. *J. Nematol.* **22**: 106-112.

Forer, L.B., Hill, N.S. and Powell, C.A. 1981. *Xiphinema rivesi* a new tomato ringspot virus vector (Abstr.). *Phytopathology* **71**: 874.

Fortuner, R. 1987. A reappraisal of Tylenchina (Nemata). 8. The family Hoplolaimidae Filipjev, 1934. *Rev. de Nematol.* **10**: 219-232.

France, R.A. and Abawi, G.S. 1994. Interaction between *Meloidogyne incognita* and *Fusarium oxysporum* f. sp. *phaseoli* on selected bean genotypes. *J. Nematol.* **26**: 467-474.

Francl, L.J., Madden, L.V., Rowe, R.C. and Riedel, R.M. 1987. Potato yield loss prediction and discrimination using preplant population densities of *Verticillium dahliae* and *Pratylenchus penetrans. Phytopathology* **77**: 579-584.

Franklin, M.T. 1950. Two species of *Aphelenchoides* associated with strawberry bud disease in Britain. *Ann. Appl. Biol.* **37**: 1-10.

Fritzsche, R. 1964. *Wiss. Z. Univ. Rostock, Math-Naturwiss. Reche* **13**: 433-347.

Fritzsche, R. 1968. *Biol. Zentralbl.* **87**: 139-146.

Fritzsche, R. and Fritzsche, R. and Thiele, S. 1979. *Nachrichtenbl. Pflanzenschutz DDR.* **33**: 103-104.

Fritzsche, R. and Kegler, H. 1964. *Naturwissenchaften* **51**: 299.

Fritzsche, R. and Kegler, H. 1968. *Tagber, Dt. Akad. Landu-Wiss. Berlin* **97**: 289-295.

Fritzsche, R., Kegler, H., Thiele, S. and Gruber, G. 1979. *Arch. Phytopathol. Pflanzenschutz* **15**: 177-180.

Fritzsche, R., Pelcz, J., Ottel, G. and Thiele, S. 1983. Interaction between *Meloidogyne incognita* and *Fusarium oxysporum* f. sp. *cucumerinum* in greenhouse cucumbers. *Tagungsbericht Akademie der Landwirschaftwissenschaften der Deutschen Demokratischen Republik* **261**: 685-690.

Fritzsche, R. and Schmelzer, K. 1967. *Naturwissenchaften* **54**: 489-499.

Garcia Gil de Bernabe, A. 1976. *La Degeneracion Infecciosa y Las Enfermedades de Virus de la Vina En La Zona Del Jerez.* Ministerio de Agricultura, Madrid, pp. 225.

Gay, C.M. and Bird, G.W. 1973. Influence of concomitant inoculation of *Pratylenchus brachyurus* and *Meloidogyne* spp. on root penetration and population dynamics. *J. Nematol.* **5**: 212-217.

Gerber, K. and Giblin-Davis, R.M. 1990. Association of the red ring nematode, *Rhadinaphelenchus cocophilus*, and other nematode species with *Rhynchophorus palmarum* (Coleoptera: Curculionidae). *J. Nematol.* **22**: 143-149.

Gerber, K., Giblin-Davis, R.M., Griffith, R., Escobar-Goyes, J. and D'Ascoli Cartaya, A. 1989. Morphometric comparisons of geographic and host isolates of the red ring nematode, *Rhadinaphelenchus cocophilus*. *Nematropica* **19**: 151-159.

Gerdemann, J.W. 1968. Vesicular *Arbuscular mycorrhizae* and plant growth. *Ann. Rev. Phytopath.* **6**: 397-418.

Gibbs, A.J and Harrison, B.D. 1964. A form of pea early browning virus found in Great Britain. *Ann. Appl. Biol.* **54**: 1-11.

Giblin-Davis, R.M. 1990. *The Red Ring Nematode and its Vectors*. Fla. Dept. Agric. & Consumer Serv., Divn. Plant Ind., Nema. Circ. No. 181.

Giblin-Davis, R.M., Gerber, K. and Griffith, R. 1989. *In vivo* and *in vitro* culture of the red ring nematode, *Rhadinaphelenchus cocophilus*. *Nematropica* **19**: 135-142.

Godfrey, G.H. 1936. The pineapple root system as affected by the root-knot nematode. *Phytopathology* **26**: 408-428.

Golden, J.K. and Van Gundy, S.D., 1972. Influence of *Meloidogyne incognita* on root rot development by *Rhizoctonia solani* and *Thielaviopsis basicola* in tomato. *J. Nematol.* **4**: 225.

Golden, J.K. and Van Gundy, S.D. 1975. A disease complex of okra and tomato involving the nematode, *Meloidogyne incognita* and the soil inhabiting fungus, *Rhizoctonia solani*. *Phytopathology* **65**: 265-273.

Gonsalves, D. 1988. Tomato ringspot virus decline. *In*: *Compendium of Grape Diseases* (R.C. Pearson and A.C. Goheen, *eds*), pp. 49-50. American Phytopathological Society Press, St Paul, Minnesota, p. 93.

Gonzalez, R. 1982. Crecimento y germinacion de chlamidosporas de *Fusarium oxysporum* f. sp. *lycopersici* en extractos radiculares de plantas de tomate infectadas con *Meloidogyne incognita*. *Agronomia Tropical* **30**: 305-313.

Good, J.M. 1964. Effect of soil application and sealing methods on the efficacy of row application of several soil nematicides for controlling root-knot nematodes, weeds and *Fusarium* wilt. *Plant Dis. Reptr.* **48**: 199-203.

Goode, M.J. and McGuire, J.M. 1967. *Phytopathology* **57**: 812.

Goodey, J.B. 1960. *Rhadinaphelenchus cocophilus* (Cobb, 1919) N. Comb., The nematode associated with "red-ring" disease of coconut. *Nematologica* **5**: 98-102.

Goodey, T. 1935. Observations on a nematode disease of yams. *J. Helminthol.* **13**: 173-190.

Goodman, R.N., Kiraly, Z. and Zaitlin, M. 1967. *The Biochemistry and Physiology of Infectious Plant Disease*. Van Nostrand, Princeton, New Jersey.

Gopinadhan, P.B., Mohandas, N. and Vasudevan, K.P. 1990. Cytoplasmic polyhedrosis virus infecting red palm weevil off coconut. *Curr. Sci.* **59**: 577-580.

Goswami, B.K. and Chenulu, V.V. 1974. Interaction of root-knot nematode, *Meloidogyne incognita* and tobacco mosaic virus in tomato. *Indian J. Nematol.* **4**: 69-80.

Goswami, B.K., Pandey, R.K., Goswami, J. and Tiwari, D.D. 2007. Management of disease complex caused by root knot nematode and root wilt fungus on pigeon pea through soil organically enriched with Vesicular Arbuscular Mycorrhiza, karanj (*Pongamia pinnata*) oilseed cake and farmyard manure. *J. Environ. Sci. Health* **42**: 899-904.

Gowdar, S.B. and Kulkarni, S. 1999. Compatibility effect of antagonists and seed dressers against *Fusarium udum*–the causal agent of pigeonpea wilt. *Karnataka J. Agric. Sci.* **12**: 197-199.

Grainger, J. and Clark, M.R.M. 1963. Interactions of *Rhizoctonia* and potato root eelworm. *European Potato J.* **6**: 131-132.

Green, C.D., Ghumra, M.F. and Salt, G.A. 1983. Interaction of root lesion nematodes, *Pratylenchus thornei* and *P. crenatus*, with the fungus *Thielaviopsis basicola* and a grey sterile fungus on roots of pea (*Pisum sativum*). *Plant Pathol.* **32**: 281-288.

Griesbach, J.A. and Maggenti, A.R. 1989. Vector capability of *Xiphinema americanum sensu lato* in California. *Rev. de Nematol.* **13**: 93-103.

Griffin, G.D. 1985. Interrelationships of *Heterodera schachtii* and *Meloidogyne hapla* on tomato. *J. Nematol.* **17**: 385-388.

Griffin, G.D. and Waite, W.W. 1982. Pathological interaction of a combination of *Heterodera schachtii* and *Meloidogyne hapla* on tomato. *J. Nematol.* **14**: 182-187.

Griffith, R. 1977. A species of bacterium pathogenic to the palm weevil, *Rhynchophorus palmarum*. *Ann. Rept.*, Red Ring Res. Divn., Ministry of Agri., Lands & Fish., Trinidad and Tabago, pp.31-35.

Griffith, R. 1987. Red ring disease of coconut palm. *Plant Dis.* **71**: 193-196.

Griffith, R. and Koshy, P.K. 1990. Nematode parasites of coconut and other palms. *In*: *Plant Parasitic Nematodes in Subtropical and Tropical Agriculture* (M. Luc, R.A. Sikora and J. Bridge, *eds.*), pp. 363-386. CAB International, Wallingford, Oxon, U.K.

Gudmestad, N.C., Huguelet, J.E. and Zink, R.T. 1978. The effect of cultural practices and straw incorporation into the soil on Rhizoctonia disease of potato. *Plant Dis. Reptr.* **62**: 985-988.

Gudmestad, N.C., Pasche, J.S. and Taylor, R.J. 2005. Infection frequency of *Colletotrichum coccodes* during the growing season. *In*: *'Proceedings of the 16th Triennial European Association for Potato Research conference'* (E. Ritter and A. Carrascal, *eds.*), pp. 765-769. Central Service of Publications of the Basque Government: Vitoria-Gasteiz, Spain.

Hafez, S.L., Al-Rehiayani, S., Thornton, M. and Sundararaju, P. 1999. Differentiation of two geographically isolated populations of *Pratylenchus neglectus* based on their parasitism of potato and interaction with *Verticillium dahliae*. *Nematropica* **29**: 25-36.

Hagley, E.A.C. 1963. The role of the palm weevil, *Rhynchophorus palmarum*, as a vector of red ring disease of coconuts. I. Results of preliminary investigations. *J. Econ. Entomol.* **36**: 375-380.

Haider, M.G., Nath, R.P., Thakur, S.C. and Ojha, K.L. 1987. Interaction of *Meloidogyne incognita* and *Pseudomonas solanacearum* on tomato plants. *Indian J. Nematol.* **17**: 174-176.

Halbrendt, J.M. and Brown, D.J.F. 1992. Morphometric evidence for three juvenile stages in some species of *Xiphinema americanum sensu lato*. *J. Nematol.* **24** 305-309.

Halbrendt, J.M. and Brown, D.J.F. 1993. Aspects of biology and development of *Xiphinema americanum* and related species. *J. Nematol.* **25**: 355-360.

Hamada, M., Hirakata, K. and Uchida, T. 1985. Influence of Southern root-knot nematode, *Meloidogyne incognita*, on the occurrence of root rot of pepper (*Piper nigrum* L.) caused by *Fusarium solani* f. sp. *piperi*. *Proc. Kanto-Tosan Plant Prot. Soc.* **32**: 236-237.

Hamm, P.B., Ingham, R.E., Jaeger, J.R., Swanson, W.H. and Volker, K.C. 2003. Soil fumigant effects on three genera of potential soilborne pathogenic fungi and their effect on yield in the Columbia Basin of Oregon. *Plant Dis.* **87**: 1449-1456.

Harley, J.L. and Smith, S.E. 1983. *Mycorrhizal Symbiosis*. Acad. Press, London.

Harrison, A.L. and Young, P.A. 1941. Effect of root-knot nematode on tomato wilt. *Phytopathology* **31**: 749-752.

Harrison, B.D. 1961. Soil-borne viruses – tobacco rattle virus. Rept. Rothamsted Exp. Sta. 1960, 118 pp.

Harrison, B.D. 1962. Rept. Rothamsted Exp. Sta. 1961, pp. 105.

Harrison, B.D. 1967. The transmission of strawberry latent ringspot virus by *Xiphinema diversicaudatum* (Nematoda). *Ann. Appl. Biol.* **60**: 405-409.

Harrison, B.D., Finch, J.T., Gibbs, A.J., Hollings, M., Shepherd, R.J., Valenta, V. and Wetter, C. 1971. Sixteen groups of plant viruses. *Virology* **45**: 356-363.

Harrison, B.D., Mowat, W.P. and Taylor, C.E. 1961. Transmission of a strain of tomato black ring virus by *Longidorus elongatus*. *Virology* **14**: 480-485.

Harrison, B.D. and Murant, A.F. 1978. Nematode transmissibility of pseudorecombinant isolates of Tomato black ring virus. *Ann. Appl. Biol.* **86**: 209-212.

Harrison, B.D., Robertson, W.M. and Taylor, C.E. 1974. Specificity of transmission of viruses by nematodes. *J. Nematol.* **6**: 155-164.

Harrison, B.D., and Robinson, D.J. 1986. Tobraviruses. *In*: *The Plant Viruses* Vol. 2: The Rod-shaped Plant Viruses (M.H.V. van Regenmortel and H. Fraenkel-Conrat, *eds.*), pp. 339-369. Plenum Press, N.Y.

Harrison, J.A.C. 1971. Association between the potato cyst nematode, *Heterodera rostochiensis* Woll., and *Verticillium dahliae* Kleb. in the early dying disease of potatoes. *Ann. Appl. Biol.* **67**: 185-193.

Hasan, A. 1985a. Synergistic interaction between *Pythium aphanidermatum* and *Rhizoctonia solani* with *Meloidogyne incognita* on chilli. *Nematologica* **31**: 210-217.

Hasan, A. 1985b. Breaking resistance in chilli to root-knot nematode by fungal pathogens. *Nematologica* **31**: 210-217.

Hasan, A. 1988. Interaction between *Pratylenchus coffeae* and *Pythium aphanidermatum* and/or *Rhizoctonia solani* on chrysanthemum. *Phytopathol. Zeist* **123**: 227-232.

Hasan, A. 1989. Efficacy of certain non-fumigant nematicides on the control of pigeonpea wilt involving *Heterodera cajani* and *Fusarium udum*. *Phytopathol. Zeist* **126**: 335-342.

Hasan, A. and Khan, M.N. 1985. The effect of *Rhizoctonia solani, Sclerotium rolfsii* and *Verticillium dahliae* on the resistance of *Meloidogyne incognita. Nematol. Medit.* **13**: 133-136.

Haseeb, A. 2003. Management of root-knot nematode and wilt complex and their interaction in vegetable crops using organic amendments and biocontrol agents. Final Technical Report of the UPCAR Sponsored Project, Lucknow, pp. 52.

Haseeb, A. and Archana. 2009. Effect of *Meloidogyne incognita* and *Macrophomina phaseolina* disease complex on plant growth and fruit yield of *Solanum melongena* cv. Nageena. *Trends in Biosciences*

Haseeb, A., Kumar, V., Shukla, P.K. and Ahmad, A. 2006. Effect of different bioinoculants, organic amendments and pesticides on the management of *Meloidogyne incognita-Fusarium solani* complex on tomato cv. K-25. *Indian J. Nematol.* **36**: 65-69.

Haseeb, A., Rahman, S. and Kumar, V. 2010. Interactive effect of *Meloidogyne incognita* and *Fusarium oxysporum* on growth and disease development of *Ocimum basilicum*. *National Conf. on Innovations in Nematological Research for Agricultural Sustainability–Challenges and A Roadmap Ahead*, Tamil Nadu Agri. Univ., Coimbatore, pp. 113.

Haseeb, A., Sharma, A. and Shukla, P.K. 2005. Studies on the management of root-knot nematode, *Meloidogyne incognita*-wilt fungus, *Fusarium oxysporum* disease complex of green gram, *Vigna radiata* cv ML-1108. *J. Zhejiang Univ. Sci.* **6B**: 736-742.

Haseeb, A. and Shukla, P.K. 2005. Wilt disease complex of pigeonpea and its management. *In*: *Plant Diseases: Biocontrol Management* (S. Nehra, *ed.*), pp. 84-96. Avishkar Publishers, Distributors, Jaipur.

Hawn, E.J. 1963. Transmission of bacterial wilt of alfalfa by *Ditylenchus dipsaci* (Kuhn). *Nematologica* **8**: 65-68.

Hawn, E.J. 1971. Mode of transmission of *Corynebacterium insidiosum* by *Ditylenchus dipsaci*. *J. Nematol.* **3**: 420-421.

Hayman, D.S. 1982. The physiology of vesicular arbuscular endomycorrhizal symbiosis. *Can. J. Bot.* **61**: 944-963.

Hazarika, B.P. and Roy, A.K. 1974. Effect of *Rhizoctonia solani* on the reproduction of *Meloidogyne incognita* on eggplant. *Indian J. Nematol.* **4**: 246-248.

Heald, C.M., Bruton, B.D. and Davis, R.M. 1989. Influence of *Glomus intraradices* and soil phosphorus on *Meloidogyne incognita* infecting *Cucumis melo*. *J. Nematol.* **21**: 69-73.

Heilmann, L.J., Nitzan, N., Johnson, D.A., Pasche, J.S., Doetkett, C. and Gudmestad, N.C. 2006. Genetic variability in the potato pathogen *Colletotrichum coccodes* as determined by AFLP and vegetative compatibility group analyses. *Phytopathology* **96**: 1097–1107.

Hendrix, F.F., Powell, W.M., Owen, J.H. and Campbell, W.A. 1965. *Phytopathology* **55**: 1061.

Hewitt, W.B., Raski, D.J. and Goheen, A.C. 1958. Nematode vector of soil-borne fanleaf virus of grapevines. *Phytopathology* **48**: 586-595.

Hide, G.A. and Corbett, D.C.M. 1973. Controlling early death of potatoes caused by *Heterodera rostochiensis* and *Verticillium dahliae*. *Ann. Appl. Biol.* **5**: 461-462.

Hide, G.A. and Corbett, D.C.M. 1974. Field experiments in the control of *Verticillium dahliae* and *Heterodera rostochiensis* on potatoes. *Ann. Appl. Biol.* **78**: 295-307.

Hide, G.A., Corbett, D.C.M. and Evans, K. 1984. Effects of soil treatments and cultivars on early dying disease of potatoes caused by *Globodera rostochiensis* and *Verticillium dahliae*. *Ann. Appl. Biol.* **104**: 277-89.

Hillocks, R.J. 1986. Localised and systemic effects of root-knot nematode on the incidence and severity of *Fusarium* wilt in cotton. *Nematologica* **32**: 202-208.

Hirano, K. and Kawamura, T. 1972. *Incidence of Complex Disease Caused by* Pratylenchus penetrans *or* P. coffeae *and* Fusarium *Wilt Fungus in Tomato Seedlings.* Tech. Bull., Faculty of Hort., Chiba Univ. **20**: 37-43.

Hogle, J. 2002. Poliovirus cell entry: Common structural themes in viral entry pathways. *Ann. Rev. Microbiol.* **56**: 677-702.

Hoof, H.A. van and Seinhorst, J.W. 1962. *Rhadinaphelenchus cocophilus* associated with little leaf of coconut and oil palm. *Tijdschrift over Plantenziekten* **68**: 521-256.

Hooker, W.G. (*ed.*). 1981. *Compendium of Potato Diseases*. American Phytopathological Society, St. Paul, MN, pp. 125.

Huang, S.P. and Chu, E.Y. 1984. Inhibitory effect of watermelon mosaic virus on *Meloidogyne javanica* (Treub) Chitwood infecting *Cucurbita pepo* L. *J. Nematol.* **16**: 109-112.

Hunt, D.T., Griffin, G.D., Murray, J.J., Pederen, M.W. and Paeaden, R.N. 1971. The effects of root-knot nematodes on bacterial wilt in tobacco. *Phytopathology* **61**: 256-259.

Husain, S.I., Khan, T.A. and Jabri, M.R.A. 1985. Studies on root-knot, root rot and pea mosaic virus complex of *Pisum sativum*. *Nematol. Medit.* **13**: 103-109.

Hussain, Z. and Bora, B.C. 2008. Integrated management of *Meloidogyne incognita* and *Ralstonia solanacearum* complex in brinjal. *Indian J. Nematol.* **38**: 159-164.

Hussain, Z. and Bora, B.C. 2009. Interrelationship of *Meloidogyne incognita* and *Ralstonia solanacearum* complex in brinjal. *Indian J. Nematol.* **39**: 41-45.

Ingham, R.E. 1988. Interactions between nematodes and vesicular-arbuscular mycorrhizae. *Agri. Ecosystems Envt.* **24**: 169-182.

Ingham, R.E., Hamm, P.B., Williams, R.E. and Swanson, W.H. 2000. Control of *Paratrichodorus allius* and corky ringspot disease of potato in the Columbia Basin of Oregon. *J. Nematol.* (Suppl.) **32**: 566-575.

Ishibashi, N. and Choi, D.R. 1991. Biological control of soil pests by mixed application of entomopathogenic and fungivorous nematodes. *J. Nematol.* **23**: 175-181.

Ismail, W., Johri, J.K., Zaidi, A.A. and Singh, B.P. 1979. Influence of root-knot nematode, tobacco mosaic virus complex on the growth and carbohydrates of *Solanum khasianum* Clarke. *Indian J. Exptl. Biol.* **17**: 1266-1267.

Iwaki, M. and Komuro, Y. 1974. Virus isolated from narcissus (*Narcissus* spp.) in Japan. 5. Arabis mosaic virus. *Ann. Phytopathol. Soc. Japan* **40**: 344-353.

Jabri, M., Khan, T.A. Husain, S.I. and Mahmood, K. 1985. Interaction of zinnia mosaic virus with root-knot nematode, *Meloidogyne incognita* on *Zinnia elegans*. *Pakistan J. Nematol.* **3**: 17-21.

Jacobsen, B.J., MacDonald, D.H. and Bissonnette, H.L. 1979. Interaction between *Meloidogyne hapla* and *Verticillium albo-atrum* in soils. *Phytopathology* **69**: 978-981.

Jaffee, B.A. 1986. Parasitism of *Xiphinema rivesi* and *X. americanum* by zoosporic fungi. *J. Nematol.* **18**: 87-93.

Jaffee, B.A. and McInnis, T. 1991. Sampling strategies for detection of density-dependent parasitism of soil-borne nematodes by nematophagous fungi. *Rev. de Nematol.* **14**: 147-150.

Jain, R.K. and Sethi, C.L. 1987. Pathogenicity of *Heterodera cajani* on cowpea as influenced by the presence of VAM fungi, *Glomus fasciculatum* or *G. epigaeus*. *Indian J. Nematol.* **17**: 165-170.

Jatala, P. 1975. Root-knot nematodes (*Meloidogyne* spp.) and their effect on potato quality. *In*: *Proceedings of the 6th Triennial Conference of the EAPR.* Wageningen, Netherlands, 194 pp.

Jatala, P. 1986. Biological control of plant parasitic nematodes. *Ann. Rev. Phytopath.* **24**: 453-489.

Jatala, P. and Martin, C. 1977a. Interactions of *Meloidogyne incognita acrita* and *Pseudomonas solanacearum* on field grown potatoes. *Proc. American Phytopathol. Soc.* **4**: 177-178.

Jatala, P. and Martin, C. 1977b. Interactions of *Meloidogyne incognita acrita* and *Pseudomonas solanacearum* on *Solanum chacoense* and *S. sparsipilum*. *Proc. American Phytopathol. Soc.* **4**: 178.

Jatala, P., Martin, C., Mendoza, H.A., Hidalgo, O.A. and Rincon, R.H. (*eds.*). 1990. Role of nematodes in the expression of *Pseudomonas solanacearum* and strategies for screening and breeding for combined resistance. *Advances en el Mejoramiento Genetico de la Papa en los Paises del cono sur.*, pp.187-189.

Jenkins, S.F. Jr. 1972. Interaction of *Pseudomonas solanacearum* and *Meloidogyne incognita* on bacterial wilt resistant and susceptible cultivars of tomato. *Phytopathology* **62**: 767.

Jenkins, W.R. 1964. A rapid centrifugal-flotation technique for separating nematodes from soil. *Plant Dis. Reptr.* **48**: 692.

Jenkins, W.R. and Coursen, B.W. 1957. The effect of root-knot nematodes, *Meloidogyne incognita acrita* and *M. hapla*, on *Fusarium* wilt of tomato. *Plant Dis. Reptr.* **41**: 182-186.

Jensen, H.J. 1978. Interrelations of nematodes and other organisms in disease complexes. *Int. Potato Centre Rept. of the Second Planning Conf. on the Devp. in the Control of Nematode Pests of Potatoes*, CIP, Lima, Peru.

Jensen, H.J. and Allen, T.C. Jr. 1964a. *Trichodorus allius*, a potential nematode vector of TRV. *Phytopathology* **54**: 1434.

Jensen, H.J. and Allen, T.C. Jr. 1964b. Transmission of tobacco rattle virus by the stubby root nematode *Trichodorus allius. Plant Dis. Reptr.* **48**: 333-334.

Jensen, H.J., Koespell, P.A. and Allen, T.C. Jr. 1974. Tobacco rattle virus and nematode vectors in Oregon. *Plant Dis. Reptr.* **58**: 269-271.

Jha, A. and Posnette, A.F. 1959. Transmission of a virus to strawberry plants by a nematode. *Nature (London)* **184**: 962-963.

Joaquim, T.R. and Rowe, R.C. 1990. Reassessment of vegetative compatibility relationships among strains of *Verticillium dahliae* using nitrate-nonutilizing mutants. *Phytopathology* **80**: 1160-1166.

Joaquim, T.R. and Rowe, R.C. 1991. Vegetative compatibility and virulence of strains of *Verticillium dahliae* from soil and potato plants. *Phytopathology* **81**: 552–558.

Joaquim, T., Smith, V.L. and Rowe, R.C. 1988. Seasonal variation and the effects of wheat rotation on populations of *Verticillium dahliae* Kleb. in Ohio potato field soils. *Ame. Potato J.* **65**: 439-447.

Johnson, A.W. and Littrell, R.H. 1969. Effect of *Meloidogyne incognita, M. hapla* and *M. javanica* on the severity of *Fusarium* wilt of chrysanthemum. *J. Nematol.* **1**: 122-125.

Johnson, A.W. and Littrell, R.H. 1970. Pathogenicity of *Pythium aphanidermatum* to chrysanthemum in combined inoculations with *Belonolaimus longicaudatus* or *Meloidogyne incognita. J. Nematol.* **2**: 255-259.

Johnson, D.A. 1994. Effect of foliar infection caused by *Colletotrichum coccodes* on yield of Russet Burbank potato. *Plant Dis.* **78**: 1075-1078.

Johnson, D.A. and Miliczky, E.R. 1993. Effects of wounding and wetting duration on infection of potato foliage by *Colletotrichum coccodes. Plant Dis.* **77**: 13-17.

Johnson, D.A., Rowe, R.C. and Cummings, T.F. 1997. Incidence of *Colletotrichum coccodes* in certified potato seed tubers planted in Washington State. *Plant Dis.* **81**: 1199-1202.

Johnson, D.A. and Santo, G.S. 2001. Development of wilt in mint in response to infection by two pathotypes of *Verticillium dahliae* and co-infection by *Pratylenchus penetrans. Plant Dis.* **85**: 1189-1192.

Johnson, H.A. and Powell, V.T. 1969. Influence of root-knot nematodes on bacterial wilt development in flue-cured tobacco. *Phytopathololology* **59**: 486-91.

Jonathan, E.I. and Rajendran, G. 1998. Interaction of *Meloidogyne incognita* and *Fusarium oxysporum* f.sp. *cubense* on banana. *Nematol. Medit.* **26**: 9-12.

Jonathan, E.I., Rajendran, G. and Arulmozhiyan, R. 1997. Interaction of *Rotylenchulus reniformis* and *Phytophthora palmivora* on betel vine. *Nematol. Medit.* **25**: 191-193.

Jonathan, E.I., Umamaheswari, R. and Bommaraju, P. 2006. Bioefficacy of native plant growth promoting rhizobacteria against *Meloidogyne incognita* and *Phytophthora capsici. Indian J. Nematol.* **36**: 230-233.

Jones, A.T., McElroy, F.D. and Brown, J.F. 1981. Tests for transmission of cherry leaf roll virus using *Longidorus, Paralongidorus* and *Xiphinema* nematodes. *Ann. Appl. Biol.* **99**: 143-150.

Jones, F.G.W. 1968. *Rep. Rothamsted Exp. Stn. for 1967*, pp. 141-162.

Jones, J.P. and Overman, A.J. 1976. Tomato wilts, nematodes and yields as affected by soil reaction and persistent contact nematicide. *Plant Dis. Reptr.* **60**: 913-917.

Kaitany, R., Melakeberhan, H., Bird G.W. and Safir, G. 2000. Association of *Phytophthora sojae* with *Heterodera glycines* and nutrient stressed soybeans. *Nematropica* **30**: 193-199.

Kaplan, D.T. and Timmer, L.W. 1982. Effects of *Pratylenchus coffeae–Tylenchulus semipenetrans* interactions on nematode population dynamics in citrus. *J. Nematol.* **14**: 368-373.

Karthikeyan, G., Duraisamy, S., Sivakumar, C.V. and Duraisamy, S. 1999. Biological control of *Pythium aphanidermatum, Meloidogyne incognita* disease complex in chilli with organic amendments. *Madras Agric. J.* **86**: 320-323.

Kegler, H., Kleinhempel, H. and Verderevskaja, T.D. 1976. *Acta Hort.* **67**: 209-213.

Khan, A.M., Saxena, S.K. and Khan, M.W. 1971. Interaction of *Rhizoctonia solani* Kuhn and *Tylenchorhynchus brassicae* Siddiqi, 1961 in pre-emergence damping-off of cauliflower seedlings. *Indian J. Nematol.* **1**: 85-86.

Khan, M.M., Khan, M.R. and Reshu. 2010. Evaluation of effectiveness of seed treatment with pesticides and biocontrol agents against the disease complex of pigeonpea in pots. *National Conf. on Innovations in Nematological Research for Agricultural Sustainability–Challenges and A Roadmap Ahead*, Tamil Nadu Agri. Univ., Coimbatore, p. 101.

Khan, M.W. 1993. Mechanisms of interactions between nematodes and other plant pathogens. *In*: *Nematode Interactions* (M.W. Khan, *ed.*), pp. 55-78. Chapman and Hall, Madras, India.

Khan, M.W. and Haider, S.R. 1991. Interaction of *Meloidogyne javanica* with different races of *Meloidogyne incognita. J. Nematol.* **23**: 298-305.

Khan, M.W. and Haq, S. 1979. Interaction between *Meloidogyne incognita* and *Tylenchorhynchus brassicae* on tomato. *Libyan J. Agri.* **8**: 181-186.

Khan, M.W. and Muller, J. 1982. Interaction between *Rhizoctonia solani* and *Meloidogyne hapla* on radish in gnotobiotic culture. *Libyan J. Agri.* **11**: 133-140.

Khan, R.M., Khan, A.M. and Khan, M.W. 1986a. Interaction between *Meloidogyne incognita, Rotylenchulus reniformis* and *Tylenchorhynchus brassicae* on tomato. *Rev. de Nematol.* **9**: 245-250.

Khan, R.M., Khan, A.M. and Khan, M.W. 1986b. Interactions of *Meloidogyne incognita, Rotylenchulus reniformis* and *Tylenchorhynchus brassicae* as coinhabitants on eggplant. *Nematol. Medit.* **14**: 201-206.

Khan, R.M., Khan, A.M. and Khan, M.W. 1987. Competitive interaction between *Meloidogyne incognita, Rotylenchulus reniformis* and *Tylenchorhynchus brassicae* on cauliflower. *Pakistan J. Nematol.* **5**: 73-83.

Khan, R.M. and Khan, M.W. 1986. Antagonistic behaviour of root-knot, reniform, and stunt nematodes in mixed infection. *Int. Nematol. Network Newsl.* **3**: 3.

Khan, R.M., Khan, M.W. and Khan, A.M.1985. Cohabitation of *Meloidogyne incognita* and *Rotylenchulus reniformis* in tomato roots and effect on multiplication and plant growth. *Nematol. Medit.* **13**: 153-159.

Khan, R.M. and Parvatha Reddy, P. 1992. Nematode pests of ornamental crops. *In*: *Nematode Pests of Crops* (D. S. Bhatti and R. K. Walia, *eds.*), pp. 250-257. CBS Publishers & Distributors, Delhi, 381 pp.

Khan, T., Khan, S.T., Fazal, M. and Siddiqi, Z.A. 1997. Biological control of *Meloidogyne incognita* and *Fusarium solani* disease complex in papaya using *Paecilomyces lilacinus* and *Trichoderma harzianum. Int. J. Nematol.* **7**: 127-132.

Khan, T.A. and Husain, S.I. 1988. Studies on the efficacy of *Paecilomyces lilacinus* as biocontrol agent against disease complex caused by the interaction of *Rotylenchulus reniformis, Meloidogyne incognita* and *Rhizoctonia solani* on cowpea. *Nematol. Medit.* **16**: 229-231.

Khan, T.A. and Hussain, S.I. 1989. Relative resistance of six cowpea cultivars as affected by concomitance of two nematodes and a fungus. *Nematol. Medit.* **17**: 39-41.

Khan, T.A. and Hussain, S.I. 1990a. Effect of interaction of various inoculum levels of *Meloidogyne incognita* and *Fusarium solani* on seedling emergence and post-emergence damping-off of papaya seedlings. *New Agric.* **1**: 17-20.

Khan, T.A. and Hussain, S.I. 1990b. Biological control of root-knot and reniform nematodes and root rot fungus on cowpea. *Bioved* **1**: 19-24.

Kheir, A.M. and Osman, A.A. 1977. Interaction of *Meloidogyne incognita* and *Rotylenchulus reniformis. Nematol. Medit.* **5**: 113-116.

Kim, J.I., Choi, D.R., Cho, C.K. and Han, H.C. 1989. Influence of plant parasitic nematodes on occurrence of *Phytophthora* blight on hot pepper and sesame. *In*: *Res. Rept. of the Rural Devp. Admn., Crop Prot.* **31**: 27-30.

Kimenju, J.W., Karanja, N.K. and Macharia, I. 1999. Plant parasitic nematodes associated with common bean in Kenya and the effect of *Meloidogyne* infection on bean nodulation. *African Crop Sci. J.* **7:** 503-510.

Kinloch, R.A. and Allen, M.W. 1972. Interaction of *Meloidogyne hapla* and *M. javanica* infecting tomato. *J. Nematol.* **4**: 7-16.

Kirkpatrick, J.D., Van Gundy, S.D. and Martin, J.P. 1965. Effects of *Xiphinema index* on growth and abscission in Carignane grape, *Vitis vinifera. Nematologica* **11**: 41.

Kishore, R., Kamalwanshi, R.S. and Pandey, M. 2005. Effect of root-knot nematode with *Fusarium solani* on damping-off of papaya. *Indian J. Nematol.* **35**: 96-97.

Kleineke-Borchers, A. and Wyss, U. 1982. Investigations on changes in susceptibility of tomato plants to *Fusarium oxysporum* f. sp. *lycopersici* after infection by *Meloidogyne incognita*. *Zeitschrift fur Pflanzenkrankheiten und Pflanzenschutz* **89**: 67-78.

Klos, E.J., Fronek, F., Knierim, J.A. and Cation, D. 1967. Peach rosette mosaic transmission and control studies. *Mich. Agric. Exp. Sta. Q. Bull.* **49**: 287-293.

Ko, M.P., Barker, K.R. and Huang, J.S. 1984. Nodulation of soybeans as affected by half-root infection with *Heterodera glycines*. *J. Nematol.* **16**: 97-105.

Kotcon, J.B., Bird, G.W., Rose, L.M. and Dimoff, K. 1985. Influence of *Glomus fasciculatum* and *Meloidogyne hapla* on *Allium cepa* in organic soils. *J. Nematol.* **17**: 55-60.

Kotcon, J.B., Rouse, D.I. and Mitchell, J.E. 1985. Interactions of *Verticillium dahliae, Colletotrichum coccodes, Rhizoctonia solan*i and *Pratylenchus penetrans* in the early dying syndrome of Russet Burbank potatoes. *Phytopathology* **75**: 68-74.

Kraaijenga, D.A. and Ouden, H. den. 1966. 'Red ring' disease in Surinam. *Tijdschrift over Plantenziekten* **72**: 20-27.

Krikun, J. and Orio, D. 1979. *Verticillium* wilt of potato: Importance and control. *Phytoparasitica* **7**: 107-116.

Krusberg, L.R. 1963. Host response to nematode infection. *Ann. Rev. Phytopath.* **1**: 219-240.

Kumar, P. and Kamalwanshi, R.S. 2009. Pathogenicity and interrelationship of *Meloidogyne incognita* and *Fusarium oxysporum* f. sp. *pisi* on pea. *Ann. Plant Prot. Sci.* **17**:

Kumar, S. and Sivakumar, C.V. 1981. Disease complex involving *Rotylenchulus reniformis* and *Rhizoctonia solani* in okra. *Indian J. Nematol.* 11:

Kumar, V. and Haseeb, A. 2009. Interactive effect of *Meloidogyne incognita* and *Rhizoctonia solani* on the growth and yield of tomato. *Indian J. Nematol.* **39**: 178-181.

Kumar, V., Haseeb, A. and Sharma, A. 2009a. Integrated management of *Meloidogyne incognita-Fusarium solani* disease complex of brinjal cv. Pusa Kranti. *Ann. Plant Prot. Sci.* **17**:

Kumar, V., Haseeb, A. and Sharma, A. 2009b. Integrated management of *Meloidogyne incognita* and *Fusarium solani* disease complex of chilli. *Indian Phytopath.* **62**: 324-327.

Labuy'ere, R.E., Den Ouden, H. and Seinhorst, J.W. 1959. *Nematologica* **4**: 336-343.

Lacey, M.S. 1936. Studies in bacteriosis. XXII. The isolation of a bacterium associated with fasciation of sweet peas, cauliflower of strawberry and leafy gall of various plants. *Ann. Appl. Biol.* **23**: 11-15.

Lacey, M.S. 1942. Studies in bacteriosis. XXV. Studies on a bacterium associated with leafy galls, fasciations and cauliflower disease of various plants. Part IV. The inoculation of strawberry plants with *Bacterium fascians* Tilford. *Ann. Appl. Biol.* **29**: 302-310.

Lamberti, F. 1981. Combating nematode vectors of viruses. *Plant Dis.* **65**: 113-117.

Lamberti, F. and Bleve-Zacheo, T. 1979. Studies on *Xiphinema americanum sensu lato* with descriptions of fifteen new species (Nematoda: Longidoridae). *Nematol. Medit.* **7**: 51-106.

Lamberti, F. and Ciancio, A. 1993. Diversity of *Xiphinema americanum*-group species and hierarchical cluster analysis of morphometrics. *J. Nematol.* **25**: 332-343.

Lamberti, F. and Roca, F. 1987. Present status of nematodes as vectors of plant viruses. *In*: *Vistas on Nematology* (J.A. Veech and D.W .Dickson, Eds), pp 321-328. Society of Nematologists, Hyattsville, Maryland. 509 pp.

LaMondia, J.A. 1999a. The effects of *Pratylenchus penetrans* on strawberry vigor and yield. *J. Nematol.* **31**: 418-423.

LaMondia, J.A. 1999b. The effects of rotation crops on the strawberry pathogens *Pratylenchus penetrans, Meloidogyne hapla,* and *Rhizoctonia fragariae. J. Nematol.* **31**:650-655.

LaMondia, J.A., Gent, M.P.N., Ferrandino, F.J., Elmer, W.H. and Stoner, K.A. 1999. Effect of compost amendment or straw mulch on potato early dying disease. *Plant Dis.* **83**: 361-366.

LaMondia, J.A. and Halbrendt, J.M. 2003. Differential host status of rotation crops to dagger, lesion and root- knot nematodes. *J. Nematol.* **35**: 349.

LaMondia, J.A. and Martin, S.B. 1989. The influence of *Pratylenchus penetrans* and temperature on black root rot of strawberry by binucleate *Rhizoctonia* spp. *Plant Dis.* **73**: 107-110.

Lanjewar, R.D. and Shukla, V.N. 1985. Parasitism and interaction between *Pythium myriotylum* and *Meloidogyne incognita* in soft rot complex of ginger. *Indian J. Nematol.* **15**: 170-173.

Lees, A.K. and Hilton, A.J. 2003. Black dot (*Colletotrichum coccodes*): an increasingly important disease of potato. *Plant Pathol.* **52**, 3-12.

Lele, V.C., Durgapal, J.C., Agarwal, D.K. and Sethi, C.L. 1978. Crown and root gall of grape (*Vitis vinefera* L.) in Andhra Pradesh. *Curr. Sci.* **47**: 280.

Libman, G. and Leach, J.G. 1962. A study of the role of some nematodes in the incidence and severity of southern bacterial wilt of tomato. *Phytopathology* **52**: 1219.

Libman, G., Leach, J.G. and Adams, R.E. 1964. Role of certain plant parasitic nematodes in infection of tomatoes by *Pseudomonas solanacearum*. *Phytopathology* **54**: 151-153.

Liburd, O.W. 1977. *Interactions Involving Root-knot Nematodes,* Fusarium oxysporum *f. sp.* lycopersici *and Tomatoes*. Ph. D. thesis, Cornell Univ., Ithaca, New York.

Lider, L.A., Olmo, H.P. and Goheen, A.C. 1988. Hybrid Grapevine Rootstock Patent No. 6166, U.S. Patent Office.

Lingaraju, S. and Mallesh, S.B. 2010. Emerging nematode diseases in two horticultural crops: Their management thro' novel approaches. *National Conf. on Innovations in Nematological Research for Agricultural Sustainability–Challenges and A Roadmap Ahead*, Tamil Nadu Agri. Univ., Coimbatore, p. 51.

Lister, R.M. 1964. Strawberry latent ringspot: A new nematode-borne virus. *Ann. Appl. Biol.* **54**: 167-176.

Littrel, R.H. and Johnson, A.W. 1969. Pathogenicity of *Pythium aphanidermatum* to chrysanthemum in combined inoculations with *Belonolaimus longicaudatus* and *Meloidogyne incognita*. *Phytopathology* **59**: 115-116.

Loos, C.A. 1959. Symptom expression of *Fusarium* wilt disease of the Gros Michel banana in the presence of *Radopholus similis* (Cobb, 1893) Thorne, 1949 and *Meloidogyne incognita acrita* Chitwood, 1949. *Proc. Helminthol. Soc. Wash.* **26**: 103-111.

Lownsberry, B.F., English, H., Noel, G.R. and Shick, F.J. 1977. Influence of Nemaguard and Lovell rootstocks and *Macroposthonia xenoplax* on bacterial canker of peach. *J. Nematol*. **9**: 221-224.

Lucas, G.B., Sasser, J.N. and Kelman, A. 1955. The relationship of root-knot nematodes to Granville wilt resistance in tobacco. *Phytopathology* **45**: 537-540.

Mace, M.E., Bell A.A. and Beckman, C.H. (*eds.*). 1981. *Fungal Wilt Diseases of Plants*. Academic Press. New York. p. 640.

MacFarlane, S.A. and Brown, D.J.F. 1995. Sequence comparison of RNA2 of nematode transmissible and nematode non-transmissible isolates of *Pea early browning virus* suggests that gene encoding the 29 kDa protein may be involved in nematode transmission. *J. Gen. Virol.* **76**: 1299-1304.

MacFarlane, S.A., Brown, D.J.F. and Bol, J.F. 1995. The transmission by nematodes of tobraviruses is not determined exclusively by the virus coat protein. *European J. Plant Pathol.* **101**: 535-539.

MacFarlane, S.A., Neilson, R. and Brown, D.J.F. 2002. Nematodes. *In*: *Advances in Botanical Research* (R.T. Plumb, *ed.*), pp. 169-198. Academic Press, San Diego, CA, USA.

MacGuidwin, A.E., Bird, G.W. and Safir, G.R. 1985. Influence of *Glomus fasciculatum* on *Meloidogyne hapla* infecting *Allium cepa*. *J. Nematol.* **17**: 389-395.

MacGuidwin, A.E. and Rouse, D.I. 1990. Role of *Pratylenchus penetrans* in the potato early dying disease of Russet Burbank potato. *Phytopathology* **59**: 1077-1082.

Mahmood, K., Qamar, S., Naqvi, A. and Alam, M.M. 1974. The influence of bottle gourd mosaic virus and brinjal mosaic virus on the development of root-knot caused by *Meloidogyne incognita* (Kofoid and White, 1919) Chitwood, 1949. *Indian Phytopath.* **27**: 610-611.

Mai, W.F. and Abawi, G.S. 1987. Interaction among root-knot nematodes and Fusarium wilt fungi on host plants. *Ann. Rev. Phytopath.* **25**: 317-338.

Mai, W.F., Bloom, J.R. and Chen, T.A. (*eds.*). 1977. *Biology and Ecology of the Plant - Parasitic Nematode* Pratylenchus penetrans. *Penn. State Univ. Bul.* 815.

Malaguti, G. 1953. "Pudrición del cogollo" de la palma de aceite africana (*E. guineensis*) en Venezuela. *Agron. Trop.* **3**: 13-31.

Malek, R.B. and Jenkins, W.R. 1964. Aspects of the host-parasite relationships of nematodes and hairy vetch. *New Jersey Agri. Expt. Sta. Bull.* **813**: p. 31.

Mali, V.R., Vanek, G. and Bojnansky, V. 1975. Transmission by nematodes of some grapevine viruses occurring in Czechoslovakia and Hungary. *Pol'nohospodarstiv A*, No. 3, p. 132.

Mallesh, S.B. and Lingaraju, S. 2009. Enhancement of plant growth and suppression of wilt complex of *Coleus forskohlii* through fluorescent pseudomonads. *Int. Conf. on Hort.*, Bangalore.

Mamiya, Y. 1984. The pine wood nematode. *In*: *Plant and Insect Nematodes* (W.R. Nickle, *ed.*), pp. 587-626. Marcel Dekker, New York.

Manoj Kumar, R., Rao, M.S. and Ramachandran, N. 2010. Biomanagement of disease complex caused by *Meloidogyne incognita* and *Phytophthora parasitica* Dastur in protected cultivation of gerbera using formulation of *Trichoderma harzianum* and *Pseudomonas fluorescens. National Conf. on Innovations in Nematological Research for Agricultural Sustainability – Challenges and A Roadmap Ahead*, Tamil Nadu Agri. Univ., Coimbatore, pp. 97-98.

Marley, P.S. and Hillocks, R.J. 1993. The role of phytoalexins in resistance to *Fusarium* wilt in pigeonpea (*Cajanus cajan*). *Plant Pathol.* **42**: 212-218.

Marley, P.S. and Hillocks, R.J. 1994. Effect of root-knot nematodes on cajanol accumulation in the vascular tissues of pigeonpea after stem inoculation with *Fusarium udum*. *Plant Pathol.* **43**: 172-176.

Mansen, A.J., Nyland, G., McElroy, F.D. and Stace-Smith, R. 1974. Origin, cause, host range and spread of cherry rasp leaf disease in North America. *Phytopathology* **64**: 721-727.

Marshall, J.W. 1989. Changes in relative abundance of two potato cyst nematode species *Globodera rostochiensis* and *G. pallida* in one generation. *Ann. Appl. Biol.* **115**: 79-87.

Martelli, G. and Taylor, C.E. 1989. Distribution of viruses and their nematode vectors. *Adv. Dis. Vector Res.* **6**: 151-189.

Martelli, G.P. and Savino, V. 1988. Fanleaf degeneration. *In*: *Compendium of Grape Diseases* (R. C. Pearson and A. C. Goheen, *eds.*), pp 48-49. American Phytopathological Society Press, St Paul, Minnesota, p. 93.

Martin, F.N. and Bull, C.T. 2002. Biological approaches for control of root pathogens of strawberry. *Phytopathology* **92**: 1356-1362.

Martin, M.J., Riedel, R.M. and Rowe, R.C. 1981. *Pratylenchus penetrans* and *Verticillium dahliae*–causal agents of early dying in *Solanum tuberosum* cv. Superior, in Ohio. *J. Nematol.* **13**: 449.

Martin, M.J., Riedel, R.M. and Rowe, R.C. 1982. *Verticillium dahliae* and *Pratylenchus penetrans*: Interactions in the early dying complex of potato in Ohio. *Phytopathology* **72**: 640-644.

Martin, S.B. 1988. Identification, isolation frequency, and pathogenicity of anastomosis groups of binucleate *Rhizoctonia* spp. from strawberry roots. *Phytopathology* **78**: 379-384.

Masefield, G.B. 1958. Some factors affecting nodulation in the tropics. *In*: *Nutrition of the Legumes* (E.G. Hallsworth, *ed.*), pp. 202-215. Butterworths Scientific Publications, London.

Mass, P.W.T. 1974. Soil disinfection to prevent transmission of tobacco rattle virus to potatoes and flower bulbs. *Agri. Envt.* **1**: 329-238.

Mayol, P.S. and Bergeson, G.B. 1970. The role of secondary invaders in *Meloidogyne incognita* infection. *J. Nematol.* **2**: 80-83.

McCarthy M.G. and Cirami, R.M. 1990. The effect of rootstocks on the performance of Chardonnay from a nematode-infested Barossa Valley vineyard. *Ame. J. Enol. and Viticulture* **41**: 126-130.

McElroy, F.D, Brown, D.J.F. and Boag, B. 1977. The virus-vector and damage potential, morphometrics and distribution of *Paralongidorus maximus*. *J. Nematol.* **9**: 122-130.

McGuire, J.M. and Good, M.J. 1970. The effect of benomyl on *Xiphinema americanum* and tobacco ring spot virus infection. *Phytopathology* **60**: 1150-1151.

McKeen, C.D. and Mountain, W.B. 1960. Synergism between *Pratylenchus penetrans* (Cobb) and *Verticillium albo-atrum* R. and B. in eggplant wilt. *Can. J. Bot.* **38**: 789-794.

McKenry, M.V. 1992. Nematodes. *In*: *Grape Pest Management* (D.L. Flaherty, L.P. Christensen, W.T. Lanini, J.J. Marois, P.A. Phillips and L.T. Wilson, *eds.*), pp. 279-293. University of California Publication 3343, Oakland, California.

McKenry, M.V. and Kretsch, J.O. 1994. Interactions of selected *Vitis* rootstocks with ectoparasitic nematodes. *Ame. J. Enol. and Viticulture* (In press).

McKenry, M.V., Viveros, M. and Teviotdale, B. 1990. *Criconema mutabile* associated with bacterial canker and Nemaguard rootstock. *Plant Dis.* **74**: 394.

McKinley, R.R.T. and Tolboys, P.W. 1979. Effects of *Pratylenchus penetrans* on development of strawberry wilt caused by *Verticillium dahliae*. *Ann. Appl. Biol.* **92**: 347-357.

Meagher, J.W. and Jenkins, P.T. 1970. Interaction of *Meloidogyne hapla* and *Verticillium dahliae* and chemical control of wilt in strawberry. *Australian J Exp. Agric. Ani. Hus.* **10**: 493-496.

Melakeberhan, H. and Ferris, H. 1988. Growth and energy demand of *Meloidogyne incognita* on susceptible and resistant *Vitis vinifera* cultivars. *J. Nematol.* **20**: 545-554.

Menge, J.A. 1982. Effect of soil fumigants and fungicides on vesicular-arbuscular fungi. *Phytopathology* **72**: 1125-1132.

Mervosh, T.L. and LaMondia, J.A. 2004. Strawberry black root rot and berry yield are not affected by use of terbacil herbicide. *Hort. Sci.* **39**: 1339-1342.

Miller, L.I. 1983. Diversity of selected taxa of *Globodera* and *Heterodera* and their interspecific and intergeneric hybrids. *In*: *Concepts in Nematode Systematics* (A.R. Stone, A.M. Platt and L.F. Khalil, *eds.*), pp. 207-220. Acad. Press, New York.

Miller, P.M. 1975. Effect of the tobacco cyst nematode, *Heterodera tobaccum,* on severity of *Verticillium* and *Fusarium* wilts of tomato. *Phytopathology* **65**: 81-82.

Miller, R.W. 1994. Estimated peach tree losses 1980 to 1992 in South Carolina: Causes and economic impact. *In*: *Proc. 6th Stone Fruit Tree Decline Workshop*, Oct. 26-28, 1992, Ft. Valley, GA, pp. 121-127.

Minz, G. and Strich-Harari, D. 1959. Inoculation experiments with a mixture of *Meloidogyne* spp. on tomato roots. *Ktavin Records Agri. Expt. Sta.* **9**: 275-279.

Misra, C. and Das, S.N. 1979. Interaction of some plant parasitic nematodes on the root-knot development in brinjal. *Indian J. Nematol.* **7**: 46-53.

Misra, S.L. and Edward, J.C. 1977. Observations on histopathology of some of the more important rootstocks for commercial citrus affected by citrus nematodes. *Allahabad Farmer* **53**: 7-18.

Mojtahedi, H., Boydston, R.A., Thomas, P.E., Crosslin, J.M and Boydston, R.A. 2002. Eliminating tobacco rattle virus from viruliferous *Paratrichodorus allius* and establishing a new virus-vector combination. *J. Nematol.* **34**: 66-69.

Mojtahedi, H. and Lownsberry, B.F. 1976. The effects of ammonia-generating fertilizer on *Criconemoides xenoplax* in pot cultures. *J. Nematol.* **8**: 306-309.

Mojtahedi, H., Lownsberry, B.F. and Moody, E.H. 1975. Ring nematodes increase development of bacterial canker in plums. *Phytopathology* **65**: 556-559.

Mojtahedi, H., Santo, G.S., Crosslin, J.M., Brown, C.R. and Thomas, P.E. 2000. Corky ringspot disease: Review of the current situation. *Proc. 39th Annual Washington State Potato Conference*. Moses Lake, WA.

Morandi, D. 1987. VA mycorrhizae, nematodes, phosphorus and phytoalexins on soybean. *In*: *Mycorrhizae in the Next Decade, Practical Applications and Research Priorities.* Proc. of the 7th N. American Conf. on Mycorrhizae (D.M. Sylvia, L.L. Hung and J.H. Graham, *eds.*), p. 212. Inst. of Food and Agri. Sci., Univ. of Fla., Gainesville.

Morsink, F. and Rich, A.E. 1968. *Phytopathology* **58**: 401.

Moura, R.M., de Echandi, E. and Powell, N.T. 1975. Interaction of *Corynebacterium michiganense* and *Meloidogyne incognita* on tomato. *Phytopathology* **65**: 1332-1335.

Mowat, W.P. and Taylor, C.E. 1962. Nematode-borne viruses. *Rept. Scot. Hort. Res. Inst.* **9**: 67-68.

Mpofu, S.I. and Hall, R. 2002. Effect of annual sequence of removing or flaming potato vines and fumigating soil on Verticillium wilt of potato. *Ame. J. Potato Res.* **79**: 1-7.

Muller, J. 1973. Zum Einfluss von *Pratylenchus penetrans* auf die *Verticillium*-Welke von *Impatiens balsamina*. *Zeitschrift fur Pflanzenkrankheiten und Pflanzenschutz* **80**: 295-311.

Munnecke, D.E., Chandler, P.A. and Starr, M.P. 1963. Hairy root (*Agrobacterium rhizogenes*) of field roses. *Phytopathology* **53**: 788-799.

Murant, A.F., Jones, A.T., Martelli, G.P. and Stace-Smith, R. 1990. Nepoviruses: Diseases and virus identification. *In*: *The Plant Viruses* (B.D. Harrison and A.F. Murant, *eds.*). Plenum Press, New York.

Murant, A.F. and Taylor, C.E. 1965. Treatment of soil with chemicals to prevent transmission of tomato black ring and raspberry ring spot virus by *Longidorus elongatus*. *Ann. Appl. Biol.* **55**: 227-237.

Mustafa, U. and Khan, M.R. 2004. Management of disease complex caused by *Fusarium oxysporum* f. sp. *gladioli* and *Meloidogyne incognita* on gladiolus. *National Symposium on Paradigms in Nematological Research for Biodynamic Farming*, Univ. of Agri. Sci., Bangalore, pp. 91-92.

Mustika, I. 1984. Effects of nematodes and fungi on the growth of pepper and yellow disease. *Pemberitan Penelitan Tanaman Industri* **8**: 28-37.

Mustika, I., Suardgat, D. and Wikanda, A. 1984. Control of pepper yellow disease with fertilizer and pesticides. *Pemberitan Penelitan Tanaman Industri* **8**: 37-43.

Muthulakshmi, M., Devarajan, K. and Jonathan, E.I. 2010. Interaction of *Meloidogyne incognita* and *Macrophomina phaseolina* on mulberry (*Morus alba* L.). *National Conf. on Innovations in Nematological Research for Agricultural Sustainability–Challenges and A Roadmap Ahead*, Tamil Nadu Agri. Univ., Coimbatore, p. 108.

Nagesh, M., Chakrabarti, S.K., Shekhawat, G.S. and Gadewar, A.V. 1997. Evaluation of potato accessions for their combined resistance to *Pseudomonas solanacearum* and *Meloidogyne incognita*. *Pest Mangmt in Hortil. Ecosystems* **3**:17-20.

Nagesh, M., Parvatha Reddy, P. and Ramachandran, N. 1998. Integrated management of *Meloidogyne incognita* and *Fusarium oxysporum* f sp. *gladioli* in gladiolus using antagonistic fungi and neem cake. *In*: *Nematology - Challenges & Opportunities in 21st Century* (Usha K. Mehta, *ed.*), pp. 263-266. Sugarcane Breeding Inst., Coimbatore.

Naik, D. 2003. *Biotechnological Approaches for the Management of Wilt Disease Complex in Capsicum (*Capsicum annuum *L.) and Egg Plant (*Solanum melongena *L.) with Special Emphasis on Biological Control*. Ph. D. thesis, Kuvempu Univ., Shimoga, p. 201.

Naik, D., Rao, M.S., Shylaja, M., Reddy, M.S. and Rahiman, B.A. 2003. Management of wilt disease complex in capsicum using *Bacillus pumilus* and *Paecilomyces lilacinus. Proc. of Int. Plant Growth Promoting Rhizobacteria Workshop*, Calicut, India, pp. 276-281.

Napiere, C.M. 1980. Varying inoculum levels of bacteria, nematodes and the severity of tomato bacterial wilt. *Ann. Trop. Res.* **2**: 129-134.

Napiere, C.M. and Quinio, A.J. 1980. Influence of root-knot nematode on bacterial wilt severity in tomato. *Ann. Trop. Res.* **2**: 29-39.

Naqvi, S.Q.A. and Alam, M.M. 1975. Influence of brinjal mosaic virus on the population of *Tylenchorhynchus brassicae* and *Rotylenchulus reniformis* around eggplant roots. *Geobios* **2**: 120-121.

Nath, R. and Kamalwanshi, R.S. 1989. Role of non-pathogenic fungi in presence of root-knot nematode in damping-off of tomato seedlings. *Indian J. Nematol.* **19**: 84-85.

Nath, R., Singh, R.K. and Haider, M.G. 1976. Combined effect of *Fusarium moniliformae* and *Hoplolaimus indicus* on sugarcane plants. *Sugarcane Pathologists' Newsl.* **17**: 24-25.

Nayar, K., Mathew, J., Kuriyan, J., Gnanamanickam, S.S. and Mahadevan, A. (*eds.*). 1988. Role and association of root-knot nematode (*Meloidogyne incognita*) in the bacterial wilt of brinjal incited by *Pseudomonas solanacearum*. *In*: *Advances in Research on Pant Pathogenic Bacteria*. Based on the proceedings of the *National Symposium on Phytobacteriology*, University of Madras, Madras, India, pp. 45-48.

Ndubizu, T.O.C. 1977. Effect of earthworms, nematodes, cultivation and host plants on *Verticillium* wilt of peach and cherry. *Ann. Appl. Biol.* **86**: 153-161.

Negron, J.A. and Acosta, N. 1989. *Fusarium oxysporum* f. sp. *coffeae - Meloidogyne incognita* complex in 'Bourbon' coffee. *Nematropica* **19**: 161-168.

NeSmith, W.C. and W.M. Dowler. 1976. Cultural practices affect cold hardiness and peach tree short life. *J. Amer. Soc. Hort. Sci.* **101**: 116-119.

Newhall, A.G. 1958. The incidence of Panama disease of banana in the presence of the root-knot and the burrowing nematodes (*Meloidogyne incognita* and *Radopholus similis*). *Plant Dis. Reptr.* **42**: 853-856.

Nickel, W.R. and Welch, H.E. 1984. History, development and importance of insect nematology. *In*: *Plant and Insect Nematodes* (W.R. Nickle, *ed.*), pp. 627-653. Marcel Dekker, New York.

Nicot, P.C. and Rouse, D.I. 1987. Precision and bias of three quantitative soil assays for *Verticillium dahliae*. *Phytopathology* **77**: 875-881.

Nigh, E.L. 1966. Incidence of crown gall infection in peach as affected by the Javanese root-knot nematode. *Phytopathology* **56**: 150.

Nirula, K.K. and Paharia, K.D. 1970. Role of root-knot nematodes in spread of brown rot in potatoes. *Indian Phytopath.* **23**: 158-159.

Nitzan, N., Cummings, T.F. and Johnson, D. 2005. Effect of seed-tuber germination, soil-borne inoculum, and azoxystrobin application on development of potato black dot caused by *Colletotrichum coccodes. Plant Dis.* **89**: 1181-1185.

Nitzan, N., Evans, M. and Johnson, D. 2006b. Colonization of potato plants after aerial infection by *Colletotrichum coccodes*, causal agent of potato black dot. *Plant Dis.* **90**: 999-1003.

Nitzan, N., Tsror (Lahkim), L. and Johnson, D. 2006a. Vegetative compatibility groups and aggressiveness of North American isolates of *Colletotrichum coccodes*, the causal agent of potato black dot. *Plant Dis.* **90**: 1287-1292.

Noguera, G.R. 1982. Alteration in the production of rishitin in roots and tyloses in stems and interaction of *Meloidogyne-Fusarium* in tomato plants. *Agronomia Tropical* **32**: 303-308.

Noguera, G.R. and Smits, B.G. 1982. Variations in the rhizosphere microflora of tomato infected with *Meloidogyne incognita. Agronomia Tropical* **32**: 147-154.

Norton, D.C. 1978. *Ecology of Plant Parasitic Nematodes*. John Wiley, New York.

Novy, R.G., Corsini, D.L., Love, L.S., Pavek, J.J. and Mosley, A.R. 2003. Alturas: a multi-purpose, russet potato cultivar with high yield and tuber specific gravity. *Ame. J. Potato Res.* **80**: 295-301.

Nyczepir, A.P. and Lewis, S.A. 1984. Incidence of *Fusarium* and *Pythium* spp. in peach feeder roots as related to dibromochloropropane application for control of *Criconemella xenoplax*. *Plant Dis.* **68**: 497-499.

Nyczepir, A.P. and Wood, B.W. 1988. Peach leaf senescence delayed by *Criconemella xenoplax*. *J. Nematol.* **20**: 585-589.

Nyland, G., Lownsbery, B.F., Lowe, S.K. and Mitchell, J.F. 1969. The transmission of cherry rasp leaf virus by *Xiphinema americanum*. *Phytopathology* **59**: 1111-1112.

O'Bannon, J.H., Inserra, R.N., Nemec, S. and Volvos, N. 1979. The influence of *Glomus mosseae* on *Tylenchulus semipenetrans*-infected and uninfected *Citrus limon* seedlings. *J. Nematol.* **11**: 247-250.

O'Bannon, J.H., Leathers, C.R. and Reynolds, H.W. 1967. *Phytopathology* **57**: 414-417.

O'Bannon, J.H. and Nemec, S. 1979. The response of *Citrus limon* seedlings to a symbiont, *Glomus etunicatus*, and a pathogen, *Radopholus similis*. *J. Nematol.* **11**: 270-275.

O'Bannon, J.H., Radewald, J.D., Tomerlin, A.T. and Inserra, R.N. 1976. Comparative influence of *Radopholus similis* and *Pratylenchus coffeae* on citrus. *J. Nematol.* **8**: 58-63.

Oehlschlager, A.C., Chinchilla, C., Castillo, G, and Gonzalez, L. 2002. Control of red ring disease by mass trapping of *Rhynchophorus palmarum* (Coleoptera: Curculionidae). *The Florida Entomologist* **85**: 507-513.

Okie, W.R., Beckman, T.G., Nyczepir, A.P., Reighard, G.L., Newall, W.C. Jr. and Zehr, E.I. 1994. BY520-9, a peach rootstock for the southeastern United States that increases scion longevity. *Hort. Sci.* **29**:705-706.

Olien, W.C., Graham, C.J., Hardin, M.E. and Bridges, W.C. Jr. 1995. Peach rootstock differences in ring nematode tolerance related to effects on tree dry weight, carbohydrate and prunasin contents. *Physiol. Plant. Path.* **94**: 117-123.

Oostenbrink, M. 1966. Major characteristics of relation between nematodes and plants. *Meded. Landb. Hoges. Ch. Wageningen*, p. 46.

Orion, D. and Krikum, J. 1976. Response of *Verticillium* resistant and *Verticillium* susceptible tomato varieties to inoculation with the nematode *Meloidogyne javanica* and with *Verticillium dahliae. Phytoparasitica* **4**: 41-44.

Orion, D., Levy, Y., Israeli, Y. and Fischer, E. 1999. Scanning electron microscope observations on spiral nematode (*Helicotylenchus multicinctus*)-infested banana roots. *Nematropica* **29**: 179-183.

Orton Williams, K.J. 1972. *Macroposthonia xenoplax*. Commonwealth Institute of Helminthology, Descriptions of Plant-parasitic Nematodes. No. 12.

Otazu, V., Gudmestad, N.C. and Zink, R.T. 1978. The role of *Colletotrichum atramentarium* in the potato wilt complex in North Dakota. *Plant Dis. Reptr.* **62**: 847-851.

Oteifa, B.A. and Salem, A.A. 1972. Biology and histopathogenesis of the reniform nematode, *Rotylenchulus reniformis,* on Egyptian cotton, *Gossypium barbadense*. *Proc. Third Cong. Medt. Phytopath. Union*, Oeiras, Portugal, pp. 299-304.

Overman, A.J. and Jones, J.P. 1970. Effect of stunt and root-knot nematodes on *Verticillium* wilt of tomato. *Phytopathology* **60**: 1306.

Oyekan, P.O. and Mitchell, J.E. 1971. Effect of *Pratylenchus penetrans* on the resistance of pea variety to *Fusarium* wilt. *Plant Dis. Reptr.* **55**: 1032-1035.

Pani, A.K. and Das, S.N. 1972. Studies on etiological complexes in plant disease. I. Association of root-knot nematodes in bacterial wilt of tomato. *J. Res. Orissa Univ. Agri. Technol.* **2**: 54-59.

Parrot, D.M., Berry, M.M. and Farell, K.M. 1976. Competition between *Globodera rostochiensis* Ro 1 and *G. pallida* Pa 3. *In*: *Nematol. Dept. Rothamsted Expt. Sta. Rept. 1975* (F.G.W. Jones, *ed.*), pp. 197-198.

Parvatha Reddy, P., Singh, D.B. and Ram Kishun. 1979. Effect of root-knot nematodes on the susceptibility of Pusa Purple Cluster brinjal to bacterial wilt. *Curr. Sci.* **48**: 915-916.

Parvatha Reddy, P., Singh, D.B. and Sharma, S.R. 1979. Interaction of *Meloidogyne incognita* and *Rhizoctonia solani* in a root rot disease complex of French bean. *Indian Phytopath.* **32**: 651-652.

Pathak, K.N. and Keshar, N. 2004. Interaction of *Meloidogyne incognita* with *Fusarium oxysporum* f. sp. *conglutinans* on cauliflower. *Indian J. Nematol.* **34**: 85-87.

Pathak, K.N., Nath, R.P. and Haider, M.G. 1985. Effect of initial inoculum level of *Meloidogyne incognita* and *Rotylenchulus reniformis* on pigeonpea and their interrelationship. *Indian J. Nematol.* **15**: 177-179.

Perkins, H.H., Giddens, J.E. and Daniell, J.W. 1979. Effect of lime and fertilizer and peach root residues on peach trees in a decline area. *Univ. Ga. Res. Bull.* 246.

Perveen, S., Ehteshamul-Haque, S. and Ghaffar, A. 1998. Efficacy of *Pseudomonas aeruginosa* and *Paecilomyces lilacinus* in the control of root rot-root knot disease complex on some vegetables. *Nematol. Medit.* **26**: 209-212.

Perveen, K., Haseeb, A. and Shukla, P.K. 1988. Influence of root-knot nematode and wilt fungus on the growth of pigeonpea. *In*: *Nematology: Challenges and Opportunities in 21st Century* (Usha K. Mehta, ed,), pp. 80-86. Sugarcane Breeding Inst., Coimbatore.

Peters, R.D., Sturz, A.V., Carter, M.R. and Sanderson, J.B. 2003. Developing disease-suppressive soils through crop rotation and tillage management practices. *Soil and Tillage Res.* **72**: 181-192.

Peters, R.D., Sturz, A.V., Carter, M.R. and Sanderson, J.B. 2004. Influence of crop rotation and soil conservation tillage practices on the severity of soilborne potato diseases in temperate humid agriculture. *Canadian J. Soil Sci.* **84**: 397-402.

Pinkerton, J.N., Ivors, K.L., Bristow, P.R. and Windom, G.E. 2002. The use of soil solarization for the management of soilborne plant pathogens in strawberry and red raspberry production. *Plant Dis.* **86**: 645-651.

Pinochet, J. and Raski, D.J. 1975. Four new species of the genus *Hemicriconemoides* (Nematoda: Criconematidae). *J. Nematol.* **7**: 263-270.

Pinochet, J., Raski, D.J. and Goheen, A.C. 1976. Effects of *Pratylenchus vulnus* and *Xiphinema index* singly and combined on vine growth of *Vitis vinifera*. *J. Nematol.* **8**: 330-335.

Pinochet, J., Raski, D.J. and Jones, N.O. 1976. Effect of *Helicotylenchus pseudorobustus* on 'Thompson Seedless' grape. *Plant Dis. Reptr.* **60**: 528-529.

Pitcher, R.S. 1963. Role of plant parasitic nematodes in bacterial diseases. *Phytopathology* **53**: 35-39.

Pitcher, R.S. 1965. Interrelationships of nematodes and other pathogens of plants. *Helminthol. Abstr.* (Series B) **34**: 1-17.

Pitcher, R.S. 1978. Interaction of nematodes with other pathogens. *In*: *Plant Nematology*, Third Edition (J.F. Southey, *ed.*), pp. 63-77. Her Majesty's Stationary Office, London.

Pitcher, R.S. and Crosse, J.E. 1956. An etiological association between nematodes (*Aphelenchoides* spp.) and *Corynebacterium fascians* in strawberry disease. *Proc. Int. Zool. Cong.*, Copenhagen, p. 376.

Pitcher, R.S. and Crosse, J.E. 1958. Studies on the relationship of eelworms and bacteria to certain plant diseases II. Further analysis of the strawberry cauliflower disease complex. *Nematologica* **3**: 244-256.

Ploeg, A.T., Robinson, D.J. and Brown, D.J.F. 1993. RNA-2 of Tobacco rattle virus encodes determinants of transmissibility by trichodorid vector nematodes. *J. Gen. Virol.* **74**: 1463-1466.

Poinar, G.O. Jr. 1975. *Entomogenous Nematodes*. E.J. Brill, Leiden, p. 317.

Poinar, G.O. Jr. and Hansen, E.C. 1986. Associations between nematodes and bacteria. *Helminthol. Abstr.* (Series B) **55**: 62-91.

Polychronopoulos, A.G., Houston, B.R. and Lownsbery, B.F. 1969. Penetration and development of *Rhizoctonia solani* in sugar beet seedlings infected with *Heterodera schachtii*. *Phytopathology* **59**: 482-485.

Powelson, M.L. and Rowe, R.C. 1993. Biology and management of early dying of potatoes. *Ann. Rev. Phytopath.* **31**: 111-126.

Prasad, K.S.K., Siddaramaiah, A.L. and Hegde, R.K. 1980. Notes on the association of *Corticium rolfsii* and *Meloidogyne incognita* in the wilt complex of *Solanum khasianum*. *Indian J. Agric. Sci.* **50**: 454.

Prince, V.E. 1966. Winter injury to peach trees in Georgia. *Proc. Amer. Soc. Hort. Sci.* **88**: 190-196.

Prince, V.E. and Horton, B.D. 1972. Influence of pruning at various dates on peach tree mortality. *J. Amer. Soc. Hort. Sci.* **97**: 303-305.

Pritts, M. and Wilcox, W. 1990. Black root rot disease of strawberry. *Cornell Small Fruits Newsletter* **5**(4): 1-2.

Putz, C. and Stocky, G. 1971. *Ann. Phytopathol.* **2**: 329-347.

Radewald, J.D. and Raski, D.J. 1962. A study of the life cycle of *Xiphinema index*. *Phytopathology* **52**: 748.

Ramachandran, N. and Sinha, P. 2000. Indian Institute of Horticultural Research, Bangalore, Karnataka. *In*: *Progress Report for the Period Sept. 1997 - Aug. 2000 of*

National Network Project on Phytophthora Diseases of Hoticultural Crops (S.R. Sarma, *ed.*), pp. 100-110. Indian Inst. of Spices Research, Calicut, Kerala.

Ramana, K.V., Eapen, S.J. and Sarma, Y.R. 1998. Effect of *Meloidogyne incognita, Pythium aphanidermatum* and *Ralstonia solanacearum* alone and in combinations in ginger. *In*: *Nematology-Challenges and Opportunities in 21st Century* (Usha K. Mehta, *ed.*), pp. 87-93. Sugarcane Breeding Institute, Coimbatore.

Ramsdell, D.C. 1988. Peach rosette mosaic virus decline. *In*: *Compendium of Grape Diseases* (R.C. Pearson and A.C. Goheen, *eds.*), pp. 51-52. American Phytopathological Society Press, St Paul, Minnesota, p. 93.

Rana, G.L. and Roca, F. 1973. 2nd *Int. Meet. Globe Artichoke, Bari*, pp. 139-140.

Rangahau, M.K. 2003. Potato early dying (Verticillium wilt) – Strategies for control. Crop and Food Research (Broad sheet) No. 126. http://www.crop.cri.nz/psp/books/bradsheets.htm.

Rankin, D.J., Hewitt, W.B. and Schmitt, R.V. 1971. Controlling fan leaf virus-dagger nematode disease complex in vineyards by soil fumigation. *Calif. Agri.* **25**: 11-14.

Rao, K.T. and Seshadri, A.R. 1981. Studies on interaction between *Meloidogyne incognita* and *Rotylenchulus reniformis* on grape seedlings. *Indian J. Nematol.* **11**: 101-102.

Rao, M.S., Naik, D., Shylaja, M. and Parvatha Reddy, P. 2002a. Prospects for the management of the root-knot nematode and bacterial wilt disease complex in capsicum by integration of fungal and bacterial biocontrol agents. *Proceedings of International Conference on Vegetables,* Bangalore, pp. 347-351.

Rao, M.S., Naik, D., Shylaja, M. and Parvatha Reddy, P. 2002b. Interaction of *Meloidogyne incognita* and *Fusarium oxysporum* f. sp. *dianthi* in the manifestation of the disease complex on tuberose. *Zonal Meeting of the Southern Chapter of Indian Phytopath. Soc.,* Bangalore.

Rao, M.S., Parvatha Reddy, P. and Walia, R.K. 2001. Biological control of nematodes in Horticultural crops. *Proc. of National Nematol. Cong.*, Indian Agri. Res. Inst., New Delhi, pp. 351-360.

Rao, M.S., Ramachandran, N., Sowmya, D.S., Ratnamma, K., Chaya, M.K. and Manoj Kumar, R. 2009. Biological control of nematode induced disease complex in certain vegetable crops. *Int. Conf. on Hort.*, Bangalore.

Rao, M.S., Shylaja, M. and Naik, D. 2003. Management of nematode induced wilt disease complex in tuberose (*Polianthes tuberosa* L.) cultivar Prajwal using *Pochonia chlamydosporia* (*Verticillium chlamydosporium*) and *Trichoderma harzianum. J. Orn. Hort.* **6**: 341-346.

Rao, Y.S. and Prasad, J.S. 1981. Interaction of *Heterodera oryzicola* and *Meloidogyne graminicola* in artificial inoculation in rice. *Indian J. Nematol.* **11**: 104.

Raski, D.J. 1952. On the morphology of *Criconemoides* Taylor 1936, with descriptions of six new species. *Proc. Helminthol. Soc. Wash.* **19**: 85-99.

Raski, D.J. 1975. Revision of the genus *Paratylenchus* Micoletzky, 1922, and descriptions of new species. Part II of three parts. *J. Nematol.* **7**: 274-295.

Raski, D.J. 1988. Dagger and needle nematodes. *In*: *Compendium of Grape Diseases* (R.C. Pearson and A.C. Goheen, *eds.*), pp 56-59. American Phytopathological Society Press, St Paul, Minnesota, p. 93.

Raski, D.J., Hewitt, W.B., Goheen, A.C., Taylor, C.E. and Taylor, R.H. 1965. Survival of *Xiphinema index* and reservoirs of fanleaf virus in fallowed vineyard soil. *Nematologica* **11**: 349-352.

Raski, D.J. and Luc, M. 1987. A reappraisal of Tylenchina (Nemata). 10. The superfamily Criconematoidea Taylor, 1936. *Rev. de Nematol.* **10**: 409-444.

Raski, D.J. and Radewald, J.D. 1958. Reproduction and symptomatology of certain ectoparasitic nematodes on roots of Thompson Seedless grape. *Plant Dis. Reptr.* **42**: 941-943.

Ravi, K., Nanje Gowda, D.N. and Parvatha Reddy, P. 2001. Integrated management of disease complex in banana involving borrowing nematode and Panama wilt–an ecofriendly approach. *In*: *IPM in Horticultural Crops: Emerging Trends in the New Millenium* (A. Verghese and P. Parvatha Reddy, *eds.*), p. 154. Association for Advancement of Pest Management in Horticultural Ecosystems, Bangalore.

Ravichandra, N.G., Krishnappa, K. and Setty, K.G.H. 1990. Interaction of *Meloidogyne javanica* and race 1, race 2, race 3 of *Meloidogyne incognita* with *Pseudomonas solanacearum* on a few brinjal lines. *Indian J. Nematol.* **20**: 138-147.

Read, P.J. and Hide, G.A. 1995. Development of black dot disease (*Colletotrichum coccodes* (Wallr.) Hughes) and its effects on the growth and yield of potato plants. *Ann. Appl. Biol.* **127**: 57-72.

Ribeiro, C.A.G. and Ferraz, S. 1983. Studies on the interaction between *Meloidogyne javanica* and *Fusarium oxysporum* f. sp. *phaseoli* on bean. *Fitopatologia Brasileira* **8**: 439-446.

Ribeiro, O.K. and Black, L.L. 1971. *Rhizoctonia fragariae:* A mycorrhizal and pathogenic fungus of strawberry plans. *Plant Dis. Reptr.* **55**: 599-603.

Riffle, J.W. 1971. Effect of nematodes on root-inhabiting fungi. *In*: *Mycorrhizae*. Proc. of the 1st N. American Conf. on Mycorrhizae (H. Hacskaylo, *ed.*), pp. 97-113. Misc. Publn., U.S.D.A., Washington D.C.

Riley, I.T., Readon, T.B. and McKay, A.C. 1988. Genetic analysis of plant pathogenic bacteria in the genus *Clavibacter* using alozyme electrophoresis. *J. Gen. Microbiol.* **134**: 3025-3030.

Rishbeth, J. 1960. *Emp. J. Exp. Agric.* **28**: 109-113.

Ritchie, D.F. 1989. Improved peach tree longevity with use of fenamiphos in peach tree short-life locations. *Plant Dis.* **73**: 160-163.

Ritchie, D.F. and Clayton, C.E. 1981. Peach tree short life: A complex of interacting factors. *Plant Dis.* **65**: 462-469.

Ritzema-bos, J. 1891. Zwei neu Nematodenkrankheiten der Erdbeerpflanze. *Zeitschrift Pflanzenkrankheiten* **1**: 1-16.

Rizvi, H., Khan, M.R. and Khan, U. 2010. Effect of inoculations with *Meloidogyne incognita* and *Rhizoctonia solani* on biochemical and morphological response of some marigold cultivars. *National Conf. on Innovations in Nematol. Res. for Agri. Sustainability – Challenges and A Roadmap Ahead*, Tamil Nadu Agri. Univ., Coimbatore, p. 103.

Robbins, R.T. and Brown, D.J.F. 1991. Comments on the taxonomy, occurrence and distribution of Longidoridae (Nematoda) in North America. *Nematologica* **37**: 395-419.

Roberts, I.M. and Brown, D.J.F. 1980. Detection of six nepoviruses in their nematode vectors by immunosorbent electron microscopy. *Ann. Appl. Biol.* **96**: 187-192.

Roberts, P.A., Magyarosy, A.C., Mathews, W. and May, D.W. 1988. Effect of metham sodium applied by drip irrigation on root-knot nematodes, *Pythium ultimum* and *Fusarium* sp. in soil and carrot and tomato roots. *Plant Dis.* **72**: 213-217.

Robinson, D. and Harrison, B.D. 1985. Unequal variation in the two genome parts of tobraviruses with evidence for the existence of three separate viruses. *J. Gen. Virol.* **66**: 171-176.

Roca, F., Martelli, G.P., Lamberti, F. and Rana, G.L. 1975. Distribution of *Longidorus attenuatus* Hooper in Apulian artichoke fields and its relationship with artichoke Italian latent virus. *Nematol. Medit.* **3**: 91-101.

Roca, F. and Rana, G.L. 1981. *Paratrichodorus tunisiensis* (Nematoda: Trichodoridae) a new vector of tobacco rattle virus in Italy. *Nematol. Medit.* **9**: 217-219.

Roca, F., Rana, G.L. and Kyriakopoulou, P.E. 1981. *Longidorus fasciatus* Roca *et* Lamberti vector of a serologically distinct strain of artichoke Italian latent virus in Greece. *Nematol. Medit.* **10**: 53-69.

Rodriguez-Kabana, R. and King, P.S. 1977. Nematicidal activity of the fungicide ethazole. *J. Nematol.* **9**: 203-206.

Romaniko, V.I. 1958. Study of the biology of the nematode *Pratylenchus pratensis* which infests the nodules of legumes and bean crops in the Chelyabinsk Region. *Akad. Nauk. USSR,* Moscow.

Rothrock, C.S. 1992. Tillage systems and plant disease. *Soil Sci.* **154**: 308-315.

Rowe, R.C. 1983. Early Dying – east and west. *Am. Veg. Grow*. **31**(3): 8-10.

Rowe, R.C., Davis, J.R., Powelson, M.L. and Rouse, D.I. 1987. Potato early dying: causal agents and management strategies. *Plant Dis.* **71**: 482-489.

Rowe, R.C. and Powelson, M.R. 2002. Potato Early Dying: Management challenges in a changing production environment. *Plant Dis.* **86**: 1184-1193.

Rowe, R.C., Ridel, R.M. and Martin, M.J. 1985. Synergistic interactions between *Verticillium dahliae* and *Pratylenchus penetrans* in potato early dying disease. *Phytopathology* **75**: 412-418.

Roy, K.W., Lawrence, G.W. Hodges, H.H., McLean, K.S. and Killebrew, J.F. 1989. Sudden death syndrome of soybean: *Fusarium solani* as incitant and relation of *Heterodera glycines* to disease severity. *Phytopathology* **79**: 191-197.

Roy, S., Ojha, P.K., Ojha, K.L., Upadhyay, J.P. and Jha, M.M. 1999. Effects of combined application of fertilizers and organic amendments on the disease intensity of wilt complexes in banana (*Musa* sp.). *J. Appl. Biol.* **9**: 84-86.

Rudel, M. 1985. Grapevine damage induced by particular virus vector combinations. *Phytopathologia Mediterranea* **24**: 183-185.

Rupe, J.C., Robbins, R.T., Becton, C.M., Sabbe, W.A. and Gbur, E.E. 1999. Vertical and temporal distribution of *Fusarium solani* and *Heterodera glycines* in fields with sudden death syndrome of soybean. *Soil Biol. Biochem.* **31**: 245-251.

Rupp, D. (1990). The leaching of nitrate and nematicide from trenched vineyards. *Wein-Wissenschaft* **45**: 135-140.

Rykbost, K.A., Ingham, R.E. and Maxwell, J. 1992. Control of nematodes and related disease in potatoes. In: *Research in the Klamath Basin*, 1991 Annual Report. Special Report 895. Agricultural Experiment Station, Oregon State University, Corvallis, OR, pp. 66-71.

Rykbost, K.A., Ingham, R.E. and Maxwell, J. 1995. Control of nematodes and related disease in potatoes. In: *Research in the Klamath Basin*, 1994 Annual Report. Special Report 949. Agricultural Experiment Station, Oregon State University, Corvallis, OR, pp. 92-108.

Saeed, I.A.M., MacGuidwin, A.E. and Rouse, D.I. 1997a. Synergism of *Pratylenchus penetrans* and *Verticillium dahliae* manifested by reduced gas exchange in potato. *Phytopathology* **87**: 435-439.

Saeed, I.A.M., MacGuidwin, A.E. and Rouse, D.I. 1997b. Disease progress based on effects of *Verticillium dahliae* and *Pratylenchus penetrans* on gas exchange in Russet Burbank potato. *Phytopathology* **87**: 440-445.

Saeed, M., Ahmad, M. and Khan, H.A. 1972. A complex disease of tomato and papaya caused by nematode fungi association in Pakistan. *Pakistan J. Sci. Industrial Res.* **15**: 312-313.

Salazar, L.F. and Chinchilla, C. 1989. Characterization morfométrica de cuatro poblaciones de *Rhadinaphelenchus* sp. obtenidos de *Cocos nucifera* y *Elaeis guineensis*. *Nematropica* **19**: 18.

Samiyappan, R. 2003. Biological control of fungal and nematode complex diseases by plant growth promoting rhizobacteria (PGPR). *Winter School on Biological Control of Plant Parasitic Nematodes*, Tamil Nadu Agri. Univ., Coimbatore, pp. 69-77.

Samuel, M. and Mathew, J. 1983. Role and association of root-knot nematode *Meloidogyne incognita* in the introduction of bacterial wilt of ginger incited by *Pseudomonas solanacearum*. *Indian Phytopath.* **36**: 398-399.

Sanger, H.L. 1961. *Proc. of 4th Potato Virus Dis.,* Braunschweig, pp. 22-28.

Santo, G.S. and Bolander, W.J. 1977a. Effect of *Macroposthonia xenoplax* on the growth of Concord grape. *J. Nematol.* **9**: 215-217.

Santo, G.S. and Bolander, W.J. 1977b. Separate and concomitant effects of *Macroposthonia xenoplax* and *Meloidogyne hapla* on Concord grapes. *J. Nematol.* **9**: 282-283.

Sarah, J.L. 1994. CIRAD-FLHOR Research actions on nematodes and black borer weevil on bananas and plantains. *In*: *Banana Nematodes and Weevil Borers in Asia and the Pacific* (R.V. Valmayor, R.G. Davide, J.M. Stanton, N.L. Treverrow and V.N. Roa, *eds.*), pp. 159-169. Serdang, Selangor, Malaysia.

Savage, E.E. and Cowart, F.F. 1954. Factors affecting peach tree longevity in Georgia. *Proc. Amer. Soc. Hort. Sci.* **64**: 81-86.

Savage, E.E., Hayden, R.A. and Ward, W.E. 1968. The effect of subsoiling on the growth and yield of peach trees. *Univ. Ga. Res. Bull.* 30.

Savitha, R.M. and Lingaraju, S. 1996. Interaction of root-knot nematode and wilt bacterium on brinjal. *Karnataka J. Agri. Sci.* **9**: 657-660.

Schindler, A.F., Stewart, R.N. and Semeniuk, P. 1961. A synergistic *Fusarium* – nematode interaction on carnations. *Phytopathology* **51**: 143-146.

Schlang, J. and Sikora, R.A. 1978. Investigations on the interaction between *Meloidogyne arenaria, Verticillium dahliae* and *Phytophthora cryptoge* on *Gerbera jomesonii. Zeitschrift fur Pflanzenkrenkheiten und Pflanzenschutz* **85**: 203-209.

Schneider, S.M., and Ferris, H. 1987. Stage-specific population development and fecundity of *Paratrichodorus minor. J. Nematol.* **19**: 267-394.

Scholte, K. 1990. Causes of differences in growth pattern, yield and quality of potatoes (*Solanum tuberosum* L.) in short rotations on sandy soil as affected by crop rotation, cultivar and application of granular nematicides. *Potato Res.* **33**: 181-190.

Scholte, K. and s'Jacob, 1989. Synergistic interactions between *Rhizoctonia solani* Kuhn, *Verticillium dahliae* Kleb., *Meloidogyne* spp. and *Pratylenchus neglectus* (Rensch) Chitwood & Oteifa, in potato. *Potato Res.* **32**: 387-395.

Schroeder, L. 1988. Strawberry black root rot complex. *The University of Connecticut Cooperative Extension Service Grower, Vegetable and Small Fruit Newsletter* **88**: 5.

Schuiling, M. and Dinther, J.B. 1981. Red ring disease in the Paricatuba Oil Palm Estate, Pará, Brazil. *Zeitschrift fur Angewandte Entomologie* **91**: 154-169.

Schuster, M.L. 1959. Relation of root-knot nematodes and irrigation water to the incidence and dissemination of bacterial wilt of bean. *Plant Dis. Reptr.* **43**: 27-32.

Seinhorst, J.W. and Oostrom, A. 1989. Reproduction of a mixture of pathotypes Ro 1 of *Globodera rostochiensis* and Pa 3 of *G. pallida* and its offspring on cvs. Cardinal and Irene. *Nematologica* **35**: 469-474.

Sellam, M.A., Rushdi, M.H. and Gendi, D.M. 1982. Interrelationship of *Meloidogyne incognita* Chitwood and *Pseudomonas solanacearum* on tomato. *Egyptian J. Phytopath.* **12**: 35-42.

Selvanathan, N. 2003. Potato early dying complex: Key factors related to disease development. *In*: *Manitoba Agronomist Conf.*, Univ. of Manitoba, pp. 198-201.

Senthamarai, M., Poornima, K. and Subramanian, S. 2006. Nematode-fungal disease complex involving *Meloidogyne incognita* and *Macrophomina phaseolina* on *Coleus forskohlii* Briq. *Indian J. Nematol.* **36**: 181-184.

Senthamarai, M., Poornima, K., Subramanian, S. and Sudheer, M.J. 2008. Nematode-fungal disease complex involving *Meloidogyne incognita* and *Macrophomina phaseolina* on medicinal *Coleus forskohlii* Briq. *Indian J. Nematol.* **38**: 30-33.

Senthil Kumar, T. and Rajendran, G. 2004. Biocontrol agents for the management of disease complex involving root-knot nematode, *Meloidogyne incognita* and *Fusarium moniliforme* on grapevine (*Vitis vinefera*). *Indian J. Nematol.* **34**: 49-51.

Seshadri, A.R. 1965. Investigations on the biology and life cycle of *Criconemella xenoplax* Raski, 1952 (Nematoda: Criconematidae). *Nematologica* **11**: 540-562.

Shafer, S.R., Rhodes, L.H. and Riedel, R.M. 1981. *In-vitro* parasitism of endomycorrhizal fungi of ericaceous plants by the mycophagous nematode *Aphelenchoides bicaudatus*. *Mycologia* **73**: 141-149.

Shahzad, S. and Gaffer, A. 1989. Use of *Paecilomyces lilacinus* in the control of root rot and root-knot complex in okra and mungbean. *Pakistan J. Nematol.* **7**: 47-53.

Sharma, J.K., Singh, I. and Chhabra, H.K. 1980. Observations on the influence of *Meloidogyne incognita* and *Rhizoctonia bataticola* on okra. *Indian J. Nematol.* **10**: 148-151.

Sharma, N.K. 1984. Effect of *Meloidogyne incognita* on nodulation and symbiotic nitrogen fixation in cowpea (Abstr.). *First Int. Cong. Nematol.*, Canada, p. 83.

Sharma, N.K. and Gill, J.S. 1979. Interaction between *Meloidogyne incognita* and *Rhizoctonia solani* on potato. *Indian Phytopath.* **32**: 293-298.

Sharma, N.K. and Sethi, C.L. 1975. Leghaemoglobin content of cowpea nodules as influenced by *Meloidogyne incognita* and *Heterodera cajani*. *Indian J. Nematol.* **5**: 113-114.

Sharma, N.K. and Sethi, C.L. 1976. Interrelationship between *Meloidogyne incognita, Heterodera cajani* and *Rhizobium* sp. on cowpea (*Vigna sinensis* (L.) Savi). *Indian J. Nematol.* **6**: 117-123.

Sharma, N.K. and Sethi, C.L. 1978. Interaction between *Meloidogyne incognita* and *Heterodera cajani* on cowpea. *Indian J. Nematol.* **6**: 1-12.

Sheela, M.S. and Venkitesan, T.S. 1981. Interrelationships of infectivity between the burrowing and root-knot nematodes in black pepper, *Piper nigrum*. *Indian J. Nematol.* **6**: 105.

Sheela, M.S. and Venkitesan, T.S. 1990. Interaction between *Meloidogyne incognita* and the fungus *Fusarium* spp. on black pepper vine. *Indian J. Nematol.* **20**: 184-188.

Shreenivasa, K.R., Krishnappa, K., Reddy, B.M.R., Ravichandra, N.G., Karuna, K. and Kantharaju, V. 2005. Integrated management of nematode complex on banana. *Indian J. Nematol.* **35**: 37-40.

Shukla, V.N. and Swarup, G. 1970. Interrelationships of the root-knot nematode, *Meloidogyne incognita* (Kofoid and White, 1919) Chitwood, 1949 and *Sclerotium rolfsii* Sacc. on the emergence of tomato seedlings. *Bull. Indian Phytopath. Soc.* **6**: 52-54.

Shylaja, M. 2004. *Studies on the Development of Bio-Intensive Disease Management Strategies in Tuberose (*Polyanthes tuberosa *L.) and Carnations (*Dianthus caryophyllus *L.) Using Biological Control Agents.* Ph. D. thesis, Kuvempu Univ., Shimoga, 132 pp.

Siddiqui, I.A. and Ehteshamul-Haque, S. 2001. Suppression of the root rot–root knot disease complex by *Pseudomonas aeruginosa* in tomato: The influence of inoculum density, nematode populations, moisture and other plant-associated bacteria. *Plant and Soil* **237**: 81-89.

Siddiqui, I.A., Ehteshamul-Haque, S. and Ghaffar, A. 1999. Root-dip treatment with *Pseudomonas aeruginosa* and *Trichoderma* spp. in the control of root rot-root knot disease complex in chilli (*Capsicum annuum* L.). *Pakistan J. Nematol.* **17**: 67-75

Siddiqui, I.A., Sher, S.A. and French, A.M. 1973. *Distribution of Plant Parasitic Nematodes in California.* California Department of Food and Agriculture, Sacramento, California, 324 pp.

Siddiqui, I.A., Qureshi, S.A., Sultana, V., Ehteshamul-Haque, S. and Ghaffar, A. 2000. Biological control of root rot-root knot disease complex of tomato. *Plant and Soil.* **227**: 163-169

Siddiqui, Z.A. and Mahmood, I. 1995. Some observations on the management of the wilt disease complex of pigeonpea by treatment with a vesicular arbuscular fungus and biocontrol agents for nematodes. *Biosource Technol.* **54**: 227-230.

Siddiqui, Z.A. and Mahmood, I. 1996. Biological control of *Heterodera cajani* and *Fusarium udum* on pigeonpea by *Glomus mosseae, Trichoderma harzianum* and *Verticillium chlamydosporium. Israel J. Plant Sci.* **44**: 49-56.

Siddiqui, Z.A. and Mahmood, I. 1999a. Biological control of *Heterodera cajani* and *Fusarium udum* on pigeonpea by *Glomus mosseae, Trichoderma harzianum* and *Verticillium chlamydosporium. Zeitschrift fur Pflanzenkrankheiten und Pflanzenschutz* **106**: 530-533.

Siddiqui, Z.A. and Mahmood, I. 1999b. Effects of *Meloidogyne incognita, Fusarium oxysporum* f. sp. *pisi, Rhizobium* sp. and different soil types on growth, chlorophyll, and carotenoid pigments of pea. *Israel J. Plant Sci.* **47**: 251-256.

Sidhu, G.S. and Webster, J.M. 1974. Genetics of resistance in the tomato to root-knot nematode-wilt fungus complex. *J. Heredity* **65**: 153-156.

Sidhu, G.S. and Webster, J.M. 1977. Predisposition of tomato to the wilt fungus (*Fusarium oxysporum lycopersici*) by the root-knot nematode (*Meloidogyne incognita*). *Nematologica* **23**: 436-442.

Sidhu, G.S. and Webster, J.M. 1981. Genetics of plant-nematode interaction. *In*: *Plant Parasitic Nematodes*, Vol. III (B.M. Zuckerman and R.A. Rohde, *eds.*), pp. 61-87. Academic Press, New York.

Sikora, R.A. 1979. Predisposition of *Meloidogyne* infection by the endotrophic mycorrhizal fungus *Glomus mosseae*. *In*: *Root-knot Nematodes* (Meloidogyne *species), Systematics, Biology and Control* (F. Lamberti and C.E. Taylor, *eds.*), pp. 399-404. Academic Press, New York.

Sikora, R.A. and Schlosser, E. 1973. Nematodes and fungi associated with root systems of banana in a state of decline in Lebanon. *Plant Dis. Reptr.* **57**: 615-618.

Singh, B., Ali, S.S., Naimuddin and Askary, T.H. 2004. Combined effect of *Fusarium udum* and *Meloidogyne javanica* on wilt resistant accessions of pigeonpea. *Ann. Pl. Prot. Sci.* **12**: 130-133.

Singh, D.B. and Parvatha Reddy, P. 1981. Influence of *Meloidogyne incognita* infestation on *Rhizobium* nodule formation in French bean. *Nematol. Medit.* **9**: 1-5.

Singh, D.B., Parvatha Reddy, P. and Sharma, S.R. 1981. Effect of the root-knot nematode, *Meloidogyne incognita* on *Fusarium* wilt of French beans. *Indian J. Nematol.* **11**: 84-85.

Singh, S., Dhawan, S.C. and Goswami, B.K. 2007. Histopathology of cowpea roots co-infected with root-knot nematode, *Meloidogyne incognita* and wilt fungus *Fusarium oxysporum*. *Indian J. Nematol.* **37**: 156-160.

Singh, S., Goswami, B.K. and Sharma, H.K. 2007. Role of reniform nematode, *Rotylenchulus reniformis* in reducing disease-complex on tomato caused by root-knot nematode, *Meloidogyne incognita* and wilt fungus *Fusarium oxysporum* f. sp. *lycopersici*. *Indian J. Nematol.* **37**: 172-175.

Singh, Y.P., Singh, R.S. and Sitaramaiah, K. 1990. Mechanism of resistance of mycorrhizal tomato against root-knot nematode. *In*: *Trends in Mycorrhizal Research*, Proc. of the National Conf. on Mycorrhizae (B.L. Jalali and H. Chand, *eds.*), pp. 96-97. Haryana Agri. Univ., Hissar.

Sitaramaiah, K. 1989. Host-parasite relationships of plant parasitic nematodes. *In*: *Perspectives in Plant Pathology* (V.P. Agnihotri, N. Singh, H.S. Chaube, U.S. Singh and T.S. Dwivedi, *eds.*), pp. 155-165. Today and Tomorrow's Printers and Publishers, New Delhi.

Sitaramaiah, K. and Parvathi Devi, G. 1994. Nematodes, soil microorganisms in relation to *Phytophthora* wilt disease in betel vine. *Indian J. Nematol.* **24**: 16-25.

Sitaramaiah, K. and Sikora, R.A. 1982. Effect of the mycorrhizal fungus *Glomus fasciculatum* on the host-parasite relationship of *Rotylenchulus reniformis* in tomato. *Nematologica* **28**: 412-419.

Sitaramaiah, K. and Sinha, S.K. 1984a. Interaction between *Meloidogyne javanica* and *Pseudomonas solanacearum* on brinjal. *Indian J. Nematol.* **14**: 1-5.

Sitaramaiah, K. and Sinha, S.K. 1984b. Histological aspects of *Pseudomonas solanacearum* and root-knot nematode wilt complex in brinjal. *Indian J. Nematol.* **14**: 175-178.

Sivaprasad, P., Jacob, A., Nair, S.K. and George, B. 1990. Influence of VA mycorrhizal colonization on root-knot nematode infestation in *Piper nigrum* L. *In*: *Trends in Mycorrhizal Research*, Proc. of the National Conf. on Mycorrhizae (B.L. Jalali and H. Chand, *eds.*), pp. 100-101. Haryana Agri. Univ., Hissar.

Smith, G.S. 1987. Interactions of nematodes with mycorrhizal fungi. *In*: *Vistas on Nematology* (J.A. Veech and D.W. Dickson, *eds.*), pp. 292-300. Soc. of Nematologists, Hayattsville.

Smith, G.S. and Kaplan, D.T. 1988. Influence of mycorrhizal fungus, phosphorus and burrowing nematode interactions on growth of rough lemon citrus seedlings. *J. Nematol.* **20**: 371-374.

Smits, B.G. and Noguera, R. 1982. Effect of *Meloidogyne incognita* on the pathogenicity of different isolates of *Fusarium oxysporum* on brinjal (*Solanum melongena* L.). *Agronomia Tropical* **32**: 284-290.

Sol, H.H. and Seinhorst, J.W. 1961. *Tijdschr. Plantenziekten* **67**: 307-309.

Son, S.H., Khan, Z., Kim, S.G. and Kim, Y.H. 2009. Plant growth-promoting rhizobacteria,

Paenibacillus polymyxa and *Paenibacillus lentimorbus* suppress disease complex caused by root-knot nematode and Fusarium wilt fungus. *J. Appl. Microbiol.* **107**: 524-532.

Sowmya, D.S., Rao, M.S., Gopalakrishnan, C. and Ramachandran, N. 2010. Biomanagement of *Meloidogyne incognita* and *Erwinia carotovora* s. sp. *carotovora* infecting carrot (*Dacus carota* L.). *National Conf. on Innovations in Nematological Research for Agricultural Sustainability–Challenges and A Roadmap Ahead*, Tamil Nadu Agri. Univ., Coimbatore, p. 97.

Srinivasan, A. 1974. *Studies on Root Lesion Nematode*, Pratylenchus delattrei. M. Sc. (Agri.) thesis, Tamil Nadu Agricultural University, Coimbatore.

Srinivasan, A. and Muthukrishnan, T.S. 1976. Control of root lesion nematode *Pratylenchus delattrei* Luc 1958 on crossandra. *Indian J. Nematol.* **6**: 111-114.

Stark, J.C., and Love, S.L. (eds.). 2003. *Potato Production Systems*. University of Idaho Extension, Moscow, ID, 426 pp.

Starr, J.L. and Mai, W.F. 1976. Efficacy of pyroxychlor in controlling *Pythium polymorphon – Meloidogyne hapla* root disease complex of celery. *Plant Dis. Reptr.* **63**: 390-392.

Stellmach, G. and Goheen, A.C. 1988. Other virus and virus-like diseases. *In*: *Compendium of Grape Diseases* (R. C. Pearson and A. C. Goheen, *eds.*), pp 53-54. American Phytopathological Society Press, St Paul, Minnesota. 93 pp.

Stephen, Z.A., El-Behadi, A.H., Al-Zaharoon, H.H., Antoon, B.G. and Georges, M.S. 1996. Control of root-knot and wilt disease complex on tomato plants. *Dirasat Series B. Pure and Applied Sciences* **23**: 13-16.

Stevenson, W.R., Green, R.J. and Bergeson, G.B. 1976. Occurrence and control of potato black dot root rot in Indiana. *Plant Dis. Reptr.* **60**: 248-251.

Stevenson, W.R., Loria, R., Franc, G.D. and Weingartner, D.P. (eds.). 2001. *Compendium of Potato Diseases*. 2nd Edition. American Phytopathological Society, St. Paul MN, 106 pp.

Stewart, R.N and Schindler, A.F. 1956. The effect of some ectoparasitic and endoparasitic nematodes in the expression of bacterial wilt in carnations. *Phytopathology* **46**: 419-422.

Storey, G.W. and Evans, K. 1987. Interactions between *Globodera pallida* juveniles, *Verticillium dahliae* and three potato cultivars, with descriptions of associated histopathologies. *Plant Pathol.* **36**: 192-200.

Stover, R.H. 1966. Fungi associated with nematode and non-nematode lesions on banana roots. *Can. J. Bot.* **44**: 1703-1710.

Strausbaugh, C.A. 1993. Assessment of vegetative compatibility and virulence of *Verticillium dahliae* isolates from Idaho potatoes and tester strains. *Phytopathology* **83**: 1253-1258.

Sugawara, K., Kobayashi, K. and Ogoshi, A. 1997. Influence of the soybean cyst nematode (*Heterodera glycines*) on the incidence of brown stem rot in soybean and adzuki bean. *Soil Biol. Biochem.* **29**: 1491-1498.

Sumner, D.R. and Johnson, A.W. 1973. Effect of root-knot nematode on *Fusarium* wilt of watermelon. *Phytopathology* **63**: 361-364.

Sundaram, R. and Tangaraju, T. 2001. Biointensive management of *Meloidogyne incognita* and *Pythium aphanidermatum* in tomato nurseries. *Natl. Cong. on Centenary of Nematol. in India–Appraisal & Future Plans*, Indian Agri. Res. Inst., New Delhi, pp. 169-170.

Sundararaju, P. and Koshy, P.K. 1982. *Indian J. Nematol.* **12**: 404-405.

Sundararaju, P. and Koshy, P.K. 1987. Separate and combined effects of *Radopholus similis* on *Cylindrocarpon obtusisporum* on areca nut seedlings. *Indian J. Nematol.* **17**: 301-305.

Sundararaju, P. and Thangavelu, R. 2009. Influence of *Pratylenchus coffeae* and *Meloidogyne incognita* on the Fusarium wilt complex of banana. *Indian J. Nematol.* **39**: 71-74.

Suresh, C.K., Bagyaraj, D.J. and Reddy, D.D.R. 1985. Effect of vesicular arbuscular mycorrhiza on survival, penetration and development of root-knot nematode in tomato. *Plant and Soil* **87**: 305-308.

Sutherland, J.R. and Fortin, J.A. 1968. Effect of the nematode *Aphelenchus avenae* on some ectotrophic, mycorrhizal fungi on a red pine mycorrhizal relationship. *Phytopathology* **58**: 519-523.

Sylvia, D.M. and Sinclair, W.A. 1983. Phenolic compounds and resistance to fungal pathogens induced in primary roots of Douglas-fir seedlings by the ectomycorrhizal fungus *Laccaris laccata*. *Phytopathology* **73**: 390-397.

Taha, A.H.Y. and Kassab, A.S. 1979. The histological reactions of *Vigna sinensis* to separate and concomitant parasitism by *Meloidogyne javanica* and *Rotylenchulus reniformis*. *J. Nematol.* **11**: 117-123.

Taha, A.H.Y. and Kassab, A.S. 1980. Interrelations between *Meloidogyne javanica, Rotylenchulus reniformis* and *Rhizobium* sp. on *Vigna sinensis*. *J. Nematol.* **12**: 57-62.

Taha, A.H.Y. and Raski, D.J. 1969. Interrelationships between root-nodule bacteria, plant-parasitic nematodes and their leguminous host. *J. Nematol.* **1**: 201-211.

Taha, A.H.Y. and Sultan, S.A. 1977. Population of *Rotylenchulus reniformis* and *Tylenchulus semipenetrans* on grape seedlings as influenced by coincident infestations. *Nematol. Medit.* **5**: 253-257.

Taylor, C.E. 1962. Transmission of raspberry ringspot virus by *Longidorus elongatus* (de Man) (Nematoda: Dorylaimoidea). *Virology* **17**: 493-494.

Taylor, C.E. 1980. Nematodes. *In*: *Vectors of Plant Pathogens* (K.F. Harris and K. Marmorosch, *eds.*), pp. 375-416. Academic Press, New York.

Taylor, C.E. 1990. Nematode interactions with other pathogens. *Ann. Appl. Biol.* **116**: 405-416.

Taylor, C.E. and Brown, B.J.F. 1981. Nematode-virus interactions. *In*: *Plant Parasitic Nematodes* (B.M. Zuckerman and R.A. Rohde, *eds.*), pp. 281-301. Acad. Press, New York.

Taylor, C.E. and Gordon, S.C. 1970. A comparison of four nematicides for the control of *Longidorus elongatus* and *Xiphinema diversicaudatum* and the virus they transmit. *Hort. Res.* **10**: 133-134.

Taylor C.E. and Robertson, W.M. 1975. Acquisition, retention and transmission of viruses by nematodes. *In*: *Nematode Vectors of Plant Viruses* (F. Lamberti, C. E. Taylor and J. W. Seinhorst, Eds), pp 253-276. Plenum Press, London, 460 pp.

Taylor, C.E. and Robertson, W.M. 1977. Virus vector relationships and mechanisms of transmission. *Proc. American Phytopath. Soc.* **4**: 20-29.

Taylor, R.J., Pasche, J.S. and Gudmestad, N.C. 2005. Influence of tillage and method of metam sodium application on distribution and survival of *Verticillium dahliae* in the soil and the development of potato early dying disease. *Ame. J. Potato Res.* **82**: 451-461.

Thakur, S.C. 1985. *Interaction of Nematode,* Meloidogyne incognita *and Bacterium Pseudomonas solanacearum on Tomato Plants*. M. Sc. thesis, Rajendra Agri. Univ., Pusa, Bihar.

Thomas, C.K.M., Hussey, R.S. and Roncadori, R.W. 1983. Interaction of vesicular arbuscular mycorrhizal fungi and phosphorus with *Meloidogyne incognita* on tomato. *J. Nematol.* **15**: 410-417.

Thomas, H.A. 1959. On *Criconemoides xenoplax* Raski with special reference to its biology under laboratory conditions. *Proc. Helminthol. Soc. Wash.* **26**: 55-59.

Thomas, R.J. and Clark, C.A. 1980. Interactions between *Meloidogyne incognita* and *Rotylenchulus reniformis* on sweet potato. *J. Nematol.* **12**: 239.

Thomas, R.J. and Clark, C.A. 1981. *Meloidogyne incognita* and *Rotylenchulus reniformis* interactions in a sweet potato field. *Phytopathology* **71**: 908.

Thomas, R.J. and Clark, C.A. 1983a. Effects of concomitant development on reproduction of *Meloidogyne incognita* and *Rotylenchulus reniformis*. *J. Nematol.* **15**: 215-221.

Thomas, R.J. and Clark, C.A. 1983b. Population dynamics of *Meloidogyne incognita* and *Rotylenchulus reniformis* alone and in combination. *J. Nematol.* **15**: 204-211.

Thomason, I.J., Erwin, D.C. and Garber, M.J. 1959. The relationship of the root-knot nematode, *Meloidogyne javanica,* to *Fusarium* wilt of cowpea. *Phytopathology* **49**: 602-606.

Timper, P. and Kaya, H.K. 1989. Role of the second-stage cuticle of entomogenous nematodes in preventing infection by nematophagous fungi. *J. Invert. Pathol.* **54**: 314-321

Triantaphyllou, A.C. 1991. Further studies on the role of polyploidy in the evolution of *Meloidogyne. J. Nematol.* **23**: 249-253.

Tribe, H.T. 1977. Pathology of cyst nematodes. *Biol. Rev.* **52**: 477-507.

Tripathi, P.K. and Singh, C.S. 2006. Effect of some compatible biocontrol agents along with mustard cake and furadan on *Meloidogyne incognita* infecting tomato plant. *Indian J. Nematol.* **36**: 309-311.

Trudgill, D.L. 1991. Resistance to and tolerance of plant parasitic nematodes in plants. *Ann. Rev. Phytopath.* **29**: 167-192.

Trudgill, D.L. and Brown, D.J.F. 1978. Ingestion, retention and transmission of two starins of raspberry ringspot virus by *Longidorus macrosoma*. *J. Nematol.* **10**: 85-89.

Tsror (Lahkim), L., Erlich, O. and Hazanovsky, M. 1999. Effect of *Colletotrichum coccodes* on potato yield, tuber quality, and stem colonization during spring and autumn. *Plant Dis.* **83**: 561-565.

Tsror (Lahkim), L. and Hazanovsky, M. 2001. Effect of coinoculation by *Verticillium dahliae* and *Colletotrichum coccodes* on disease symptoms and fungal colonization in four potato cultivars. *Plant Pathol.* **50**: 483-488.

Umesh, K.C., Krishnappa, K. and Bagyaraj, D.J. 1988. Interaction of burrowing nematode, *Radopholus similis* and VA mycorrhiza, *Glomus fasciculatum* in banana. *Indian J. Nematol.* **18**: 6-11.

Valdez, R.B. 1972. Transmission of raspberry ringspot virus by *Longidorus caespiticola, L. leptocephalus* and *Xiphinema diversicaudatum* and arabis mosaic virus by *L. caespiticola* and *X. diversicaudatum. Ann. Appl. Biol.* **71**: 229-234.

Van Gundy, S.D. and Kirkpatrick, J.D. 1975. Nematode-nematode interactions on tomato. *J. Nematol.* **7**: 330.

Van Gundy, S.D., Kirkpatrick, J.D. and Golden, J. 1977. The nature and role of metabolic leakage from root-knot nematode galls and infection by *Rhizoctonia solani*. *J. Nematol.* **9**: 113-121.

Van Gundy, S.D. and Rackham, R.L. 1961. Studies on the biology and pathogenicity of *Hemicycliophora arenaria. Phytopathology* **51**: 393-397.

Van Gundy, S.D. and Tsao, P.H. 1963. Growth reduction of citrus seedlings by *Fusarium solani* as influenced by the citrus nematode and other soil factors. *Phytopathology* **53**: 488-489.

van Hoof, H.A. 1962. *Trichodorus pachydermus* and *T. teres* vectors of the early browning virus of peas. *Tijdschr. Plantenziekten* **68**: 391-396.

van Hoof, H.A. 1968. Transmission of tobacco rattle virus by *Trichodorus* species. *Nematologica* **14**: 20-24.

van Hoof, H.A., Maat, D.Z. and Seinhorst. J.W. 1966. Viruses of the tobacco rattle virus group in Northern Italy: Their vectors and serological relationships. *Neth. J. Plant Pathol.* **72**: 253-258.

Vargas, M.T., Zavaleta Mejia, E. and Hernandez, A.M. 1996. Breaking of resistance to *Phytophthora capsici* Leo in chilli (*Capsicum annuum* L.) Serrano CM-34 by *Nacobbus aberrans* Thorne and Allen. *Nematropica* **26**: 159–166.

Varshney, V.P., Khan, A.M. and Saxena, S.K. 1987. Interaction between *Meloidogyne incognita, Rhizoctonia solani* and *Rhizobium* species on cowpea. *Int. Nematol. Network Newsl.* **4**(3): 11-14.

Vats, R. and Dalal, M.R. 1988. Interrelationships between *Rotylenchulus reniformis* and *Rhizobium leguminosarum* on *Pisum sativum*. *In*: *Nematology-Challenges and Opportunities in 21st Century* (Usha K. Mehta, *ed.*), pp. 98-101. Sugarcane Breeding Institute, Coimbatore.

Vats, R. and Dalal, M.R. 1997. Interaction between *Rotylenchulus reniformis* and *Fusarium oxysporum* f. sp. *pisi* on pea (*Pisum sativum* L.). *Ann. Biol.* **13**: 239-42.

Vellios, E., Duncan, G., Brown, D. and MacFarlane, S. 2002. Immunogold localization of tobravirus 2b nematode transmission helper protein associated with virus particles. *Virology* **300**: 118-124.

Venkata Rao, A., Narasimhan, V., Rajagopalan, P., Vidyasekaran, P. and Muthukrishnan, T.S. 1973. Efficacy of nematicides for the control of betel vine wilt. *Indian J. Agric. Sci.* **43**: 1072-1075.

Verdejo, M.S., Green, C.D. and Podder, A.K. 1988. Influence of *Meloidogyne incognita* on nodulation and growth of pea and black bean. *Nematologica* **34**: 88-97.

Verma, K.K., Goel, S.R. and Paruthi, I.J. 2010. Studies on biomanagement of root-knot and root-rot complex in cowpea (*Vigna unguiculata*) by fungal antagonists. *National Conf. on Innovations in Nematological Research for Agricultural Sustainability–Challenges and A Roadmap Ahead*, Tamil Nadu Agri. Univ., Coimbatore, p. 83.

Vishwadhar and Chaudhary, R.G. 2000. Disease resistance in pulse crops–Current status and future approaches. *In*: *Proc. of Nat. Symp. on Role of Resistance in Intensive Agri.*, Directorate of Wheat Res., Karnal, pp. 144-157.

Visser, P.B. and Bol, J.F. 1999. Nonstructural proteins of tobacco rattle virus which have a role in nematode-transmission: Expression pattern and interaction with viral coat protein. *J. Gen. Virol.* **80**: 3272-3280.

Vovlas, N., Inserra, R.N. and Greco, N. 1992. *Schistonchus caprifici* parasitizing caprifig (*Ficus caricasylvestris*) florets and the relationship with its vector fig wasp (*Blastophaga psenes*). *Nematologica* **38**: 215-226.

Vrain, T.C. 1993. Restriction fragment length polymorphism separates species of the *Xiphinema americanum* group. *J. Nematol.* **25**: 361-364.

Walia, K.K. and Gupta, D.C. 1986a. Antagonism between *Heterodera cajani* and *Rhizoctonia solani* on cowpea (*Vigna unguiculata*). *Indian J. Nematol.* **16**:41-43.

Walia, K.K. and Gupta, D.C. 1986b. Effect of the fungus *Rhizoctonia solani* on the population development of pigeon-pea cyst nematode, *Heterodera cajani* on cowpea (*Vigna unguiculata*). *Indian J. Nematol.* **16**: 131-132.

Walker, G. E. 1997. Effects of *Meloidogyne* spp. and *Rhizoctonia solani* on the growth of grapevine rootings. *J. Nematol.* **29**: 190-198.

Walker, M.A., Wolpert, J.A. and Weber, E. 1994. Field screening of grape rootstock selections for resistance to fanleaf degeneration. *Plant Dis.* (In press).

Walkinshaw, C.H., Griffin, G.D. and Larson, R.H. 1961. *Trichodorus christiei* as a vector of potato corky ringspot (tobacco rattle) virus. *Phytopathology* **51**: 806-808.

Wallace, H.R. 1978. The diagnosis of plant diseases of complex etiology. *Ann. Rev. Phytopath.* **16**: 379-342.

Walter, I.C. 1965. Host resistance as relates to root pathogens and soil microorganisms. *In*: *Ecology of Soil-borne Plant Pathogens* (K.F. Barker and W.C. Snyder, *eds.*), pp. 314-320. Univ. Calif. Press, Berkeley.

Wang, L.H. 1973. *Biochemical and Physiological Changes in Root Exudates, Xylem Sap and Cell Permeability of Tomato Plants Infected with* Meloidogyne incognita. Ph. D. thesis, Purdue Univ., Indiana.

Webster, J.M. 1975. Aspects of the host parasite relationship of plant parasitic nematodes. *Advanced Parasitol.* **13**: 225-230.

Webster, J.M. 1985. Interaction of *Meloidogyne* with fungi on crop plants. *In*: *An Advanced Treatise on Meloidogyne, Vol. I: Biology and Control* (J.N. Sasser and C.C. Carter, *eds.*), pp. 183-192. North Carolina State Univ. Graphics, Raleigh.

Weingartner, D.P., and Shumaker, J.R. 1990. Effects of soil fumigants and aldicarb on corky ringspot disease and trichodorid nematodes in potato. *J. Nematol.* (Suppl.) **22**: 775-778.

Weingartner, D.P., Shumaker, J.R. and Smart, G.C. Jr. 1975. Incidence of corky ringspot disease on potato as affected by different fumigation rates, cultivars, and harvest dates. *Proc. Soil and Crop Sci. Soc. of Florida* **34**:194-196.

Weischer, B. 1975. Further studies on the population development of *Ditylenchus dipsaci* and *Aphelenchoides ritzemabosi* in virus-infected and virus-free tobacco. *Nematologica* **21**: 213-218.

Weller, D.M., Cook, R.J., MacNish, G., Bassett, E.N., Powelson, R.L. and Peterson, R.R. 1986. *Rhizoctonia* root rot of small grains favored by reduced tillage in the Pacific Northwest. *Plant Dis.* **70**: 70-73

Welsch, J. and Sikora, R.A. 1987. Influence of endotrophic mycorrhizal fungi on *Globodera rostochiensis* hatch and early root infection on potato. *Revue de Nematol.*

Wescott, S.W. and Barker, K.R. 1976. Interaction of *Acrobeloides buetschlii* and *Rhizobium leguminosarum* on Wando pea. *Phytopathology* **66**: 468-472.

Wescott, S.W. and Burrows, P.M. 1991. Degree-day models for predicting egg hatch and population increase of *Criconemella xenoplax*. *J. Nematol.* **23**: 386-392.

Wheeler, T.A., Madden, L.V., Rowe, R.C. and Riedel, R.M. 1992. Modelling of yield loss in potato early dying caused by *Pratylenchus penetrans* and *Verticillium dahliae*. *J. Nematol.* **24**: 99–102.

Wilhelm, S. 1950. Vertical distribution of *Verticillium albo-atrum* in soils. *Phytopathology* **40**: 368-376.

Winoto, S. and Lim, T.K. 1979. Interaction of *Meloidogyne incognita* and *Rotylenchulus reniformis* on tomato. *Malaysian Agri. Res.* **1**: 6-13.

Wolfe, D., Hartman, J. R., Brown, G. R. and Strang, J. 1990. The influence of soil fumigation on strawberry yield and economics in black root rot infested fields. *Appl. Agr. Res.* **5**:17-20.

Yagita, H. and Komuro, Y. 1972. Transmission of mulberry ringspot virus by *Longidorus martini* Merny. *Ann. Phytopathol. Soc. Japan* **38**: 275-283.

Yen, J.H., Huang, J.W., Lin, C.Y., Chen, D.Y. and Tsay, T.T. 1998. Interaction between *Fusarium oxysporum* f. sp. *niveum* in watermelon roots. *Plant Pathol. Bull.* **7**: 201.204.

Yousif, G.M. 1979. Histological responses of four leguminous crops infected with *Meloidogyne incognita. J. Nematol.* **11**:395-401.

Yuen, G.Y., Craig, M.L. and Geisler, L.J. 1994. Biological control of *Rhizoctonia solani* on tall fescue using fungal antagonists. *Plant Dis.* **78**:118-123.

Yuen, G.Y., Schroth, M.N., Weinhold, A.R. and Hancock, J.G. 1991. Effect of soil fumigation with methyl bromide and chloropicrin on root health and yield of strawberry. *Plant Dis.* **75**: 416-420.

Zadoks, J.C. and Sachein, R.D. 1979. *Epidemiology and Plant Disease Management*. Oxford Univ. Press, Oxford.

Zambolin, L. and Oliveira, A.A.R. 1986. Interacao entre *Glomus etunicatum* and *Meloidogyne javanica* em feijao (*Phaseolus vulgaris* L.). *Fitopathologia Brasileira* **11**: 217-216.

Zehr, E. I., Miller, R. W. and Smith, E. H. 1976. Soil fumigation and peach rootstocks for protection against peach tree short life. *Phytopathology* **66**: 689-694.

Zeller, S. M. 1932. A strawberry disease caused by *Rhizoctonia. Ore. Agr. Expt. Sta. Bul.* **295**: 1-22.

Zentmyer, G.A. 1961. Chemotaxis of zoospores for root exudates. *Science* **133**: 1595-1596.

Zutra, D. and Orion, D. 1982. Crown gall bacteria (*Agrobacterium radiobacter* var. *tumefaciens*) on cotton in Israel. *Plant Dis.* **66**: 1200-1201.

SUBJECT INDEX

S

T

V

W

X

Z